VOLUME TWO

ADVANCES IN NEUROTOXICOLOGY

Linking Environmental Exposure to Neurodevelopmental Disorders

VOLUME TWO

ADVANCES IN NEUROTOXICOLOGY

Linking Environmental Exposure to Neurodevelopmental Disorders

Edited by

MICHAEL ASCHNER

Albert Einstein College of Medicine, Bronx, NY, United States

LUCIO G. COSTA

University of Washington, Seattle, WA, United States; University of Parma, Parma PR, Italy

ELSEVIER

ACADEMIC PRESS

An imprint of Elsevier

Academic Press is an imprint of Elsevier
50 Hampshire Street, 5th Floor, Cambridge, MA 02139, United States
525 B Street, Suite 1650, San Diego, CA 92101, United States
The Boulevard, Langford Lane, Kidlington, Oxford OX5 1GB, United Kingdom
125 London Wall, London, EC2Y 5AS, United Kingdom

First edition 2018

ISBN: 978-0-12-815551-6
ISSN: 2468-7480

Publisher: Zoe Kruze
Acquisition Editor: Sam Mahfoudh
Editorial Project Manager: Alina Cleju
Production Project Manager: Vignesh Tamil
Cover Designer: Alan Studholme

Typeset by SPi Global, India

CONTENTS

CONTRIBUTORS

Navatha Alugubelly
Center for Environmental Health Sciences, Department of Basic Sciences, College of Veterinary Medicine, Mississippi State University, Mississippi State, MS, USA

Daniel M.P. Ardisson-Araújo
Programa de Pós-graduação em Ciências Biológicas: Bioquímica Toxicológica; Laboratório de Virologia de Insetos, Departamento de Bioquímica e Biologia Molecular, Centro de Ciências Naturais e Exatas, Universidade Federal de Santa Maria, Santa Maria, RS, Brazil

Michael Aschner
Department of Molecular Pharmacology, Albert Einstein College of Medicine, Bronx, New York, NY, USA

David C. Bellinger
Department of Neurology, Boston Children's Hospital, Harvard Medical School, Boston, MA, United States

Aaron B. Bowman
Vanderbilt University, Vanderbilt University Medical Center, Department of Pediatrics; Vanderbilt University, Vanderbilt University Medical Center, Department of Neurology; Vanderbilt University, Vanderbilt University Medical Center, Department of Biochemistry, 1161 21st Avenue S, D-4105 MCN, Nashville, TN, United States

Gemma Calamandrei
Center for Behavioral Sciences and Mental Health, Istituto Superiore di Sanità, Viale Regina Elena 299, I-00161 Rome, Italy

Russell L. Carr
Center for Environmental Health Sciences, Department of Basic Sciences, College of Veterinary Medicine, Mississippi State University, Mississippi State, MS, USA

Deborah A. Cory-Slechta
Department of Environmental Medicine, University of Rochester School of Medicine, Rochester, NY 14642, United States

Randall L. Davis
Department of Pharmacology and Physiology, Center for Health Sciences, Oklahoma State University, Tulsa, OK, United States

José G. Dórea
Professor Emeritus, Faculdade de Ciências da Saúde, Universidade de Brasília, Brasília, DF, Brazil

Marina Guizzetti
Department of Behavioral Neuroscience, Oregon Health & Science University and VA Portland Health Care System, Portland, OR, USA

Joel G. Hashimoto
Department of Behavioral Neuroscience, Oregon Health & Science University and VA Portland Health Care System, Portland, OR, USA

Edward D. Levin
Department of Psychiatry and Behavioral Sciences, Duke University Medical Center, Box 104790 DUMC, Durham, NC 27710, USA

Ashley Malin
Department of Preventive Medicine, Icahn School of Medicine at Mount Sinai, New York, NY, United States

Afzaal N. Mohammed
Center for Environmental Health Sciences, Department of Basic Sciences, College of Veterinary Medicine, Mississippi State University, Mississippi State, MS, USA

Rachana Nitin
Vanderbilt University, Vanderbilt University Medical Center, Department of Pediatrics; Vanderbilt University, Vanderbilt University Medical Center, Department of Neurology; Vanderbilt University, Vanderbilt University Medical Center, Department of Biochemistry, 1161 21st Avenue S, D-4105 MCN, Nashville, TN, United States

Pablo A. Nogara
Programa de Pós-graduação em Ciências Biológicas: Bioquímica Toxicológica, Centro de Ciências Naturais e Exatas, Universidade Federal de Santa Maria, Santa Maria, RS, Brazil

Cláudia S. Oliveira
Programa de Pós-graduação em Ciências Biológicas: Bioquímica Toxicológica, Centro de Ciências Naturais e Exatas, Universidade Federal de Santa Maria, Santa Maria, RS, Brazil

Laura Ricceri
Center for Behavioral Sciences and Mental Health, Istituto Superiore di Sanità, Viale Regina Elena 299, I-00161 Rome, Italy

João B.T. Rocha
Departamento de Bioquímica e Biologia Molecular, Centro de Ciências Naturais e Exatas, Universidade Federal de Santa Maria, Santa Maria, RS, Brazil

Robert O. Wright
Department of Preventive Medicine, Icahn School of Medicine at Mount Sinai, New York, NY, United States

Xiaolu Zhang
Department of Pharmaceutical Science, Northern Jiangsu People's Hospital, Yangzhou, China; Department of Behavioral Neuroscience, Oregon Health & Science University and VA Portland Health Care System, Portland, OR, USA

PREFACE

The second volume of *Advances in Neurotoxicology* is dedicated to environmental triggers on neurodevelopmental disorders. It is a timely addition to the literature as the prevalence of neurodevelopmental disorders, such as autism spectrum disorders and attention-deficit hyperactivity disorder, is skyrocketing. Thus, further insight into etiologic factors and risk factors associated with these, and other, disorders is meritorious. The volume focuses on several environmental agents with known fetal and childhood developmental sequelae, and where appropriate genetic risks for diseases and their interactions with the environmental cause are addressed. This volume highlights the relationship between neurodevelopmental disorders and environmental toxicants, addressing environmental factors that affect normal neurodevelopment. The various chapters are authored by internationally recognized scientists, and we are indebted to all of them for their great contributions. Among environmental factors discussed in the chapters are well-known ones, such as methylmercury, lead, or alcohol, as well as others which are less in the limelight (e.g., manganese, cocaine). Also of relevance are the discussions on the role of stress and neuroinflammation in neurodevelopmental disorders.

This series fills a gap in the growing field of neurotoxicology, and the volumes should be of interest to academic and industrial researchers, graduate students, and practicing toxicologists. The series editors will be happy to entertain proposals for specific topics to be the subject of future volumes, and for guest editing.

MICHAEL ASCHNER
LUCIO G. COSTA

CHAPTER ONE

The Neurodevelopmental Toxicity of Lead: History, Epidemiology, and Public Health Implications

David C. Bellinger*, Ashley Malin†, Robert O. Wright†,[1]
*Department of Neurology, Boston Children's Hospital, Harvard Medical School, Boston, MA, United States
†Department of Preventive Medicine, Icahn School of Medicine at Mount Sinai, New York, NY, United States
[1]Corresponding author: e-mail address: robert.wright@mssm.edu

Contents

Although the neurotoxicity of lead was recognized more than two millennia ago (Bellinger, 2018), research conducted in the last two decades has significantly enhanced our understanding of the ways in which it harms human health. This, in turn, has led to important changes in the conclusions of risk assessments and thus changes in regulatory standards. In 2012 the US Centers for Disease Control and Prevention acknowledged that there is no level of lead that can be considered "safe," replacing its action level with a reference level that, rather than being health-based, corresponds simply to the 97.5 percentile of the distribution of blood lead concentrations in young children (US CDC, 2012). The reference level can thus change as characteristics of the distribution change. In 2011, the WHO/FAO Joint Expert Committee on Food Additives withdrew its Provisional Tolerable Weekly Intake for

Advances in Neurotoxicology, Volume 2
ISSN 2468-7480
https://doi.org/10.1016/bs.ant.2018.03.009

lead (25 μg/kg body weight) because it concluded that it was not possible to identify an intake that would adequately protect children from IQ loss (JECFA, 2011).

The global scope of the health outcomes associated with lead toxicity has also become more apparent. In the most recent estimation of the global burden of disease, WHO ranked risk factors within different regions in terms of their contributions to the total number of disability-adjusted life-years (DALYs). Depending on the region, lead ranked was the 17th to 30th most important contributor to the DALY total (Lim et al., 2012), and was responsible for 12.4% of the burden associated with idiopathic intellectual disability (Institute for Health Metrics and Evaluation, 2015; https://vizhub.healthdata.org/gbd-compare/). In a comparative analysis of children's exposures to neurotoxic chemicals, the Environmental Protection Agency of Denmark concluded that, "...lead constitutes by far the greatest risk of chronic neurotoxic effects."

The goal of this chapter is to provide an overview of the research that prompted these conclusions. The topics covered include which aspects of neurodevelopment appear to be most vulnerable to lead, properties of the dose–effect/response relationships and the time course of the relationships, the late structural and functional brain correlates of childhood lead exposure, the factors that modify lead's neurotoxicity, and critical windows of vulnerability, during which lead is most toxic to neurodevelopment.

1. NEUROPSYCHOLOGICAL FUNCTION

Much of the impetus for the recent sea change in the assessment of lead risks came from a set of analyses of IQ data collected in seven prospective studies ($N = 1333$) conducted in several countries (Lanphear et al., 2005). The motivation for pooling the data from these studies was to increase the precision with which the shape of the dose–effect relationship for blood lead concentrations <10 μg/dL could be ascertained. The relationships between IQ, measured at age 5–10 years, and four indices of lead exposure were evaluated: the blood lead concentration measured the closest in time to the IQ assessment (concurrent), the maximum blood lead concentration prior to IQ measurement (peak), the mean blood lead concentration between 6 months of age and IQ measurement (lifetime average), and the mean level between 6 and 24 months of age (early childhood). A variety of statistical models were compared in terms of goodness-of-fit to the data, adjusting for 10 covariates, such as maternal IQ and home quality. A log-linear model involving concurrent

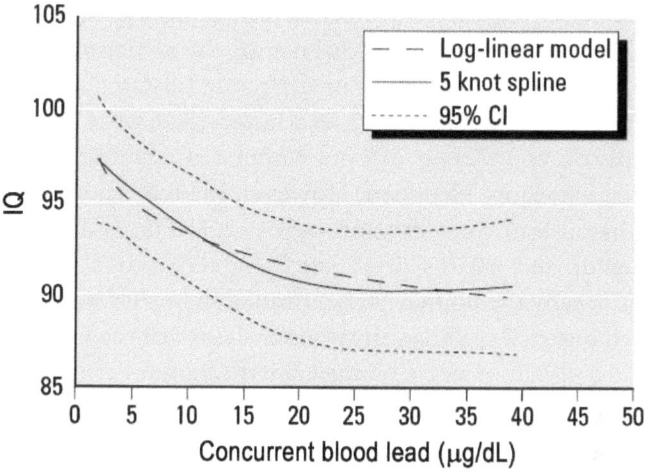

Fig. 1 Restricted cubic splines and log-linear model for concurrent blood lead concentration and Full-Scale IQ from the pooled analyses of seven prospective studies (Lanphear et al., 2005). *Reproduced From Environmental Health Perspectives (https://www.ncbi.nlm.nih.gov/pmc/article/PMC12576521).*

blood lead concentration provided the best fit (i.e., the largest R^2) and indicated that an increase in blood lead from 2.4 to 30 µg/dL (the 5th and 95th percentiles of the blood lead distribution in the pooled dataset) was associated with an IQ reduction of 6.9 points (95% CI: 4.2, 9.4) (Fig. 1). One of the most important findings was that the association appeared to be non-linear, insofar as much of the overall reduction, 3.9 points (95% CI: 2.4, 5.3) occurred over the range of 2.4–10 µg/dL. The increase from 10 to 20 µg/dL was associated with a further reduction of 1.9 (95% CI: 1.2, 2.6) points, and the increase from 20 to 30 µg/dL with an additional reduction of 1.1 (95% CI: 0.7, 1.5) points. The analyses provided no evidence of a blood lead concentration below which no adverse impact on IQ is evident, i.e., a threshold or "safe" concentration. Sensitivity analyses indicated that the results were not unduly influenced by the data from any of the seven studies. The change in the coefficient for concurrent blood lead concentration produced by excluding the data for one of the seven studies ranged from −2.6% to +8.6%.

Although the mechanism(s) by which the proportional loss in IQ are greater at lower than at higher blood lead concentrations remain unknown, similar supra-linear relationships between blood lead and other cognitive (e.g., Kordas et al., 2006; Téllez-Rojo et al., 2006) and educational achievement outcomes (Evens et al., 2015; Skerfving et al., 2015) were subsequently reported. Some speculated that the supra-linear relationship was a statistical artifact created by the combination of the log-normal distribution of blood lead concentration and the normal distribution of IQ scores (Bowers and Beck, 2006). Whether or not this is true, the criticism is not germane in this case. In additional analyses of the pooled dataset, piecewise linear models were fit to different blood lead ranges that were defined a priori (e.g., <7.5 μg/dL vs. ≥7.5 μg/dL) and within neither range were significant deviations from linearity identified. However, the regression coefficients for concurrent blood lead concentration were −2.94 (95% CI: −5.16, −0.71) below 7.5 μg/dL and −0.16 (95% CI: −2.4, −0.8) at or above 7.5 μg/dL.

One reason why the pooled analyses had such impact was that the results demonstrated more clearly than any prior analyses that the CDC action level at the time, 10 μg/dL, is not a biologically meaningful criterion. Although the CDC never claimed that it was and promoted it solely as a risk management guideline, many in the medical community seemed to interpret it as a threshold for neurotoxicity. In the pooled analyses, adverse impacts on IQ were clearly evident at concentrations below 10 μg/dL and, furthermore, the decline in IQ for each μg/dL increase was substantially greater below 10 than above 10. Although the regulations promulgated over the preceding decades have been a resounding success, producing a 90% decline in the mean blood lead concentration in children, and by one estimate, a four- to five-point increase in the mean IQ of US adults (Kaufman et al., 2014), the pooled analyses showed that large numbers of children nevertheless remain at risk of lead-associated IQ deficits.

Although lead's impact on children's cognitive development is often characterized in the literature as "irreversible," this term is ambiguous and could create scenarios in which lead-exposed children are treated as though developmental delay in inevitable, which can itself be detrimental to their subsequent neurodevelopment. While childhood lead poisoning tends to peak in the early postnatal years, there are many years of development that follow during which brain plasticity is still operative. Therefore, interventions might be successful, although apart from chelation, few interventions have been tried. Therefore, while from a toxicodynamic perspective the biological injury that lead causes at a target might be immutable, it does not necessarily follow that the functional sequelae of the injury cannot be mitigated by means of an intervention. Observational prospective studies show that the adverse impacts of lead are persistent, but these studies did not include the implementation of any interventions that might have altered the long-term impact of early exposure. These might include developmental supports to help children find strategies for working around the deficits, and/or interventions (e.g., environmental, nutritional) to reduce future lead exposure that might contribute to the maintenance of deficits associated with early-life exposure. Nevertheless, it is becoming clear that absent such interventions, children who experience greater childhood exposure to lead tend to be less successful in adulthood. In a subset of participants in the Boston Lead Study, Mazumdar et al. (2011) found that IQ at about age 30 years was inversely related to childhood blood lead levels, in particular those measured between the ages of 4 and 10. In the large Dunedin Multidisciplinary Health and Development Study, Reuben et al. (2017) found that blood lead concentration measured at 11 years of age was inversely related to IQ at age 38 years, even after adjustment for IQ at age 11 years. Early lead exposure was also inversely related to socioeconomic status at age 38, suggesting that children who were more highly exposed enjoyed lower upward social mobility and, in fact, failed to match the socioeconomic achievements of their parents.

One unfamiliar with lead research might wonder why it took approximately 40 years to reach the conclusion that a threshold for lead neurotoxicity does not exist. Instead, the concentration identified as "acceptable" underwent gradual reduction, from 60 μg/dL in the 1960s, to 40 in 1971, 30 in 1975, 25 in 1985, and 10 in 1991, to be replaced, in 2012, by the nonhealth-based reference value of 5 μg/dL. Why did this process move by gradual steps rather than all at once? One explanation is that the questions a researcher could ask were limited by the distribution of children's blood lead distribution at the time that a study was initiated. In the late 1970s, 88% of US children had a blood lead concentration >10 μg/dL, which made it difficult to implement a study designed to determine whether adverse effects occur at concentrations below 10. It simply would have been difficult to find a sufficient number of children with such concentrations. As an illustration of the challenges created by the ubiquity and severity of lead exposure during this period, in one study published in 1972, the mean blood lead concentration of children in the "control group" was 38 μg/dL (the mean of "exposed" children was 81 μg/dL). Each time the "acceptable" blood lead concentration was lowered, new studies were initiated to determine whether the new value provided children with an adequate margin of safety. Each time, the weight of evidence suggested that it did not because adverse impacts could be reliably identified at concentrations below it. We are presently at the logical terminus of this process, as the consensus now is that no concentration greater than zero provides an adequate margin of safety. Thus, somewhat ironically, it is the dramatic success of regulations put in place in the latter decades of the 20th century to reduce population exposures that created the conditions that allowed investigators to ask whether blood lead concentrations of 10 or 5 μg/dL are neurotoxic.

An argument frequently advanced by those skeptical about the importance of lead is that its impact on the neurodevelopment of a child is so modest that it is clinically insignificant. While it is true that a reduction of a few IQ points would not be immediately apparent, this perspective fails to consider the issue in the context of population health. Effect estimates from epidemiologic data are in essence a population average effect and should be interpreted in the context of a population and not at the individual level as some individuals will be resistant and some will be more sensitive. Moreover, the impact of a factor at the population level depends not only on the effect size but also on the distribution of the factor or, in the case of a dichotomous factor, its incidence or prevalence. In a set of comparative analyses of pediatric disease and events, such as brain tumors, congenital heart disease, traumatic brain injury, iron deficiency, and lead exposure, the total number of IQ points lost among US children 1–5 years of age associated with each disease or event was estimated (Bellinger, 2012). The estimate for the loss associated with lead exposure was nearly 23 million IQ points, exceeded only by preterm birth. Among the reasons for this is the absence of a threshold for its inverse relationship with IQ and the fact that virtually every child has a blood lead concentration above the detection limit. As a result, and in contrast to most other diseases and events, every child contributes to the IQ loss associated with lead exposure, with the greatest contributions to the total contributed by the very large proportion of children with blood lead concentrations at the low end of the distribution.

1.1 Other Neuropsychological Endpoints

Epidemiological studies have generally included assessment of other aspects of children's neuropsychological functioning. IQ is an apical measure, integrating children's performance in diverse verbal and nonverbal domains. If the effects of a neurotoxicant are focal, limited to only certain domains, one would expect that tests that focus more narrowly on the vulnerable domains would be more strongly related to lead biomarkers than is IQ. It is somewhat surprising, therefore, that, IQ is more consistently associated with lead biomarkers than performance on tests that focus on more narrowly-defined domains (e.g., memory, language). This might provide some insight into the mechanism of lead-associated neurotoxicity, suggesting diffuse rather than focal neuronal and or white matter injury, thereby affecting global higher cortical functioning. It is also possible that the domains most affected vary across studies because of the joint effect of cohort differences in the timing of lead exposure and differences across domains in the timing of greatest susceptibility. The domains vulnerable to lead exposure also depend, to some extent, on cohort characteristics or context specific aspects such as genetic susceptibility, social environment, mixed chemical exposures.

2. ACADEMIC ACHIEVEMENT

In one of the seminal papers in the lead literature, Byers and Lord (1943) described the neurological outcomes of 20 children who presented with sub-encephalopathic lead poisoning. Their observations presage many of the findings of later studies. Byers and Lord noted that, "after recovery from their lead poisoning, these…children made an extremely poor record in competition with their fellows," insofar as, "with one definite and a second possible exception, none of the 20 children succeeded in school" (p. 479). Numerous subsequent studies have reported significant inverse associations, in the general population of children with much more modest lead exposure and reduced success in school, expressed as grade retention, receipt of special education services, failure to complete qualifications, lower scores on standardized tests, etc. (e.g., Amato et al., 2012; Fergusson et al., 1997; Magzamen et al., 2015; Needleman et al., 1990). In ecologic analyses, Nevin (2009) found an inverse relationship between preschool blood lead concentrations and SAT scores, lagged by 17 years. Evens et al. (2015) evaluated the relationship between blood lead concentration and performance on the Illinois Standard Achievement Test in a cohort of 47,168 Chicago schoolchildren. Adjusting for covariates, they found dose-dependent reductions, extending below 10 μg/dL, in both reading and math scores. Each 5 μg/dL increase in blood lead was associated with an increase of 1.3 in the risk of reading and math scores considered to represent "failure." A non-linear dose–response relationship was found for failure, with the slope steeper for blood lead concentrations <10 μg/dL compared to >10, although this was significant only for reading.

In a similar study, Miranda et al. (2009) combined a North Carolina blood lead surveillance database with a database containing scores on a reading test given to school children in the state at the end of fourth grade. Among the 57,678 children included in the analyses, blood lead concentration ranged from 1 to 16 μg/dL, with a 75th percentile of 6 μg/dL. A significant dose–response effect relationship was found between blood lead and test score, without apparent threshold. Quantile regression analyses, furthermore, indicated that the adverse impact of lead on test scores was more pronounced at the lower end of the test score distribution than at the higher end. In other words, the impact of lead is disproportionally greater on children who are already at academic risk than on children who are at low academic risk. This might be attributable to contextual factors that influence a child's resilience or susceptibility to lead exposure, such as nutrition, social environment, and other chemical exposures. Magzamen et al. (2015) replicated this finding using data collected in a cohort of Wisconsin schoolchildren.

Skerfving et al. (2015) evaluated the performance of 3176 Swedish children on an examination taken at the end of compulsory schooling (age 16 years) in relation to blood lead concentration measured when the children were in primary school (age 7–12 years). The distribution of blood lead concentrations was very similar to that of the participants in the Miranda et al. (2009) study: range 0.6–16.2 μg/dL, 90th percentile 6 μg/dL. They found an inverse non-linear association between blood lead concentration and school performance, with a steeper slope at concentrations below 5 μg/dL than above 5. They estimated that the adverse impact of an increase in blood lead concentration from 2.5 to 5 μg/dL was similar in magnitude to the impact of having a mother with a university vs. a primary school education.

Surkan et al. (2007) reported that children with a blood lead concentration of 5–10 μg/dL, compared to children with a concentration of 1–2 μg/dL, had significantly lower reading and mathematics scores, even after adjustment for Full-Scale IQ score. This suggests that children with greater exposures were not achieving at a level commensurate with their ability.

3. ADHD

Prominent in the presentation of the children in the case series of Byers and Lord (1943) were severe behavioral pathologies that included hyperactivity, reduced impulse control, and aggression. In the 1970s, case–control and chelation challenge studies suggested that children diagnosed with hyperactivity had greater lead burdens (e.g., David, 1974). Beginning around 1980, epidemiological studies supported the hypothesis that even among children who were not clinically lead poisoned and did not meet diagnostic criteria for hyperactivity, a greater lead burden was associated with increased distractibility, reduced ability to follow directions, disorganization, daydreaming, and lack of task persistence (e.g., Needleman et al., 1979). Numerous studies in subsequent decades evaluated the associations between early childhood lead burden and behaviors indicative of attention problems. A meta-analysis of 33 such studies conducted between 1972 and 2010, involving >10,000 children, found significant effect sizes linking greater exposure and dimensional measures of both inattentive and hyperactive/impulsive symptoms (Goodlad et al., 2013).

A direct association between lead burden and behaviors associated with attention problems does not necessarily establish, however, that early childhood lead exposure is associated with the clinical syndrome of ADHD. In was in 2006 that the association between blood lead concentration and risk of the diagnosis of ADHD was first evaluated in a general population cohort. Using data from NHANES 1999–2002, Braun et al. (2006) found that the adjusted odds ratio for parent-reported ADHD among 6–16-year-old children with a blood lead concentration in the fifth quintile (>2 μg/dL) was 4.1 (95% CI: 1.2, 4.0), compared to children with a concentration <0.8 μg/dL. A dose–response relationship was observed, with an adjusted odds ratio of 1.1 (95% CI: 0.4, 3.4) among children in the second quintile, 2.1 (95% CI: 0.7, 6.8) among children in the third quintile, and 2.7 (95% CI: 0.9, 8.4) among children in the fourth quintile. Among the limitations of this study are the absence of a clinician-confirmed diagnosis of ADHD and the fact that data on important covariates, such as family history of ADHD, were not available. A study of 8–15 year olds by Froehlich et al. (2009), using NHANES 2001–2004 data, addressed the first issue. The Diagnostic Interview Schedule for Children, a clinician-administered diagnostic interview based on DSM-IV criteria, was used to confirm a diagnosis of ADHD. Children with a blood lead concentration in the upper tertile (>1.3 μg/dL) were 2.3 (95% CI: 1.5, 3.8) times more likely to meet diagnostic criteria than children with a concentration in the lowest tertile.

This association has been replicated in several subsequent case–control studies. In a study of 4–12-year-old Chinese children that included 630 cases of DSM-IV criteria for ADHD and 630 controls, conditional logistic regression analyses, adjusting for covariates found the risk of ADHD was 6.0 (95% CI: 4.1, 8.8) among children with a blood lead concentration >10 μg/dL, and 4.9 (95% CI: 3.5, 7.0) among children with a concentration of 5–10 μg/dL (Wang et al., 2008). The referent group consisted of children with a blood lead concentration <5 μg/dL. In another case–control study, a blood lead concentration >2.3 μg/dL was associated with an odds ratio of 2.5 (95% CI: 1.1, 5.9) (Park et al., 2016). Choi et al. (2016) found that blood lead concentration was significantly greater among incident cases of ADHD identified over a 2-year follow-up period. The relative risk for children with a blood lead concentration >2.17 μg/dL was 1.6 (95% CI: 1.0, 2.4).

Nigg et al. (2008) provided some clues as to the underlying behavioral mechanisms of the association between lead and ADHD symptoms. Children 8–17 years old who met rigorous criteria for the diagnosis of ADHD had a significantly higher blood lead concentration than controls, even though concentration for all participants ranged only from 0.4 to 3.47 μg/dL, mean of 1.03. A significant relationship was found between blood lead concentration and total ADHD symptoms. In this study, IQ was measured and a Stop task was administered, providing assessments of a child's ability to suppress a prepared response (stop signal reaction time) and variability of reaction time on the "go" trials (response variability, readiness, and control). Mediation analyses exploring the relationships among blood lead, IQ, hyperactivity–impulsivity, and Stop task scores suggested that lead exposure might increase a child's risk of ADHD by impairing cognitive control abilities. Analyses also suggested that the association between blood lead and IQ was mediated by the association between blood lead and hyperactivity–impulsivity, not vice versa. Nigg et al. argued that the plausibility of the link between lead and ADHD is supported by the evidence that lead disrupts midbrain dopamine circuitry (striatum and fronto-striatal networks), the same circuitry that is thought to underlie ADHD.

Nigg et al. (2010) replicated the associations between blood lead concentration and ADHD symptoms in a larger sample, including adjustment for additional covariates, and at even lower blood lead concentrations (range 0.3–2.20 μg/dL, mean of 0.73).

4. ANTI-SOCIAL BEHAVIOR

In Byers and Lord's (1943) case series, some children exhibited aggression that was so disruptive that they were excluded from school. These behaviors included fire-setting, biting others, stealing supplies, repeatedly dancing on desks, and sticking a fork into a classmate's face. Other problem behaviors noted included cruelty to animals and lack of response to punishment. This observation was not systematically followed up, however, until Needleman et al. (1996) reported that children with high bone lead concentrations were rated by both parents and teachers as higher than controls on the aggression and delinquency scales of the Child Behavior Checklist. They also self-reported engaging more frequently in delinquent behaviors.

In a subsequent study, 12–18-year-old youths who had been arrested and adjudicated as delinquent were found to have significantly higher bone lead concentrations (Needleman et al., 2002). Lead in different bone compartments has a half-life of 10 or more years in adults, so it is possible that these levels were due in part to early childhood exposures. However, the kinetics of bone lead might be quite different in children due to active bone remodeling. Several ecological studies subsequently reported associations, at the areal or national level, between crime rates and exposure metrics such as lead emissions or blood lead concentrations (e.g., Boutrell et al., 2016; Nevin, 2000, 2007; Reyes, 2007; Taylor et al., 2016). For example, using national data on preschool blood lead concentrations and crime rate for nine countries, Nevin (2007) found that the best-fitting model included a two-decade lag between blood lead measurement and crime, implicating childhood lead exposure as a risk factor. Analyses reported by Taylor et al. (2016) produced consistent results. Air lead concentrations accounted for nearly 30% of the variance in assault rates 21 years later.

The strongest evidence of a potential link has come from prospective studies that followed participants from birth to young adulthood. In the Christchurch Health and Development Study (Otago, New Zealand), children with greater lead concentrations in the dentine layer of teeth at ages 6–9 years had modestly, but significantly higher rates of violence/property convictions and self-reported violence/property offenses at ages 14–21, even after adjustment for confounding (Fergusson et al., 2008). The authors suggested that this association, which accounted for only about 1% of the variance in crime, was mediated by the lower educational achievement among participants with greater dentine lead concentrations. In the Cincinnati Prospective Lead Study, county criminal justice records were accessed for information on the number of times a member of the birth cohort had been arrested after age 19 years (Wright et al., 2008). Covariate-adjusted rate ratios for arrests were calculated for each 5 μg/dL increase in blood lead concentration in different intervals during childhood. The rate ratios ranged from 1.07 to 1.48, and most were statistically significant.

The evidence that childhood lead exposure increases the risk of criminal behavior in adulthood has been met with some skepticism. However, viewed as one possible endpoint of a developmental cascade, it is easier to conceive of how some lead-exposed children might grow into such adults. Children with greater early lead exposure arrive at school less well prepared in terms of readiness skills (McLaine et al., 2013) and remain behind their peers in achievement (Amato et al., 2012), more frequently requiring remedial services (Miranda et al., 2010), and more frequently being retained in grade (Fergusson et al., 1997). They experience greater difficulty maintaining attention, planning, organizing, and executing their activities (Chandramouli et al., 2009), and conforming to behavioral expectations (Amato et al., 2013). For some, the absence of satisfaction in academic pursuits might lead to early termination of education, reducing job opportunities and limiting income and other resources. Many other risk factors for criminal behavior track with lead poisoning, such as poverty, family violence and other hardships. While frequently thought of as potential confounders, it is equally probable that exposures to these multiple risk factors act synergistically during development to produce violent or criminal behaviors in young adults. Reduced impulse control and inability to delay gratification might lead some to make decisions that reduce their long-term prospects for success. Indeed, some studies have reported association between greater blood lead concentrations and behaviors that are associated with executive dysfunction, including drug use (Dietrich et al., 2001), teen pregnancy (Reyes, 2014), sexually transmitted diseases and injuries (Kincl et al., 2006). Presented with opportunity, some individuals might have difficulty avoiding the commission of unlawful acts. Of course, this is by no means inevitable, and only some lead-exposed children would be expected to follow this trajectory. Whether a child makes the transition from one state to the next in this cascade depends on many other factors. Each transition point thus provides an opportunity for timely interventions to reduce the likelihood that a child will progress to the next stage in the cascade.

5. NEUROIMAGING

The neural bases of lead neurotoxicity remain somewhat obscure. Byers and Lord (1943) laid the foundation for understanding this issue from a developmental perspective, however, observing, with reference to their case series of clinically poisoned children, that, "It seems likely…that the lead in the circulation of an infant in some way interferes with the changes normally occurring in the cortex and in a high percentage of cases prevents the normal growth and development of the cortex" (p. 483). Since their report, only a few neuroimaging studies have explored the associations between lead exposure history and brain structure and function among individuals from the general population. The most comprehensive data come from the Cincinnati Prospective Lead Study, in which a variety of neuroimaging studies have been reported when the participants reached young adulthood (19–24 years of age). Volumetric imaging revealed significant inverse associations between annual mean blood lead concentration between 3 and 6 years of age and gray matter volume, particularly in the frontal regions of the brain, including the anterior cingulate cortex and the ventrolateral prefrontal cortex, areas usually considered to be related to executive functions, mood regulation, and decision-making (Cecil et al., 2008). These associations were stronger for males than females. White matter abnormalities, specifically reduced fractional anisotropy and axial diffusivity on diffusion tensor imaging, were also associated with greater childhood lead exposure (Brubaker et al., 2009). These findings suggest impaired myelination and reduced axonal integrity in regions that regulate executive functions.

Two methods have been used to investigate associations brain function in lead-exposed cohorts. Cecil et al. (2011) measured levels of brain metabolites, such as *N*-acetyl aspartate and creatine and phosphocreatine, using proton magnetic resonance spectroscopy, and found that childhood blood lead concentration was associated with reductions in several metabolites throughout the brain, including gray matter (basal ganglia, cerebellum) and white matter (left frontal, left parietal). Yuan et al. (2006) measured brain activation patterns while participants performed a verb generation task. Individuals with greater blood lead concentrations in childhood showed dose-dependent changes in the left frontal cortex and the left middle temporal gyrus. Studies of adult lead-workers show correlations between brain volume in certain areas and performance on neuropsychological tests (Schwartz et al., 2007).

In an fMRI study that involved participants completing the Wisconsin Card Sorting Test, an assessment of executive functioning, Seo et al. (2014) found that neural activation in the left dorsolateral prefrontal cortex was greater in individuals with lesser lead exposures, and, moreover, cortical activation was inversely associated with blood lead concentration. The same group (Seo et al., 2014) performed fMRI scanning while participants performed on another task that is sensitive to working memory, the n-back test. Individuals with greater lead exposure performed worse on the trials that imposed greater memory load (2-back trial) and showed reduced activation in the dorsolateral prefrontal cortex, ventrolateral prefrontal cortex, pre-supplementary motor areas, and inferior parietal cortex. These findings suggest that exposure to lead produces impairments in the fronto-parietal working memory network.

6. POTENTIAL EFFECT MODIFIERS OF LEAD NEUROTOXICITY

It is important, from the standpoint of prevention and intervention, to understand if there are other factors that modify the neurotoxicity of a child's lead exposure. While the evidence is limited, there are some clues as to why some children are more vulnerable than others. For example, some but not all studies suggest that children who are iron deficient experience greater cognitive adversity from a given level of lead than do iron sufficient children (Shah-Kulkarni et al., 2016; Solon et al., 2008) or fail to improve as much following a decline in blood lead concentration (Ruff et al., 1996). Other studies show that iron deficiency increases the risk of greater blood lead levels, perhaps through increased absorption rates of environmental lead (Wright et al., 2003a,b). Folate deficiency has also been reported to increase lead's neurotoxic impact (Solon et al., 2008). Co-exposure to other neurotoxicants, such as second-hand tobacco smoke (Froehlich et al., 2009; Joo et al., 2017), manganese (Claus Henn et al., 2012; Lin et al., 2013), and cadmium (Kim et al., 2013) are reported to potentiate the effects of lead. The evidence is inconsistent with regard to possible sex differences in vulnerability. In most studies, no sex differences are found, while, as some cohorts, boys seem to be more vulnerable (e.g., Froehlich et al., 2007; Jedrychowski et al., 2009) and, in others, girls (e.g., Tong et al., 2000). Raising lead-exposed rats in an "enriched" environment (i.e., larger cage, materials to manipulate, other rats) blunts the effects of lead on water-maze learning (Guilarte et al., 2003), but systematic evidence demonstrating the effectiveness of such environmental manipulations in assisting lead-exposed children is lacking. However, early findings from a prospective study indicated that higher HOME scores and high maternal IQ were associated with greater recovery from lead-associated neurodevelopmental deficits (Bellinger et al., 1990). Permissive parenting (Hubbs-Tait et al., 2009) or being raised in a single-parent family (Choi et al., 2016) are risk factors for greater lead-associated neurodevelopmental deficits. Finally, lead-exposed children of mothers with greater self-esteem manifested better neurodevelopment than their peers with similar exposures but mothers with lower self-esteem (Surkan et al., 2008) and, broadly consistent with rodent studies (Weston et al., 2014), children born to women who experienced greater stress during pregnancy had greater lead-associated neurodevelopmental deficits (Tamayo et al., 2017).

A surprisingly small number of genes have been linked to the risk of lead poisoning, although this is largely due to the relatively limited research on the topic. Most studies have used a candidate gene approach rather than a genomics approach, focusing on genes with a priori biological plausibility rather than scanning the genome to discover genes that interact with lead. In studies applying the candidate gene approach, the vitamin D receptor (VDR) gene, the hemochromatosis gene (HFE), and the aminolevulinic acid dehydratase gene (ALAD) have received the greatest attention, and several variants have been associated with variations in blood lead levels. For these three genes, variants appear to predict differences in children's blood lead levels, with a few studies also suggesting differences in the severity of toxicity. The results are often mixed, however, likely due in part to differences in environmental exposure, differences in the age of the population under study, and differences in covariance structure of the population (SES, nutrition, social environment). Being possible effect modifiers themselves, these factors might impede the identification of the contributions of genetics.

Some genetic variants could impact lead toxicity without impacting internal dose. For example, APOE is a cytoplasmic and serum transport protein for cholesterol that binds to APOE receptors at nerve endings (Dietschy and Turley, 2001). APOE is the main apolipoprotein produced and secreted within the brain parenchyma, where it is involved in the redistribution of lipids among cells and in the regulation of cholesterol homeostasis (Barres and Smith, 2001). Glial-secreted cholesterol and the protein apolipoprotein E appear to be critical to the formation and efficient function of synapses (Göritz et al., 2002). Mauch et al. (2001) demonstrated that cholesterol secreted by glial cells and bound to APOE is the primary signaling molecule for synapse formation and mediates the synaptogenic effects of glial cells. In analyses of a cohort of Mexican children, we found that certain APOE4 isoforms predict improved neurodevelopment and may modify the neurodevelopmental impact of lead (Wright et al., 2003a,b). After adjustment for covariates, APOE4 carriers had 24-month Mental Development Index (MDI) scores that were 4.4 points (95% confidence interval: 0.1–8.7) higher than that of non-carriers. In stratified regression models, the negative association between umbilical cord blood lead on MDI was approximately fourfold greater among APOE3/APOE2 carriers than among APOE4 carriers. The APOE4 variant might therefore modify the toxic effects of heavy metals via its role in regulating synaptic architecture in the brain.

With respect to genomic research on lead toxicity, one recent study using a genome-wide interaction analysis in two birth cohorts (Mexico, Bangladesh) identified a locus on chromosome 8, containing UNC5D, a gene related to synaptic organization. A pathway consisting of a network of 52 genes enriched with oxidative stress genes and neurodevelopmental genes was also identified.

7. DEVELOPMENTAL ORIGINS OF NEURODEGENERATIVE DISEASE

One aspect of lead neurotoxicity that receives little attention is the possible role of early-life exposure to lead as, itself, an effect modifier of later events and exposures, or even processes associated with normal aging, by increasing their adverse impacts. For example, Weiss et al. (2002) speculated that if early-life exposure to a neurotoxicant increases the annual rate of neuronal loss by $<1\%$, clinical signs of neurodegeneration would appear several years earlier than they would in the absence of such an exposure. Limited animal studies support the hypothesis that early-life exposure is a risk factor for less optimal resilience in adulthood in response to an insult. Rats exposed to lead early in development but not thereafter were less successful than control rats in recovering function (beam walking, limb placing) after a laser-induced stroke in the somatosensory cortex in adulthood (Schneider and DeKamp, 2007). Early-life exposure to lead might also influence function in adulthood by altering the trajectory of processes associated with aging. Provocative studies in rodents and non-human primates suggest that exposure in infancy initiates epigenetic processes, perhaps involving altered patterns of DNA methylation, that result in adult-onset overexpression of proteins involved in the neurodegenerative processes characteristic of Alzheimer's Disease (e.g., increased deposition of β-amyloid, increased hyper-phosphorylation of tau protein) (Gąssowska et al., 2016).

8. CRITICAL WINDOWS OF VULNERABILITY

It has been difficult to obtain a clear answer to the question of the period (or developmental life-stage) during which children are most vulnerable to lead's neurotoxic effects. One reason is the fact that, in most cohorts studied, exposure tends to be more-or-less stable over developmental epochs (Baghurst et al., 1992; Dietrich et al., 1993). A high degree of intra-individual stability, or tracking, in blood lead concentrations will result in a confounding of timing of exposure by the chronicity of exposure. An investigator is largely limited to comparing their magnitude and statistical significance associated with the inverse associations between neurodevelopmental outcomes and lead biomarkers that represent different exposure averaging times. For example, in the pooled analyses reported by Lanphear et al. (2005), the four indices of blood lead variables, which were highly correlated with one another (0.74–0.96), were all significantly related to children's Full-Scale IQ scores, with the adjusted estimates ranging from −2.04 (−3.27 to −0.81) for early childhood to −3.04 (−4.33 to −1.75) for lifetime average. Concurrent blood lead concentration was selected as the primary index in additional analyses because it, "...exhibited the strongest relationship with IQ, as measured by R^2" (p. 896). However, the differences among the competing indices were modest, with substantial overlap of the confidence intervals.

Another obstacle is the difficulty of measuring exposure repeatedly in short time intervals in order to capture exposures that coincide with critical windows. Brain development in utero is extraordinarily rapid with entire brain regions formed in a matter of days. Most epidemiologic studies measure lead biomarkers at most two or three times over the course of pregnancy. This means that a blood lead level on a single day of pregnancy might be taken to represent an entire trimester, and in some cases, the entire pregnancy. The probability of missing a critical window due to exposure misclassification in this scenario, particularly in very early life, is quite high. The recent development of a tooth-based lead biomarker that can reconstruct both the relative dose of lead exposure in pregnancy and childhood as well as the date of exposure holds considerable promise for identifying possible time-dependent effects of lead (Arora et al., 2014).

Despite the rather equivocal nature of the findings to date, it has become commonplace to find statements in the literature suggesting that exposure during middle childhood is the most vulnerable time with regard to lead neurotoxicity. It is important to distinguish between this statistical association and its toxicological interpretation. Because lead is an accumulative toxicant, a biomarker concentration measured at a given time reflects exposure over an interval that will vary depending on its exposure averaging time. Although strictly speaking, the half-life of lead in blood is about 30 days, in practical terms it is much longer due to the complex kinetics of lead and its distribution and movement among different body pools with blood lead level remaining elevated over time due, in part, to redistribution from tissues to blood. The half-life of lead in blood is likely to be much longer, especially in children with unusually long duration of exposures in which three-compartment kinetics would dominate due to the long half-life of lead in bone. For example, Roberts et al. (2001) reported that, among children with a baseline blood lead concentration of 25–29 μg/dL it took, on average, 2 years for the concentration to decline below 10 μg/dL even after being placed in a case management program that would have reduced ongoing exposure.

Therefore, in any study, these considerations complicate the interpretation of any observed association in terms of a critical window because the fact of the association is not necessarily informative about when the underlying damage to the neural substrate occurred. The biomarker bearing the association to the outcome might simply be correlated with the exposure that is responsible. In other words, a statistical association should not necessarily be interpreted as having toxicological importance in this regard.

It is unlikely that there is a single time at which children are most vulnerable to lead and that the neurobehavioral domains affected by lead vary depending on the developmental processes most active at the time of exposure. One result would be the absence of a single "signature" phenotype for lead exposure. Rather, lead's effects are likely to be endpoint-specific and depend on the timing of the development of the neural substrate on which it depends. Thus, for example, one would expect that the critical windows of vulnerability for sensory functions differ from those for higher cortical function, probably occurring much earlier in development. It is plausible to hypothesize that lead exposure in early and middle childhood might be particularly harmful with regard to the latter, as expressed by children's performance on an IQ test. Lead is known to interfere with neurotransmitter release and uptake (Bressler et al., 1999), which would be expected to impair not only synaptic functioning but the selection of synapses that will survive the extensive pruning that occurs during this period and thus establish the micro-circuitry of the brain (White et al., 2007). Lead also impairs myelination and, thus, the development of white matter tracts that link different brain regions, facilitating their integration and coordination (Brubaker et al., 2009).

9. CONCLUSION

The 120-year history of childhood lead poisoning should be understood as a cautionary tale. Understanding it requires consideration not only of its scientific and medical aspects, but also its political, sociological, economic contexts. If the lessons this history affords are recognized, they can be applied to other neurotoxicant exposures, with the result that children can be provided more timely protection from their adverse impacts on neurodevelopment. Examining only the science of lead toxicology, it seems somewhat puzzling that it took more than a century to reach the conclusion that even blood lead concentrations below 5 μg/dL cause are hazardous to children, the conclusion reached by the National Toxicology Program in 2012. Three decades earlier, Herbert Needleman reasoned that if a child is at risk of encephalopathy at a blood lead concentration of 80–100 μg/dL, it would not be too surprising to learn that a concentration one-tenth as high also poses some risk to brain development (personal communication). After all, a 10-fold uncertainty factor is often applied in risk assessments to take account of inter-individual variability in susceptibility. The following paragraphs briefly discuss some of the reasons for the protracted delay in recognizing the dimensions of the problem of so-called "low-level lead toxicity" and the deliberate pace of efforts to eliminate it.

Having been mined for >6000 years, lead is familiar and ubiquitous in modern life. It appears in many common idioms and phrases, such as "leaden sky," "get the lead out," "went over like lead balloon," "it's a lead pipe cinch," "fill him full of lead," etc. Familiarity with lead likely bred complacency regarding its dangers due to a cognitive bias that produced a fallacy in reasoning. Between 1966 and 1970, 43% of the children screened in Baltimore had a blood lead concentration >60 μg/dL (Lin-Fu, 1973), perilously close to the concentration associated with the severe brain injuries resulting from edema and hemorrhage. In New York City, >45% of the children screened in 1969 had a concentration of 40 μg/dL or greater (Lin-Fu, 1972). Because such concentrations were so prevalent among children, they were considered not to pose risk. In other words, the central tendency of the distribution of blood lead concentrations in children was apparently equated with the blood lead concentration that is "physiologically normal." This bias persisted even with the more or less continuous downward shift in population blood lead concentration that occurred during the 1970s and 1980s. When 88% of US children had a blood lead concentration of 10 μg/dL or greater (Mahaffey et al., 1982), it was difficult to arouse concern that a concentration of 10 was unhealthy.

At different times, the effort to reduce lead exposure and its toxicities has clashed with competing economic and social goals, and these other goals have won out, at least in the short-term. For example, in the 1920s, priority was given to applying technological advances to meet the needs of a rapidly growing economy. It was learned that the more powerful automobile engines developed by the automobile industry ran more efficiently if tetraethyl lead was added to gasoline (Rosner and Markowitz, 1985). In the 1980s in the United States, the Supreme Court ruled that policies that excluded women of reproductive age from access to jobs that involved lead exposure violated Title VII of the 1964 Civil Rights Act (as amended by the Pregnancy Discrimination Act of 1978), thereby judging that eliminating gender discrimination in access to employment is more important than fetal protection (Bertin et al., 1992). Sociological factors have also played a role in the importance accorded to lead toxicity. As long as it was viewed as a problem affecting primarily politically and economically disenfranchised subgroups of the population, i.e., poor, minority children, the appropriation of resources to reduce the problem was an uphill battle. Indeed, this is likely one reason for the slow response to the lead contamination of water that occurred recently in Flint, Michigan (Bellinger, 2016).

Lead toxicity can be prevented, and the necessary elements of a national strategy to do so are straightforward (Bellinger et al., 2017). As always, the determining factor will be whether the benefits of achieving this goal are viewed as sufficiently important to warrant investment of the resources that will be required.

REFERENCES

Amato, M.S., Moore, C.F., Magzamen, S., Imm, P., Havlena, J.A., Anderson, H.A., Kanarek, M.S., 2012. Lead exposure and educational proficiency: moderate lead exposure and educational proficiency on end-of-grade examinations. Ann. Epidemiol. 22 (10), 738–743.

Amato, M.S., Magzamen, S., Imm, P., Havlena, J.A., Anderson, H.A., Kanarek, M.S., Moore, C.F., 2013. Early lead exposure (<3 years old) prospectively predicts fourth grade school suspension in Milwaukee, Wisconsin (USA). Environ. Res. 126, 60–65.

Arora, M., Austin, C., Sarrafpour, B., Hernández-Ávila, M., Hu, H., Wright, R.O., Tellez-Rojo, M.M., 2014. Determining prenatal, early childhood and cumulative long-term lead exposure using micro-spatial deciduous dentine levels. PLoS One 9, e97805.

Baghurst, P.A., Tong, S.L., McMichael, A.J., Robertson, E.F., Wigg, N.R., Vimpani, G.V., 1992. Determinants of blood lead concentrations to age 5 years in a birth cohort study of children living in the lead smelting city of Port Pirie and surrounding areas. Arch. Environ. Health 47 (3), 203–210.

Barres, B.A., Smith, S.J., 2001. Neurobiology. Cholesterol-making or breaking the synapse. Science 294 (5545), 1296–1297.

Bellinger, D.C., 2012. A strategy for comparing the contributions of environmental chemicals and other risk factors to neurodevelopment of children. Environ. Health Perspect. 120, 501–507.

Bellinger, D.C., 2016. Lead contamination in Flint: an abject failure to protect public health. N. Engl. J. Med. 374 (12), 1101–1103.

Bellinger, D.C., 2018. Tetraethyl lead, paints, pipes, and other lead exposure routes: the impact on human health. In: Encyclopedia of the Anthropocene.
Bellinger, D., Leviton, A., Sloman, J., 1990. Antecedents and correlates of improved cognitive performance in children exposed in utero to low levels of lead. Environ. Health Perspect. 89, 5–11.
Bellinger, D.C., Chen, A., Lanphear, B.P., 2017. Establishing and achieving national goals for preventing lead toxicity and exposure in children. JAMA Pediatr. 171 (7), 616–618.
Bertin, J.E., Ashford, N.A., Bellinger, D., Landrigan, P.J., Legator, M.S., Mattison, D.R., McBeath, W.H., Rosen, J.F., Stellman, J.M., 1992. The goal: safety and equality. Am. J. Ind. Med. 21 (4), 463–465.
Boutrell, B.B., Nelson, E.J., Emo, B., Vaughn, M.G., Schootman, M., Rosenfeld, R., Lewis, R., 2016. The intersection of aggregate-level lead exposure and crime. Environ. Res. 148, 79–85.
Bowers, T.S., Beck, B.D., 2006. What is the meaning of non-linear dose-response relationships between blood lead concentrations and IQ? Neurotoxicology 27 (4), 520–524.
Bressler, J., Kim, K.A., Chakraborti, T., Goldstein, G., 1999. Molecular mechanisms of lead neurotoxicity. Neurochem. Res. 24, 595–600.
Brubaker, C.J., Schmithorst, V.J., Haynes, E.N., Dietrich, K.N., Egelhoff, J.C., Lindquist, D.M., Lanphear, B.P., Cecil, K.M., 2009. Altered myelination and axonal integrity in adults with childhood lead exposure: a diffusion tensor imaging study. Neurotoxicology 30 (6), 867–875.
Byers, R.K., Lord, E.E., 1943. Late effects of lead poisoning on mental development. Am. J. Dis. Child. 66 (5), 471–494.
Braun, J.M., Kahn, R.S., Froehlich, T., Auinger, P., Lanphear, B.P., 2006. Exposures to environmental toxicants and attention deficit hyperactivity disorder in U.S. children. Environ. Health Perspect. 114 (12), 1904–1909.
Cecil, K.M., Brubaker, C.J., Adler, C.M., Dietrich, K.N., Altaye, M., Egelhoff, J.C., Wessel, S., Elangovan, I., Hornung, R., Jarvis, K., Lanphear, B.P., 2008. Decreased brain volume in adults with childhood lead exposure. PLoS Med. 5 (5), e112.
Cecil, K.M., Dietrich, K.N., Altaye, M., Egelhoff, J.C., Lindquist, D.M., Brubaker, C.J., Lanphear, B.P., 2011. Proton magnetic resonance spectroscopy in adults with childhood lead exposure. Environ. Health Perspect. 119 (3), 403–408.
Chandramouli, K., Steer, C.D., Ellis, M., Emond, A.M., 2009. Effects of early childhood lead exposure on academic performance and behaviour of school age children. Arch. Dis. Child. 94 (11), 844–848.
Choi, W.J., Kwon, H.J., Lim, M.H., Lim, J.A., Ha, M., 2016. Blood lead, parental marital status and the risk of attention-deficit/hyperactivity disorder in elementary school children: a longitudinal study. Psychiatry Res. 236, 42–46.
Claus Henn, B., Schnaas, L., Ettinger, A.S., Schwartz, J., Lamadrid-Figueroa, H., Hernández-Avila, M., Amarasiriwardena, C., Hu, H., Bellinger, D.C., Wright, R.O., Téllez-Rojo, M.M., 2012. Associations of early childhood manganese and lead coexposure with neurodevelopment. Environ. Health Perspect. 120 (1), 126–131.
David, O.J., 1974. Association between lower level lead concentrations and hyperactivity in children. Environ. Health Perspect. 7, 17–25.
Dietrich, K.N., Berger, O.G., Succop, P.A., Hammond, P.B., Bornschein, R.L., 1993. The developmental consequences of low to moderate prenatal and postnatal lead exposure: intellectual attainment in the Cincinnati Lead Study Cohort following school entry. Neurotoxicol. Teratol. 15 (1), 37–44.
Dietrich, K.N., Ris, M.D., Succop, P.A., Berger, O.G., Bornschein, R.L., 2001. Early exposure to lead and juvenile delinquency. Neurotoxicol. Teratol. 23 (6), 511–518.
Dietschy, J.M., Turley, S.D., 2001. Cholesterol metabolism in the brain. Curr. Opin. Lipidol. 12 (2), 105–112.

Evens, A., Hryhorczuk, D., Lanphear, B.P., Rankin, K.M., Lewis, D.A., Forst, L., Rosenberg, D., 2015. The impact of low-level lead toxicity on school performance among children in the Chicago public schools: a population-based retrospective cohort study. Environ. Health 14, 21.

Fergusson, D.M., Horwood, L.J., Lynskey, M.T., 1997. Early dentine lead levels and educational outcomes at 18 years. J. Child Psychol. Psychiatry 38, 471–478.

Fergusson, D.M., Boden, J.M., Horwood, L.J., 2008. Dentine lead levels in childhood and criminal behavior in late adolescence and early adulthood. J. Epidemiol. Community Health 62 (12), 1045–1050.

Froehlich, T.E., Lanphear, B.P., Dietrich, K.N., Cory-Slechta, D.A., Wang, N., Kahn, R.S., 2007. Interactive effects of a DRD4 polymorphism, lead, and sex on executive functions in children. Biol. Psychiatry 62 (3), 243–249.

Froehlich, T.E., Lanphear, B.P., Auinger, P., Hornung, R., Epstein, J.N., Braun, J., Kahn, R.S., 2009. Association of tobacco and lead exposures with attention-deficit/hyperactivity disorder. Pediatrics 124 (6), e1054–63.

Gąssowska, M., Baranowska-Bosiacka, I., Moczydłowska, J., Tarnowski, M., Pilutin, A., Gutowska, I., Strużyńska, L., Chlubek, D., Adamczyk, A., 2016. Perinatal exposure to lead (Pb) promotes Tau phosphorylation in the rat brain in a GSK-3β and CDK5 dependent manner: relevance to neurological disorders. Toxicology 10, 347–349.

Goodlad, J.K., Marcus, D.K., Fulton, J.J., 2013. Lead and attention-deficit/hyperactivity disorder (ADHD) symptoms: a meta-analysis. Clin. Psychol. Rev. 33 (3), 417–425.

Göritz, C., Mauch, D.H., Nägler, K., Pfrieger, F.W., 2002. Role of glia-derived cholesterol in synaptogenesis: new revelations in the synapse-glia affair. J. Physiol. Paris 96 (3–4), 257–263.

Guilarte, T.R., Toscano, C.D., McGlothan, J.L., Weaver, S.A., 2003. Environmental enrichment reverses cognitive and molecular deficits induced by developmental lead exposure. Ann. Neurol. 53 (1), 50–56.

Hubbs-Tait, L., Mulugeta, A., Bogale, A., Kennedy, T.S., Baker, E.R., Stoecker, B.J., 2009. Main and interaction effects of iron, zinc, lead, and parenting on children's cognitive outcomes. Dev. Neuropsychol. 34 (2), 175–195.

Institute for Health Metrics and Evaluation, 2015. GBD Compare, 2015. https://vizhub.healthdata.org/gbd-compare/. Accessed 6 July 2017.

JECFA, 2011. Seventy-third report of the Joint WHO/FAO Expert Committee on Food Additives and Contaminants. In: Technical Report Series No. 960. World Health Organization, Geneva. http://apps.who.int/iris/bitstream/10665/44515/1/WHO_TRS_960_eng.pdf. Accessed 3 July 2017.

Jedrychowski, W., Perera, F., Jankowski, J., Mrozek-Budzyn, D., Mroz, E., Flak, E., Edwards, S., Skarupa, A., Lisowska-Miszczyk, I., 2009. Gender specific differences in neurodevelopmental effects of prenatal exposure to very low lead levels: the prospective cohort study in three-year-olds. Early Hum. Dev. 85 (8), 503–510.

Joo, H., Lim, M.H., Ha, M., Kwon, H.J., Yoo, S.J., Choi, K.H., Paik, K.C., 2017. Secondhand smoke exposure and low blood lead levels in association with attention-deficit hyperactivity disorder and its symptom domain in children: a community-based case-control study. Nicotine Tob. Res. 19 (1), 94–101.

Kaufman, A.S., Zhou, X., Reynolds, M.R., Kaufman, N.L., Green, G.P., Weiss, L.G., 2014. The possible societal impact of the decrease in U.S. blood lead levels on adult IQ. Environ. Res. 132, 413–420.

Kim, Y., Ha, E.H., Park, H., Ha, M., Kim, Y., Hong, Y.C., Kim, E.J., Kim, B.N., 2013. Prenatal lead and cadmium co-exposure and infant neurodevelopment at 6 months of age: the Mothers and Children's Environmental Health (MOCEH) study. Neurotoxicology 35, 15–22.

Kincl, L.D., Dietrich, K.N., Bhattacharya, A., 2006. Injury trends for adolescents with early childhood lead exposure. J. Adolesc. Health 39 (4), 604–606.
Kordas, K., Canfield, R.L., López, P., Rosado, J.L., Vargas, G.G., Cebrián, M.E., Rico, J.A., Ronquillo, D., Stoltzfus, R.J., 2006. Deficits in cognitive function and achievement in Mexican first-graders with low blood lead concentrations. Environ. Res. 100 (3), 371–386.
Lanphear, B.P., Hornung, R., Khoury, J., Yolton, K., Baghurst, P., Bellinger, D.C., Canfield, R.L., Dietrich, K.N., Bornschein, R., Greene, T., Rothenberg, S.J., Needleman, H.L., Schnaas, L., Wasserman, G., Graziano, J., Roberts, R., 2005. Low-level environmental lead exposure and children's intellectual function: an international pooled analysis. Environ. Health Perspect. 113 (7), 894–899.
Lim, S.S., Vos, T., Flaxman, A.D., Danaei, G., Shibuya, K., Adair-Rohani, H., et al., 2012. A comparative risk assessment of burden of disease and injury attributable to 67 risk factors and risk factor clusters in 21 regions, 1990-2010: a systematic analysis for the Global Burden of Disease Study 2010. Lancet 380 (9859), 2224–2260.
Lin, C.C., Chen, Y.C., Su, F.C., Lin, C.M., Liao, H.F., Hwang, Y.H., Hsieh, W.S., Jeng, S.F., Su, Y.N., Chen, P.C., 2013. In utero exposure to environmental lead and manganese and neurodevelopment at 2 years of age. Environ. Res. 123, 52–57.
Lin-Fu, J.S., 1972. Undue absorption of lead among children—a new look at an old problem. N. Engl. J. Med. 286 (13), 702–710.
Lin-Fu, J.S., 1973. Vulnerability of children to lead exposure and toxicity. N. Engl. J. Med. 289 (23), 1229–1233.
Magzamen, S., Amato, M.S., Imm, P., Havlena, J.A., Coons, M.J., Anderson, H.A., Kanarek, M.S., Moore, C.F., 2015. Quantile regression in environmental health: early life lead exposure and end-of-grade exams. Environ. Res. 137, 108–119.
Mahaffey, K.R., Annest, J.L., Roberts, J., Murphy, R.S., 1982. National estimates of blood lead levels: United States, 1976-1980: association with selected demographic and socioeconomic factors. N. Engl. J. Med. 307 (10), 573–579.
Mauch, D.H., Nägler, K., Schumacher, S., Göritz, C., Müller, E.C., Otto, A., Pfrieger, F.W., 2001. CNS synaptogenesis promoted by glia-derived cholesterol. Science 294 (5545), 1354–1357.
Mazumdar, M., Bellinger, D.C., Gregas, M., Abanilla, K., Bacic, J., Needleman, H.L., 2011. Low-level environmental lead exposure in childhood and adult intellectual function: a follow-up study. Environ. Health 10, 24.
McLaine, P., Navas-Acien, A., Lee, R., Simon, P., Diener-West, M., Agnew, J., 2013. Elevated blood lead levels and reading readiness at the start of kindergarten. Pediatrics 131 (6), 1081–1089.
Miranda, M.L., Kim, D., Reiter, J., Galeano, M.A.O., Maxson, P., 2009. Environmental contributors to the achievement gap. Neurotoxicology 30 (6), 1019–1024.
Miranda, M.L., Maxson, P., Kim, D., 2010. Early childhood lead exposure and exceptionality designations for students. Int. J. Child Health Hum. Dev. 3, 77–84.
Needleman, H.L., Gunnoe, C., Leviton, A., Reed, R., Peresie, H., Maher, C., Barrett, P., 1979. Deficits in psychologic and classroom performance of children with elevated dentine lead levels. N. Engl. J. Med. 300 (13), 689–695.
Needleman, H.L., Schell, A., Bellinger, D., Leviton, A., Allred, E.N., 1990. The long-term effects of exposure to low doses of lead in childhood. An 11-year follow-up report. N. Engl. J. Med. 322 (2), 83–88.
Needleman, H.L., Riess, J.A., Tobin, M.J., Biesecker, G.E., Greenhouse, J.B., 1996. Bone lead levels and delinquent behavior. J. Am. Med. Assoc. 275 (5), 363–369.
Needleman, H.L., McFarland, C., Ness, R.B., Fienberg, S.E., Tobin, M.J., 2002. Bone lead levels in adjudicated delinquents. A case control study. Neurotoxicol. Teratol. 24 (6), 711–717.

Nevin, R., 2000. How lead exposure relates to temporal changes in IQ, violent crime, and unwed pregnancy. Environ. Res. 83 (1), 1–22.
Nevin, R., 2007. Understanding international crime trends: the legacy of preschool lead exposure. Environ. Res. 104 (3), 315–336.
Nevin, R., 2009. Trends in preschool lead exposure, mental retardation, and scholastic achievement: association or causation? Environ. Res. 109 (3), 301–310.
Nigg, J.T., Knottnerus, G.M., Martel, M.M., Nikolas, M., Cavanagh, K., Karmaus, W., Rappley, M.D., 2008. Low blood lead levels associated with clinically diagnosed attention-deficit/hyperactivity disorder and mediated by weak cognitive control. Biol. Psychiatry 63, 325–331.
Nigg, J.T., Nikolas, M., Knottnerus, G.M., Cavanagh, K., Friderici, K., 2010. Confirmation and extension of association of blood lead with attention-deficit/hyperactivity disorder (ADHD) and ADHD symptom domains at population-typical exposure levels. J. Child Psychol. Psychiatry 51, 58–65.
Park, J.H., Seo, J.H., Hong, Y.S., Kim, Y.M., Kang, J.W., Yoo, J.H., Chueh, H.W., Lee, J.H., Kwak, M.J., Kim, J., Woo, H.D., Kim, D.W., Bang, Y.R., Choe, B.M., 2016. Blood lead concentrations and attention deficit hyperactivity disorder in Korean children: a hospital-based case control study. BMC Pediatr. 16 (1), 156.
Reuben, A., Caspi, A., Belsky, D.W., Broadbent, J., Harrington, H., Sugden, K., Houts, R.M., Ramrakha, S., Poulton, R., Moffitt, T.E., 2017. Association of childhood blood lead levels with cognitive function and socioeconomic status at age 38 years and with IQ change and socioeconomic mobility between childhood and adulthood. J. Am. Med. Assoc. 317 (12), 1244–1251.
Reyes, J.W., 2007. Environmental policy as social policy: the impact of childhood lead exposure on crime. In: National Bureau of Economic Research Working Paper 13097. http://www.nber.org/papers/w13097. Accessed 5 July 2017.
Reyes, J.W., 2014. Lead exposure and behavior: effects on antisocial and risk behaviors among children and adolescents. In: National Bureau of Economic Research Working Paper 20366. http://www.nber.org/papers/w20366. Accessed 5 July 2017.
Roberts, J.R., Reigart, J.R., Ebeling, M., Hulsey, T.C., 2001. Time required for blood lead levels to decline in nonchelated children. J. Toxicol. Clin. Toxicol. 39 (2), 153–160.
Rosner, D., Markowitz, G., 1985. A 'gift of God'?: the public health controversy over leaded gasoline during the 1920's. Am. J. Public Health 75 (4), 344–352.
Ruff, H.A., Markowitz, M.E., Bijur, P.E., Rosen, J.F., 1996. Relationships among blood lead levels, iron deficiency, and cognitive development in two-year-old children. Environ. Health Perspect. 104 (2), 180–185.
Schneider, J.S., DeKamp, E., 2007. Postnatal lead poisoning impairs behavioral recovery following brain damage. Neurotoxicology 28 (6), 1153–1157.
Schwartz, B.S., Chen, S., Caffo, B., Stewart, W.F., Bolla, K.I., Yousem, D., Davatzikos, C., 2007. Relations of brain volumes with cognitive function in males 45 years and older with past lead exposure. Neuroimage 37 (2), 633–641.
Seo, J., Lee, B.K., Jin, S.U., Park, J.W., Kim, Y.T., Ryeom, H.K., Lee, J., Suh, K.J., Kim, S.H., Park, S.J., Jeong, K.S., Ham, J.O., Kim, Y., Chang, Y., 2014. Lead-induced impairments in the neural processes related to working memory function. PLoS One 9 (8), e105308.
Shah-Kulkarni, S., Ha, M., Kim, B.M., Kim, E., Hong, Y.C., Park, H., Kim, Y., Kim, B.N., Chang, N., Oh, S.Y., Kim, Y.J., Lee, B., Ha, E.H., 2016. Neurodevelopment in early childhood affected by prenatal lead exposure and iron intake. Medicine (Baltimore) 95 (4), e2508.
Skerfving, S., Lofmark, L., Lundh, T., Mikoczy, Z., Stromberg, U., 2015. Late effects of low blood lead concentrations in children on school performance and cognitive functions. Neurotoxicology 49, 114–120.

Solon, O., Riddell, T.J., Quimbo, S.A., Butrick, E., Aylward, G.P., Lou Bacate, M., Peabody, J.W., 2008. Associations between cognitive function, blood lead concentration, and nutrition among children in the central Philippines. J. Pediatr. 152 (2), 237–243.
Surkan, P.J., Zhang, A., Trachtenberg, F., Daniel, D.B., McKinlay, S., Bellinger, D.C., 2007. Neuropsychological function in children with blood lead levels <10 microg/dL. Neurotoxicology 28 (6), 1170–1177.
Surkan, P.J., Schnaas, L., Wright, R.J., Téllez-Rojo, M.M., Lamadrid-Figueroa, H., Hu, H., Hernández-Avila, M., Bellinger, D.C., Schwartz, J., Perroni, E., Wright, R.O., 2008. Maternal self-esteem, exposure to lead, and child neurodevelopment. Neurotoxicology 29 (2), 278–285.
Tamayo, Y.O.M., Téllez-Rojo, M.M., Trejo-Valdivia, B., Schnaas, L., Osorio-Valencia, E., Coull, B., Bellinger, D., Wright, R.J., Wright, R.O., 2017. Maternal stress modifies the effect of exposure to lead during pregnancy and 24-month old children's neurodevelopment. Environ. Int. 98, 191–197.
Taylor, M.P., Forbes, M.K., Opeskin, B., Parr, N., Lanphear, B.P., 2016. The relationship between atmospheric lead emissions and aggressive crime: an ecological study. Environ. Health 15, 23.
Téllez-Rojo, M.M., Bellinger, D.C., Arroyo-Quiroz, C., Lamadrid-Figueroa, H., Mercado-García, A., Schnaas-Arrieta, L., Wright, R.O., Hernández-Avila, M., Hu, H., 2006. Longitudinal associations between blood lead concentrations lower than 10 microg/dL and neurobehavioral development in environmentally exposed children in Mexico City. Pediatrics 118 (2), e323–30.
Tong, S., McMichael, A.J., Baghurst, P.A., 2000. Interactions between environmental lead exposure and sociodemographic factors on cognitive development. Arch. Environ. Health 55 (5), 330–335.
US Centers for Disease Control and Prevention, 2012. Low Level Lead Exposure Harms Children: A Renewed Call for Primary Prevention. https://www.cdc.gov/nceh/lead/acclpp/final_document_030712.pdf. Accessed 3 July 2017.
Wang, H.L., Chen, X.T., Yang, B., Ma, F.L., Wang, S., Tang, M.L., Hao, M.G., Ruan, D.Y., 2008. Case-control study of blood lead levels and attention deficit hyperactivity disorder in Chinese children. Environ. Health Perspect. 116 (10), 1401–1406.
Weiss, B., Clarkson, T.W., Simon, W., 2002. Silent latency periods in methylmercury poisoning and in neurodegenerative disease. Environ. Health Perspect. 110 (Suppl. 5), 851–854.
Weston, H.I., Weston, D.D., Allen, J.L., Cory-Slechta, D.A., 2014. Sex-dependent impacts of low-level lead exposure and prenatal stress on impulsive choice behavior and associated biochemical and neurochemical manifestations. Neurotoxicology 44, 169–183.
White, L.D., Cory-Slechta, D.A., Gilbert, M.E., Tiffany-Castiglioni, E., Zawia, N.H., Virgolini, M., Rossi-George, A., Lasley, S.M., Qian, Y.C., Basha, M.R., 2007. New and evolving concepts in the neurotoxicology of lead. Toxicol. Appl. Pharmacol. 225, 1–27.
Wright, J.P., Dietrich, K.N., Ris, M.D., Hornung, R.W., Wessel, S.D., Lanphear, B.P., Ho, M., Rae, M.N., 2008. Association of prenatal and childhood blood lead concentrations with criminal arrests in early adulthood. PLoS Med. 5 (5), e101.
Wright, R.O., Tsaih, S.W., Schwartz, J., Wright, R.J., Hu, H., 2003a. Association between iron deficiency and blood lead level in a longitudinal analysis of children followed in an urban primary care clinic. J. Pediatr. 142 (1), 9–14.
Wright, R.O., Hu, H., Silverman, E.K., Tsaih, S.W., Schwartz, J., Bellinger, D., Palazuelos, E., Weiss, S.T., Hernandez-Avila, M., 2003b. Apolipoprotein E genotype predicts 24-month Bayley scales infant development score. Pediatr. Res. 54 (6), 819–825.
Yuan, W., Holland, S.K., Cecil, K.M., Dietrich, K.N., Wessel, S.D., Altaye, M., Hornung, R.W., Ris, M.D., Egelhoff, J.C., Lanphear, B.P., 2006. The impact of early childhood lead exposure on brain organization: a functional magnetic resonance imaging study of language function. Pediatrics 118 (3), 971–977.

FURTHER READING

Bellinger, D.C., Matthews-Bellinger, J.A., Kordas, K., 2016. A developmental perspective on early-life exposure to neurotoxicants. Environ. Int. 94, 103–112.

Environmental Protection Agency of Denmark, 2017. Exposure of children and unborn children to selected chemical substances. In: Survey of Chemical Substances in Consumer Products No. 158, April 2017. http://www2.mst.dk/Udgiv/publications/2017/04/978-87-93529-84-7.pdf. Accessed 5 July 2017.

Haynes, E.N., Kalkwarf, H.J., Hornung, R., Wenstrup, R., Dietrich, K., Lanphear, B.P., 2003. Vitamin D receptor Fok1 polymorphism and blood lead concentration in children. Environ. Health Perspect. 111 (13), 1665–1669.

Hopkins, M.R., Ettinger, A.S., Hernández-Avila, M., Schwartz, J., Téllez-Rojo, M.M., Lamadrid-Figueroa, H., Bellinger, D., Hu, H., Wright, R.O., 2008. Variants in iron metabolism genes predict higher blood lead levels in young children. Environ. Health Perspect. 116 (9), 1261–1266.

Karwowski, M.P., Just, A.C., Bellinger, D.C., Jim, R., Hatley, E.L., Ettinger, A.S., Hu, H., Wright, R.O., 2014. Maternal iron metabolism gene variants modify umbilical cord blood lead levels by gene-environment interaction: a birth cohort study. Environ. Health 13, 77.

Kelada, S.N., Shelton, E., Kaufmann, R.B., Khoury, M.J., 2001. Delta-aminolevulinic acid dehydratase genotype and lead toxicity: a HuGE review. Am. J. Epidemiol. 154 (1), 1–13.

Kotok, D., 1972. Development of children with elevated blood lead levels: a controlled study. J. Pediatr. 80 (1), 57–61.

Leggett, R.W., 1993. An age-specific kinetic model of lead metabolism in humans. Environ. Health Perspect. 101 (7), 598–616.

O'Flaherty, E.J., 1998. A physiologically based kinetic model for lead in children and adults. Environ. Health Perspect. 106 (Suppl. 6), 1495–1503.

Wang, Z., Claus Henn, B., Wang, C., Wei, Y., Su, L., Sun, R., Chen, H., Wagner, P.J., Lu, Q., Lin, X., Wright, R., Bellinger, D., Kile, M., Mazumdar, M., Tellez-Rojo, M.M., Schnaas, L., Christiani, D.C., 2017. Genome-wide gene by lead exposure interaction analysis identifies UNC5D as a candidate gene for neurodevelopment. Environ. Health 16 (1), 81.

CHAPTER TWO

Neurodevelopmental Effects of Mercury

Cláudia S. Oliveira*,[1], Pablo A. Nogara*, Daniel M.P. Ardisson-Araújo*,†, Michael Aschner‡, João B.T. Rocha§, José G. Dórea¶

*Programa de Pós-graduação em Ciências Biológicas: Bioquímica Toxicológica, Centro de Ciências Naturais e Exatas, Universidade Federal de Santa Maria, Santa Maria, RS, Brazil
†Laboratório de Virologia de Insetos, Departamento de Bioquímica e Biologia Molecular, Centro de Ciências Naturais e Exatas, Universidade Federal de Santa Maria, Santa Maria, RS, Brazil
‡Department of Molecular Pharmacology, Albert Einstein College of Medicine, Bronx, New York, NY, USA
§Departamento de Bioquímica e Biologia Molecular, Centro de Ciências Naturais e Exatas, Universidade Federal de Santa Maria, Santa Maria, RS, Brazil
¶Professor Emeritus, Faculdade de Ciências da Saúde, Universidade de Brasília, Brasília, DF, Brazil
[1]Corresponding author: e-mail address: claudia.bioquimica@yahoo.com.br

Contents

1. INTRODUCTION

Mercury is a metal with a unique chemical characteristic, namely, it is silvery liquid at room temperature. Its physical properties, including low viscosity, high density, excellent electrical conductance, and reflective surface characteristics, drew the attention of researchers and industries to the application of Hg (Clarkson and Magos, 2006). In this context, there are records of Hg usage as early as ancient China, where cinnabar (HgS) was used to

Advances in Neurotoxicology, Volume 2
ISSN 2468-7480
https://doi.org/10.1016/bs.ant.2018.03.005

make a red dye. On the American continent, Hg was brought by the Spanish Galleons to be employed in the process of gold and silver extraction (Clarkson and Magos, 2006; Oliveira et al., 2017a). Nowadays, Hg is used as part of some medical devices, as a preservative of vaccines, and as a catalyst in caustic soda and chloride factories. In many parts of the world, Hg is still a constituent of dental amalgams (Horowitz et al., 2014). Of particular environmental concern, in countries, such as Brazil, Venezuela, Ecuador, Indonesia, Tanzania, and Ghana, Hg is still used in artisanal small-scale gold mining (Kristensen et al., 2014), and this activity has apparently increased in the last year (Branco et al., 2017). Another common practice in the latin and African countries is the use of Hg in ritualistic and spiritual religions, the adepts use Hg in amulets, oral ingestion, Santerias, sprinkling it on the floor of a home or vehicle, burning it in a candle, or even mixing Hg with soaps and detergents (Masur, 2011). The environmental levels of Hg increased considerably with anthropogenic intervention (Fig. 1). In fact, it is estimated that the atmospheric burden of Hg nowadays is 300–500% higher than that in the precolonial period (AMAP/UNEP, 2013; De Simone et al., 2016).

In this context, the toxicology of Hg has been extensively studied, and it has been established that Hg does not have biochemical and/or physiologic functions in living organisms (WHO, 2007). In fact, exposure to even low levels of Hg can be toxic, causing molecular, cellular, developmental and

Fig. 1 Natural and anthropogenic sources of Hg emission in the environment. The volcanic and erosion processes are natural sources of Hg. The process of gold extraction, the uncontrolled release of waste from industry and medical devices increase the emission/levels of Hg in the environment almost exponentially. The red dotted line represents the emission of Hg.

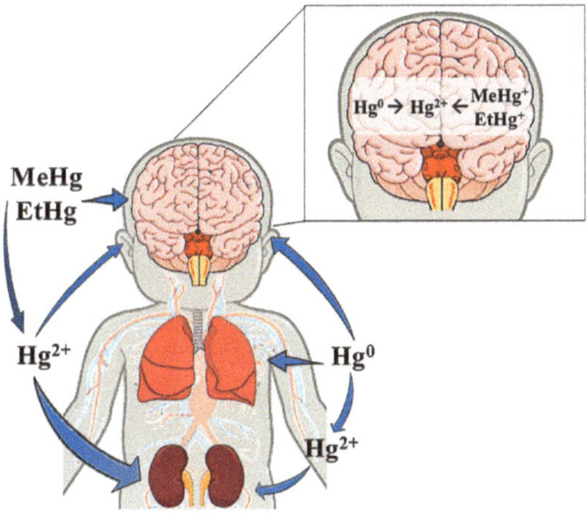

Fig. 2 Main targets of Hg different chemical species. The main interconversions among the Hg species are demonstrated. The thickness of the arrows is proportional to the amount of Hg that reaches a given organ.

behavioral alterations in different organisms (Chang et al., 2017; Chauhan et al., 2017; Fujimura and Usuki, 2017; Mora-Zamorano et al., 2017), including humans (Choi et al., 2017; Junior et al., 2017). The toxic effects are directly influenced by the chemical form of Hg: inorganic elemental Hg (Hg^0), ionic inorganic mercury (Hg^{2+}) and organic mercury (Hg atom bound to one carbon atom, for example, CH_3Hg^+ and $C_2H_5Hg^+$) (Fig. 2).

Although immediately after intoxication, Hg chemical species accumulate in different target organs, multiple studies have demonstrated that they can reach the brain causing the accumulation of Hg^{2+} (Fig. 2). The neurotoxic effects of Hg are usually much more pronounced in the immature nervous system. This review will briefly discuss the neurodevelopmental toxic effects caused by mercury exposure.

2. INORGANIC ELEMENTAL MERCURY

Inorganic elemental Hg (Hg^0) is found as a liquid at room temperature, but due to its low vapor pressure (0.2613 Pa at 25 °C), it volatilizes slowly and increasingly with rise in temperature (Gonzalez-Raymat et al., 2017; Huber et al., 2006; Lee et al., 2009; Risher et al., 2003). As a vapor, it can be inhaled and the first target is the respiratory tract (Oz et al., 2012) followed by the nervous and renal systems (Oliveira et al., 2017a; Reinhardt, 1992). Hg^0 can easily cross plasma membranes, reaching the intracellular milieu due to its high diffusibility and lipid solubility (Clarkson, 1972). In the blood, Hg^0

can be oxidized to Hg^{2+} in erythrocytes by the enzyme catalase (Halbach and Clarkson, 1978). The remaining Hg^0 can reach the brain and once inside the brain cells, it slowly is oxidized to Hg^{2+} (Eide and Syversen, 1983). The oxidized form (Hg^{2+}) has a great affinity for protein thiol (–SH) and selenol (–SeH) groups, causing several alterations that will be further discussed in Section 6.

Hg^0 is found in medical devices such as thermometers, barometers, sphygmomanometers (Risher et al., 2003), and in dental amalgams (Reinhardt, 1992); it is also commonly used in fluorescent lamps (Hu and Cheng, 2012). The main anthropogenic source of Hg^0 is artisanal small-scale gold mining, and data from UNEP (United Nations Environment Programme) estimate that 727 tons of mercury were released by gold extraction in 2010 (AMAP/UNEP, 2013). Occupational exposure to elemental Hg is common among workers that inappropriately handled the elemental mercury. Contaminated individuals presented symptoms such as skin rash, pruritus, myalgia, sleep disturbance fatigue, insomnia, nightmares, anxiety disorder and elevated hair, urine and blood Hg levels (Do et al., 2017; Mostafazadeh et al., 2013; Noble et al., 2016; Schuurs, 1999). In addition, there are cases of accidental or even intentional contamination with Hg^0 (Florentine and Sanfilippo, 1991; Gao et al., 2017; Oz et al., 2012).

Intrauterine exposure to Hg^0 has been reported to cause Hg accumulation and neurobehavioral changes in metallothionein-null mice (Yoshida et al., 2002, 2005) and in rats (Danielsson et al., 1993; Morgan et al., 2002). Metallothioneins are low molecular mass proteins rich in cysteine. These proteins are responsible by body essential metal homeostasis, as well as they are involved in the process of toxic metals detoxification. The cysteines present in the protein bind with the toxic metals, for example, mercury, forming a less reactive complex (Shen et al., 2005; Weeb and Cain, 1982).

Moreover, Vimy et al. (1990) used pregnant sheep as the experimental model and demonstrated that the Hg^0 released from dental amalgam fillings crossed the placental barrier and accumulated in the fetus' organs. Despite these experimental findings, Watson et al. (2012, 2013) did not find a correlation between mothers' dental amalgam and the development of their children at ages 9 and 30 months and 5 years in the Seychelles cohort.

3. IONIC INORGANIC MERCURY

The ionic inorganic form of Hg, Hg^{2+}, can be formed from the oxidation of elemental mercury or by the cleavage of the carbon–mercury bound in the organic mercury molecules. The process of carbon–mercury bond cleavage is very slow for $MeHg^{+}$, but it is much faster for $EtHg^{+}$ (Burbacher et al., 2005; Dórea et al., 2013). These chemical conversions occur in both the environment and in the body (Oliveira et al., 2017a).

Presently, populations are exposed to Hg^{2+} mainly through the ingestion of contaminated water and food (Park and Zheng, 2012). Another reported means of Hg^{2+} exposure is the addition of Hg^{2+} to cosmetic creams for skin lightening and acne treatments (Chan, 2011; Copan et al., 2015). Although it is known that Hg^{2+} is poorly absorbed by the intestine and skin cells, the toxic effects of the exposure to inorganic Hg^{2+} cannot be neglected.

The primary target of Hg^{2+} is the kidney (Zalups, 2000). Some studies have demonstrated that a few hours after intravenous Hg^{2+} injection, approximately 50% of the injected dose is found in the kidneys (Oliveira et al., 2017b; Zalups, 1993; Zalups and Barfuss, 1995). Hg cations are not found free in the bloodstream; rather, they are bound to thiol (–SH)-containing molecules such as albumin, cysteine, and glutathione (GSH) (Bridges and Zalups, 2017; Section 6). The high affinity of –SH-containing molecules for Hg^{2+} (Fuhr and Rabenstein, 1973; Rabenstein and Isab, 1982), particularly for cysteine, which is abundant in the blood, will direct Hg-cysteine complexes to the kidneys (Section 6). Interestingly, the fate of Hg^{2+} in the soil is also dictated by the thiol present in the dissolved organic matter (Leterme et al., 2014; Skyllberg, 2008). Hg bound to two cysteines (Cys-Hg-Cys) gets entry into the kidney's proximal tubular cells as a mimic of the amino acid cystine (Cys-Cys) by System b^{0+}, OAT1 and OAT3 transporters (Bridges and Zalups, 2017; Bridges et al., 2004; Zalups and Ahmad, 2004). Due to its electrical polarity, Hg^{2+} poorly crosses the plasma membrane or the blood–brain and placental barriers. However, some studies with experimental animals demonstrated that some of Hg^{2+} crosses the placental barrier and accumulates in the fetus's organs, which can be harmful to the animal's development (Bernard et al., 1992; Danielsson et al., 1984; Oliveira et al., 2012, 2015). Moreover, several studies have demonstrated the neurotoxic effects of inorganic mercury in developing mammalian brains (Franciscato et al., 2009; Moraes-Silva et al., 2014; Peixoto et al., 2007; Sahin et al., 2017). Although there are studies showing that in small quantities Hg^{2+} crosses the placental and blood–brain barriers (BBBs) (Alves et al., 2017), the mechanism by which it crosses these barriers is not well understood. Amino acid transporters, similar to those that transport Cys-Hg-Cys complexes in the kidneys, are probably involved.

Of toxicological significance, the access of Hg^{2+} into the immature mammalian brain is higher that observed in the adult brain. In early life, the BBB is not completely developed and is leakier to charged toxic metals (Johansson, 1990). Rat pups exposed subcutaneously to Hg^{2+} ($HgCl_2$) during the second week of postnatal life showed behavioral alterations such as loss of motor coordination and delay in the development of the geotaxis reflex (Moraes-Silva et al., 2014; Peixoto et al., 2007). These alterations were observed even after the cessation of $HgCl_2$ exposure (Franciscato et al., 2009). Huang et al. (2014) demonstrated that mice pups exposed to Hg^{2+} accumulate this metal in the cerebral neuron bodies.

Cases of human Hg^{2+} direct contamination causing neurotoxic effects do not exist. However, there is evidence that other chemical species of mercury (organic and elemental) are converted to Hg^{2+} and trapped within the brain (Burbacher et al., 2005; Ishitobi et al., 2010; Yamamoto et al., 1986). Based on a systematic review, Rooney (2014) concluded that the half-life of Hg^{2+} in the brain varies from years to decades. Of particular neurotoxicological significance, data have been published linking Hg accumulation in the brain with neurodegenerative diseases, namely, Alzheimer's disease (Chakraborty, 2017; Mutter et al., 2004; Walach et al., 2014), amyotrophic lateral sclerosis (Johnson and Atchison, 2009; Pamphlett and Jew, 2013, 2014) and autism spectrum disorders (ASDs) (Bjørklund et al., 2017a,b), as will be discussed in Section 5.

4. ORGANIC MERCURY

The organic mercury form is represented by a Hg atom bound to a C atom. The major concern of the human population on the subject of the organic mercury species exposure is related to methylmercury (CH_3Hg^+; $MeHg^+$) and ethylmercury ($C_2H_5Hg^+$; $EtHg^+$). $MeHg^+$ and $EtHg^+$ are potent neurotoxic agents. These organic mercury molecules form complexes with low molecular mass thiols (LMM-SH, e.g., cysteine and GSH, see Section 6 for more details) that can easily cross the plasma membrane and the blood–brain barrier as well as the placental barrier (Dórea et al., 2013; Farina et al., 2011a,b).

Exposure to $MeHg^+$ is related to eating habits, i.e. populations consuming a diet based on seafood/fish from contaminated sites are more likely to be exposed to MeHg (Bosch et al., 2016; Burger et al., 2005; Gochfeld and Burger, 2005; Montero-Alvarez et al., 2014; Schmidt et al., 2013). Another source of dietary MeHg is rice from contaminated areas (Brombach et al., 2017; Cui et al., 2017; Li et al., 2016a; Wu et al., 2018). In this context, longitudinal studies carried out with the Seychelles (Davidson et al., 1998, 2006,

2008a,b, 2010; Myers et al., 2009) and the Faroe Islands (Debes et al., 2006; Grandjean et al., 1997, 2003, 2005) populations confirmed the finds about the developmental deficts in subjects exposed to MeHg during periods of intense brain growth. Based on it, expert panel from National Academy of Sciences recommended that Hg levels should not exceed 5.0 μg/L in whole blood and 1.0 μg/g in hair, which corresponds to a reference-dose of 0.1 μg/kg body-weight/day (Martinez-Finley et al., 2015).

Regarding $EtHg^+$, the main source of exposure is thimerosal, a preservative in vaccines and other medications. After the immunization with thimerosal-containing vaccines (TCVs), this molecule is metabolized, releasing $EtHg^+$ (Dórea, 2017a,b; Dórea et al., 2013). There are several controversies about the use of thimerosal to preserve vaccines. On the one hand, there are scientists from the pharmaceutical industry supporting the use of thimerosal, claiming that the levels used are within safe limits. On the other hand, there are concerns that the cumulative number of shots during the critical periods of brain development may cause moderate to high dose exposure, and indeed these concerns have led rich countries to ban its use for young children. Independently, studies have questioned the double standard of using thimerosal in pediatric vaccines for children in less developed countries (Dórea, 2017a,b). From a toxicological point of view, the intentional exposure of humans or animals to any form of mercury is illogical.

The neurotoxic effects of the organic forms of mercury became evident after two outbreaks of human contamination. The first case is known as Minamata disease, which took place in Minamata Bay, Japan. For approximately 20 years (mid 1950s to mid 1970s) a factory released large amounts of Hg^{2+} into Minamata Bay (Hachiya, 2012; Harada, 1995; Kudo and Turner, 1999). Hg^{2+} was used as the catalyst in the synthesis of acetaldehyde, but a small amount of $MeHg^+$ (about 1%) was formed as byproduct, which may have increased the extent of human and environmental contamination (Kudo and Turner, 1999). Once in the aquatic environment, Hg is methylated by bacteria and incorporated in the food chain, leading to the biomagnification and bioaccumulation processes, i.e., predatory old fish accumulated more $MeHg^+$ than herbivorous or predatory juvenile fish (Oliveira et al., 2017a). As the main source of protein for the Minamata population was fish, they were consequently exposed to high levels of $MeHg^+$. Although the adults showed low to moderate neurologic symptoms, the infants, who were exposed indirectly to $MeHg^+$ (in utero or via breast milk), presented severe learning and locomotor deficits (Ekino et al., 2007; Harada, 1995).

The second major case of organic Hg contamination happened in Iraq (1971–72), where farmers and their families, inadvertently prepared and ate homemade bread made with seeds that had been treated with a fungicide containing organic Hg ($MeHg^+$ and $EtHg^+$). Approximately 6500 people were poisoned and had severe neurological symptoms, and about 500 died. As in Minamata, infants were much more incapacitated neurologically than adults (Bakir et al., 1973; Shikara and Al Dabbagh, 1973).

However, human contamination with $EtHg^+$ (in TCVs) is silent, when compared with the poisonous outbreaks of organic Hg. As mentioned before, infants and children in less developed countries are constantly exposed to $EtHg^+$ after immunization with TCVs during vulnerable periods of brain development (Dórea, 2017a,b). Based on this, several epidemiological studies have indicated adverse effects of TCVs in the normal development of children (Geier and Geier, 2006; Marques et al., 2014, 2016a; Mrozek-Budzyn et al., 2012).

5. NEURODEVELOPMENTAL TOXICITY: Hg VERSUS ASD

The brain is a complex organ that presents non-uniform development, i.e. it occurs in multiple stages. The cortex, hippocampus, and cerebellum are examples of brain structures that show late development (Andersen, 2003). Consequently, the vulnerability of different brain areas to neurotoxic insults will be different depending on the developmental period of exposure. In addition, symptoms/effects that are not present during the exposure period may appear later in life. For example, epidemiological evidence has been accumulated supporting the proposal that environmental insults in utero and/or in early postnatal life in association with a genetic predisposition may facilitate the development of neurodegenerative diseases in adult life (Grandjean and Landrigan, 2006; Modgil et al., 2014).

Among the environmental insults, metals such as cadmium, lead, manganese, aluminum, and mercury are known to be important neurotoxicants. Neurodevelopmental exposure to mercury, in utero and/or in breast milk or even in vaccines, has been associated with learning/locomotor deficits, low IQ, attention-deficit disorder, mental retardation and ASD (Debes et al., 2016; Drum, 2009; Geier et al., 2018a; Jeong et al., 2017; Johansson et al., 2007).

ASD is a group of heterogeneous dysfunctions in social interaction and communication that lead to the appearance of unusual and repetitive patterns of behavior (American Psychiatric Association, 2013). The causes of ASD are usually attributed to complex genetic modifications, but the exact genes involved in ASD development are not fully known. A recent review from Fakhoury (2018) highlighted that mutations in genes involved in the pre- and post-synaptic binding proteins, neuronal cell adhesion, and GABAA receptor, among others, may be involved in the development of ASD. However, to quote Dr. Fakhoury, "the characterization of the neurobiological mechanisms by which common genetic risk variants might operate to give rise to ASD symptomatology has proven to be far more difficult than expected."

As mentioned above, Hg has emerged as a putative environmental associated factor for ASD development. In fact, several studies have demonstrated that ASD patients presented elevated levels of Hg in blood and hair when compared to the same age control subjects (see Table 1 and the elegant review by Kern et al., 2016). The difference in the mercury values in the studies mentioned in Table 1 can probably be explained by different protocols used to quantify mercury, dietary habits of the subjects, and the occurrence of environmental mercury, among others.

Hg is a neurotoxic agent, and an immature organism can be exposed in utero and/or via breast milk (if the mother is/was exposed to metal) or via TCVs, food (fish and rice), and from environmental contamination (for example, children that live close to artisanal small-scale gold mining areas (Ruggieri et al., 2017)). Once Hg reaches the brain, it causes several alterations, including oxidative stress, enzyme inhibition, depletion of thiol molecules, DNA and gene expression alteration, excitotoxicity, and cell apoptosis. These alteration can permanently change the normal brain connectivity either by changing the migration/connection between cells or by decreasing the number of cells (Ceccatelli et al., 2010; Farina et al., 2011a,b; Garrecht and Austin, 2011; Oliveira et al., 2017a). Hg exposure can also activate the immune system (Brenden et al., 2001; Hultman et al., 1998), and several papers have indicated that the maternal and/or fetal activation of the brain immune system can be interconnected with the neurodevelopmental deficit observed in ASD (Bilbo et al., 2018; Estes and McAllister, 2016; Fernández de Cossío et al., 2017; Meltzer and Van De Water, 2017).

Further support for the potential role Hg in the etiology of ADS is the fact that males are more affected than females (Loomes et al., 2017; Palmer et al., 2017). Several studies have demonstrated that males are

Table 1 Levels of Hg in ASD patients: examples of case–control studies.

Hg levels	Sample	Country	References
Control: 0.30 μg/g Autism: 4.50 μg/g	Hair	Kuwait	Fido and Al-Saad (2005)
Control: 0.37 μg/g Autism: 1.61 μg/g	Hair	India	Priya and Geetha (2011)
Control: 2.87 μg/g Autism: 4.02 μg/g	Nail	India	Priya and Geetha (2011)
Control: 0.07 μg/g Autism: 0.15 μg/g	Baby teeth	United States	Adams et al. (2007)
Control: 0.25 μg/g Autism: 0.39 μg/g	Hair	Egypt	Mohamed et al. (2015)
Control: 0.12 μg/g Autism: 0.79 μg/g	Hair	Egypt	El-baz et al. (2010)
Control: 2.71 μg/g Autism: 3.66 μg/g	Red blood cells	Saudi Arabia	El-Ansary et al. (2017)
Control: 13.47 μg/L Autism: 55.59 μg/L	Blood	China	Li et al. (2018)
Control: 12.08 μg/L Autism: 32.90 μg/L	Blood	Egypt	Khaled et al. (2016)
Control: 0.60 μg/g Autism: 7.00 μg/g	Hair	Oman	Hodgson et al. (2014)

more sensitive than females to mercury exposure (Ekstrand et al., 2010; Gimenez-Llort et al., 2001; Malagutti et al., 2009; Marques et al., 2015; Mckeown-Eyssen et al., 1983; Rossi et al., 1997), which corroborates the hypothesis that Hg may be involved in ASD etiology. The differential susceptibility of males to toxic agents, when compared to females, is elegantly discussed by McCarthy and Wright (2017). They hypothesized that during the process of brain masculinization (exposure to high levels of testosterone) the brain is more susceptible to immune activation and/or neurotoxins.

6. POTENTIAL MOLECULAR TARGET OF Hg AND MECHANISM OF NEUROTOXICITY

6.1 Thiol (–SH) and Selenol (–SeH) Groups

The electrophilic forms of mercury (E^+Hg, i.e., Hg^{2+}, $MeHg^+$ and $EtHg^+$) can interact with different ligands (functional groups) found in the cell milieu and in the natural organic matter found in the environment (Table 2). However, the affinity of E^+Hg for soft nucleophiles (SNu^-) is orders of magnitude higher than that for hard nucleophiles (HNu^-) (Table 2). Thus, the fate of Hg^0 and E^+Hg forms in the environment and inside living cells is greatly influenced by the interaction of Hg^0 or E^+Hg with thiol/thiolate groups (–SH/S^-) ubiquitously found in complex biological systems (Farina et al., 2011a,b, 2017; Gu et al., 2011; LoPachin et al., 2012; Loux, 1998; Skyllberg et al., 2006). Another important functional group that can bind E^+Hg with high affinity

Table 2 Formation constants of $MeHg^+$ or Hg^{2+} with nucleophilic centers found in biomacromolecules.

Functional group	$MeHg^+$	Hg^{2+}
R-COOH	≈2.5–3.0	≈4–7
R-$(COOH)_2$	–	≈5–9
R-$(COOH)_3$ (citrate)	–	≈11
R-NH_2	≈7–8	≈4–5
R-$(NH_2)_2$	–	≈11
R_2-NH	≈6.7	≈4–5
R-S-R	≈7–8	≈6–12
R-SH	≈15–26	≈40–50
R-SeH	≈16–18	≈50–60

The nucleophilic centers can be categorized in hard (HNu^-) or soft (SNu^-) centers. –SeH is the softest of the SNu^- centers found in living cells, followed by –SH. The –NH and –COOH-containing molecules are HNu^- and the centers containing O are harder than those containing N (Hancock and Martell, 1996; Martin, 2002).
The values represent the log of the constant and a high value indicates that a given complex has high thermodynamic stability. The values were obtained from Stricks and Kolthoff (1953), Rabenstein et al. (1974), Rabenstein and Fairhurst, (1975), Rabenstein (1978), Jawaid and Ingman (1981), Arnold and Canty, (1983), Arnold et al. (1986), Berthon (1995), Seby et al. (2001), Anderegg et al. (2003), Mousavi (2011), Liem-Nguyen et al. (2017).

is the selenol/selenolate (–SeH/–Se^-) groups found in a small number of selenoproteins (Labunskyy et al., 2014). Although selenol/selenolate groups have a stronger affinity for E^+Hg than their sulfur-containing analogs (Bjørklund et al., 2017c; Farina et al., 2011a,b), the number of studies analyzing the constant of formation are limited (Table 2). Despite this limitation, the affinity of –SeH/–Se^- for E^+Hg is possibly one to four orders of magnitude higher than that of –SH/–S^- analogs (Farina et al., 2011a,b; Khan and Wang, 2009; Rabenstein et al., 1986; Sugiura et al., 1976).

Exposure to $MeHg^+$ occurs via fish consumption, whereas $EtHg^+$ is intramuscularly injected in the chemical form of thimerosal in vaccines. Exposure to Hg^{2+} can occur after the absorption of Hg^0 or occasionally via ingestion of contaminated food and water. For instance, about 2–10% of the total Hg found in fish meat can be present as inorganic Hg (Montero-Alvarez et al., 2014).

The strong affinity of E^+Hg for –SH and –SeH groups implies that nearly all the E^+Hg will be bound to one of these groups. Since –SH is much more abundant than –SeH (Rocha et al., 2017), the majority of E^+Hg forms will be complexed with –SH groups. In fact, $MeHg^+$ can bind to low molecular mass thiol molecules (LMM-SH, e.g., cysteine, GSH) or to cysteinyl residues in proteins or high molecular mass thiol molecules (HMM-SH). Some important complexes of E^+Hg forms with abundant LMM-SH molecules found in the cellular environment are shown in Fig. 3.

Exposure to $MeHg^+$ usually occurs via fish consumption (Bosch et al., 2016; Burger et al., 2005; Gochfeld and Burger, 2005; Montero-Alvarez et al., 2014; Schmidt et al., 2013), where the majority of $MeHg^+$ is in the complexed form with cysteine (MeHg-S-Cys) (Kuwabara et al., 2007). After the digestion of fish proteins containing the MeHg-S-Cys complex, the mercurial part can be absorbed by distinct mechanisms that are not yet well characterized. The interaction of E^+Hg with LMM-SH is critical for its distribution in the body of vertebrates and determined by exchange reactions of the type:

$$R_1\text{-S-HgE}^+ + R_2SH \leftrightarrow R_1SH + R_2\text{-S-HgE}^+$$

where R_1-SH and R_2-SH can be hypothetically either LMM-SH or HMM-SH (for a detailed discussion see Farina et al., 2011a,b, 2017; Rabenstein, 1978; Rabenstein and Fairhurst, 1975; Rabenstein and Isab, 1982; Rabenstein and Reid, 1984). A schematic representation of the fate of dietary $MeHg^+$ or

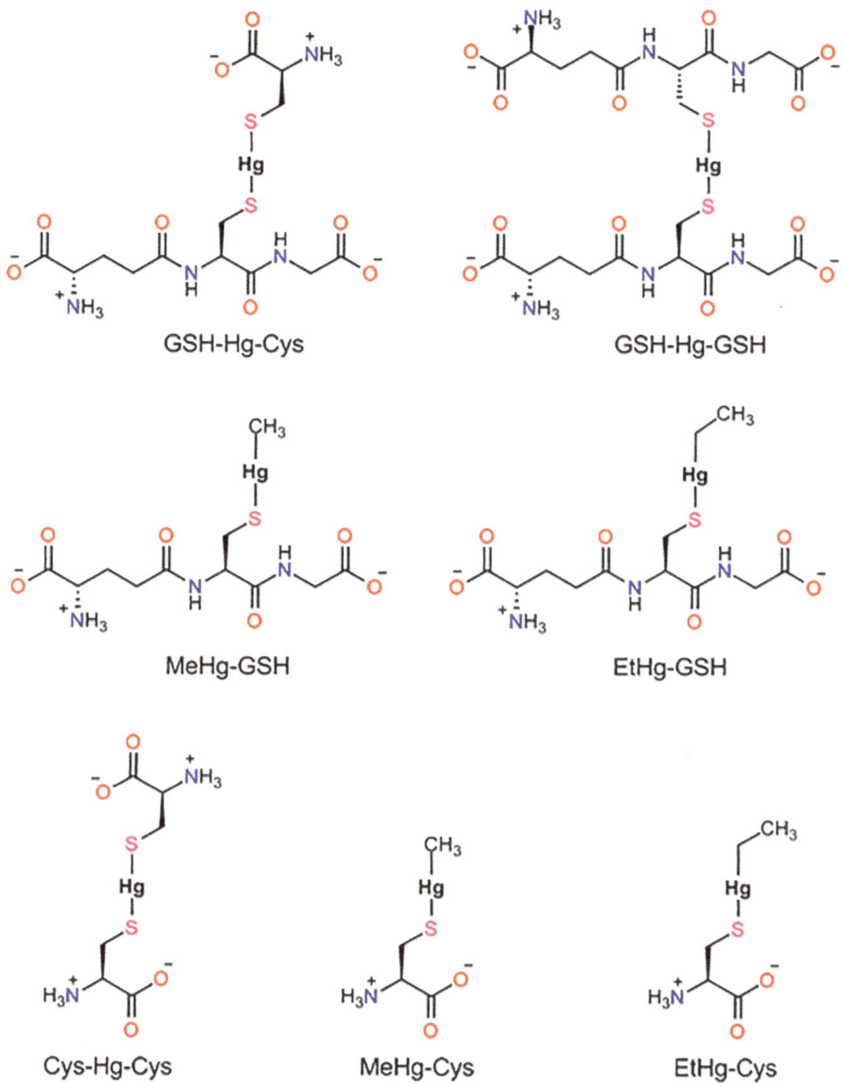

Fig. 3 Low molecular mass (LMM-SH) complexes of cysteine and/or glutathione with electrophile forms of mercury (E^+Hg, i.e., $MeHg^+$; $EtHg^+$ or Hg^{2+}). The formation of the LMM-S-E^+Hg complexes depicted here is of crucial importance for the preferential distribution of E^+Hg forms to their target organs (Bridges and Zalups, 2005, 2010, 2017; Farina et al., 2011a,b; Zalups, 2000). The distribution of $EtHg^+$–S– complexes has been rarely studied in the literature (Dórea et al., 2013).

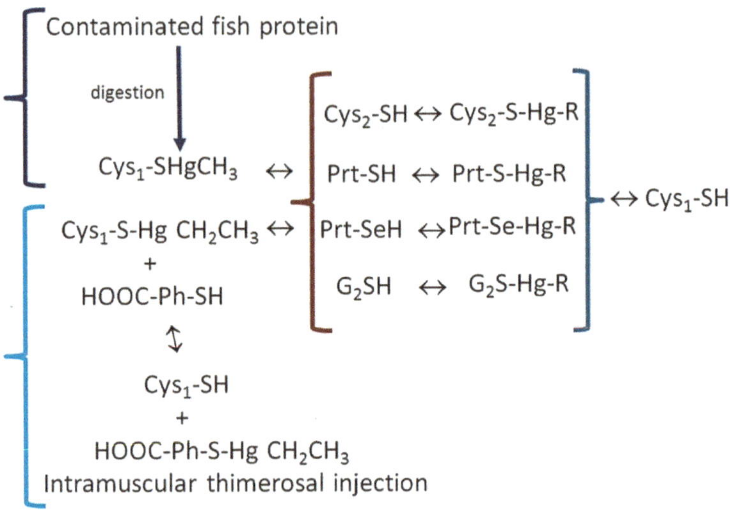

Fig. 4 The fate of methylmercury ($MeHg^+$) from the diet and ethylmercury (thimerosal) intramuscularly injected. The methylmercury-cysteine complex is represented by Cys_1-$SHgCH_3$ and thimerosal (ethylmercury) is represented by HOOC–Ph–S–Hg–CH_2–CH_3. In the figure, the first complex of $MeHg^+$ (Cys_1-$SHgCH_3$) will enter the body and will be exchanged with a second cysteine (Cys_2-SH) to form a new complex (Cys_2–S–Hg–R). Thimerosal (HOOC–Ph–S–Hg–CH_2CH_3), which is the complex of Ethylmercury ($EtHg^+$) with thiobenzoic acid (HOOC–Ph–SH) will exchange with a free cysteine or Cys_1-SH to release the free thiobenzoic acid (HOOC–Ph–SH) and the complex of ethylmercury with the Cys_1-SH (Cys_1–S–Hg–CH_2CH_3). To simplify, both the complexes of ethyl- and methyl-mercury are represented as R-Hg-Cys in the second set of reactions (inside the {}). The reaction of either MeHg Cys_1 or EtHg-Cys_1 complexes with different free thiols in LMM-SH (Cys_2–SH or G_2SH) or HMM-SH molecules (Prt-SH or Prt-SeH) will form new complexes of E^+Hg with target and non-target molecules. The symbol "↔" indicates that the exchange reactions and the reactions can occur in both directions. Here it is important to emphasize that although the constant of formation of Hg–S bond is extremely high and, consequently, thermodynamically favorable, the Hg–S can rapidly exchange with another similar thiolate group (R–S^-).

intramuscularly administered $EtHg^+$ is depicted in Fig. 4. Hg^{2+}'s fate will also depend on the binding to LMM-SH groups, and here the chemistry is a little more complicated because the exchange reactions can involve one or two thiol groups. The binding of Hg^{2+} to two cysteine molecules is critical for its movement in the body and, particularly, for its renal toxicity (Bridges and Zalups, 2005, 2010, 2017; Zalups, 2000; Fig. 5).

More recently, $MeHg^+$ and other E^+Hg forms have been described in rice (Brombach et al., 2017; Cui et al., 2017; Li et al., 2016a; Wu et al., 2018), and the exposure to E^+Hg forms via rice ingestion can cause neurodevelopmental

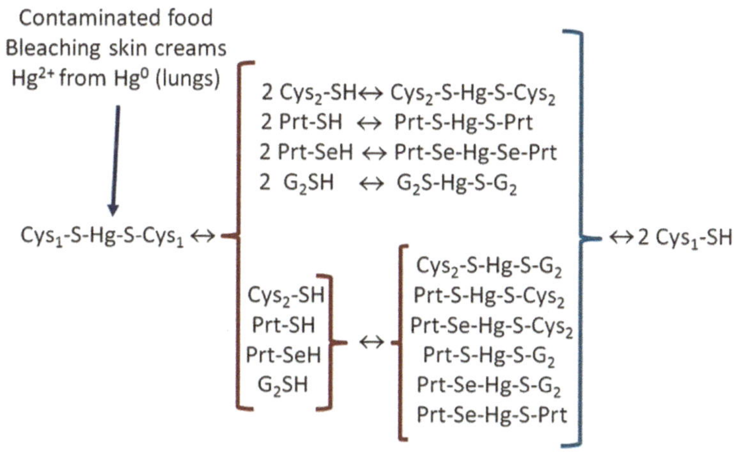

Fig. 5 The fate of Hg^{2+} from the diet, bleaching skin creams and those oxidized from Hg^0 absorbed in the lungs. After entering the body, any free Hg^{2+} will rapidly react with free thiol groups. Here the low molecular mass thiol-containing molecules (LMM-SH molecules) are represented by cysteine (Cys_1-SH, Cys_2-SH or complexed with Hg^{2+} and other organic groups represented by G_2S-, Prt-S- or Prt-Se-). The complex Cys_1–S–Hg–S–Cys_1 can be absorbed from the intestinal tract and transported in the blood to the main target tissue (i.e., the kidney). (The molecular basis of why the kidney is the target organ of Hg^{2+} is elegantly reviewed in Zalups, 2000.) The reactions represented here can involve only one new thiol group, for instance, only one Cys_2-SH can react with the complex Cys_1–S–Hg–S–Cys_1 forming Cys_2–S–Hg–S–Cys_1 and Cys_1-SH as products and this applies to all the other thiols and complexes depicted in Fig. 3.

toxicity in humans (Rothenberg et al., 2016). Little is known about the speciation of E^+Hg forms in rice (Wu et al., 2018), but in view of their high affinity for –SH groups, we can presume that they will be found forming stable complexes with both LMM-SH and HMM-SH molecules. Accordingly, the metalloproteomic analysis of rice experimentally exposed to E^+Hg forms has indicated that E^+Hg can bind to HMM-SH molecules (Li et al., 2016b).

Thimerosal is a complex $EtHg^+$ with 2-mercaptobenzoic acid or thiosalicylic acid (Fig. 6) and is usually administered intramuscularly (as part of the TCV). The exposure to free Hg^{2+} is rare, but the presence of Hg^{2+} in contaminated food and whitening or bleaching skin creams has been reported frequently (Agrawal and Sharma, 2017; Sun et al., 2017). In food, low and intermediate levels of Hg^{2+} (from picomolar to low micromolar concentrations) will be complexed with low molecular mass containing thiol molecules (LMM-SH, e.g., cysteine and GSH as complexes of the type LMM-S-Hg-S-LMM; Fig. 3). In the case of contamination via skin creams,

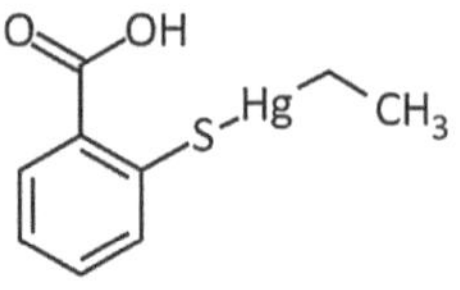

((2-carboxyphenyl)thio)(ethyl)mercury

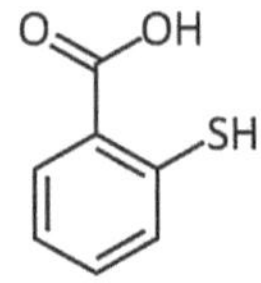

2-mercaptobenzoic acid

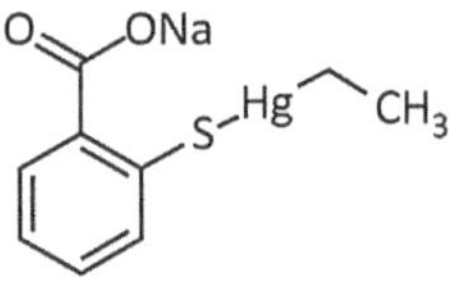

sodium ((2-carboxylatophenyl)thio)(ethyl)mercury

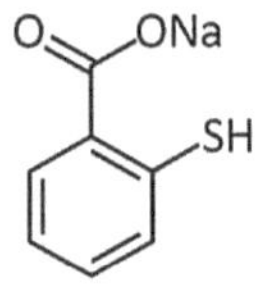

sodium 2-mercaptobenzoate

Fig. 6 Thimerosal is a complex of mercaptobenzoate with ethylmercury (Et-Hg^+)—the structures of thimerosal (either as an acid or a sodium salt) and mercaptobenzoic or salicylic acid are depicted in the figure.

absorption will involve binding to either LMM-SH or to thiol-containing proteins (HMM-SH). Hg^0 is normally absorbed by the lungs, and in the tissues and in the blood it can be oxidized to Hg^{2+}. The free Hg^{2+} formed will rapidly react with free thiol groups of LMM-SH and/or HMM-SH molecules found in the blood or tissues (Fig. 3).

As described in Fig. 4 for organic mercury forms, the basic molecular events involved in the absorption, transport, distribution and toxicity of Hg^{2+} are determined by exchange reactions of the original complex with one or two free cysteines (Cys_2-SH) to form a new complex (Cys_1–S–Hg–S–Cys_2 or Cys_2–S–Hg–S–Cys_2; for more details, see the caption of Fig. 3). The exchange reactions can occur with any free –SH group, and they are depicted in the scheme by the high molecular mass thiol- or selenol-containing proteins (HMM-SH or HMM-SeH) molecules or by LMM-SH molecules (Cys-SH or G_2-SH). The reaction of Cys-Hg complexes with other thiol-containing molecules can generate a diverse number of R–S–Hg–S–R complexes either with different LMM-SH free thiols (Cys-SH or GSH) or HMM-SH/HMM-SeH molecules (Prt-SH or Prt-SeH), which can be both target or non-target molecules of E^+Hg forms.

In view of the strong affinity of thiol/thiolate and selenol/selenolate groups for electrophilic forms of mercury (E^+Hg), the interaction of E^+Hg forms with –SH or –SeH groups will be critical for initiating their toxicity. Nevertheless, our knowledge about the preferential targets of

E^+Hg is still incomplete and here we will discuss some of the caveats when deciphering the primary targets of E^+Hg. The first seems to be related to the absorption of E^+Hg forms, which will depend greatly on the type and level of exposure. Here we will not discuss the exposure to high levels of E^+Hg forms, but only to environmentally or occupationally relevant levels. As a rule, exposure to low (pico- or nanomolar) or moderate levels (low micromolar concentrations) of E^+Hg forms does not saturate the E^+Hg binding sites (i.e., the thiol groups of non-target and target proteins) because the concentration of –SH groups is in the millimolar range in vertebrate fluids and tissues.

The question of potential saturation of –SeH groups found in a few selenoproteins, which are present at a much lower concentration than the thiol groups, is rather complex because we have practically no information about the distribution and concentration of selenoproteins in different cells and subcellular compartments. Indeed, our knowledge about the physiological function of the roughly two dozen selenoproteins found in vertebrates is incomplete, and only about one-half of them have a defined biochemical role in living cells (Fomenko et al., 2008; Labunskyy et al., 2014). Despite the limited number of studies on the interaction of $MeHg^+$ with selenoproteins, data in the literature have indicated that important selenoenzymes might be targets for both in vitro and in vivo $MeHg^+$ exposures (Branco et al., 2011, 2012, 2014, 2017; Carvalho et al., 2008, 2011; Dalla Corte et al., 2013; Wagner et al., 2010). Studies on the effects of $EtHg^+$ on selenoenzymes are scarce, but analogous to Hg^{2+} and $MeHg^+$, $EtHg^+$ is also a potent inhibitor of TrxR (Carvalho et al., 2008, 2011; Rodrigues et al., 2015).

There is a large number of studies corroborating the inhibitory effect of E^+Hg in thiol-containing proteins (Farina et al., 2017; some examples are listed in Table 3). Among them, are neurotransmitter receptors, various enzymes and cation channels (Farina et al., 2017). However, the question of concentration, distribution and reactivity of cysteinyl residues in thiol-containing proteins is also a critical point and serious limitation in understanding the neurotoxicity of E^+Hg forms. As in the case of selenoproteins, we have only limited knowledge on the concentration, reactivity, and distribution of the majority of the thousands of proteins containing –SH groups (Dröse et al., 2014; Ferrer-Sueta et al., 2011; Heppner et al., 2018; Pereira et al., 2015; Reisz et al., 2013; Riemer et al., 2015; Rudyk and Eaton, 2014; Short et al., 2016; Weerapana et al., 2010; Wible and Sutter, 2017).

Table 3 Examples of human thiol-containing proteins involved in the regulation of brain physiology that are targets of $E^{+}Hg$ forms.

Protein	Gene	N° CXXC	N° CXC	N° CC	N° C	N° aa	Function	Localization	Hg effects	References
Part A: enzymes										
5′-Nucleotidase	NT5E	1	0	0	10	574	Nucleotide binding	Plasma membrane	Inhibition	Sood et al. (1995)
Acetylcholinesterase	ACHE	0	0	0	8	614	Neurotransmitter degradation	Plasma membrane	Inhibition	Franciscato et al. (2009); Moraes-Silva et al. (2014)
Caspase-8	CASP8	1	0	1	13	479	Apoptosis	Cytoplasm	Activation	Shenker et al. (2002)
Complex I (NADH–ubiquinone oxidoreductase 75 kDa subunit)	NDUFS1	3	0	0	18	727	Electron transport	Mitochondrion	Inhibition	Glaser et al. (2010a)
Complex II (succinate dehydrogenase flavoprotein subunit)	SDHA	0	0	2	18	664	Electron transport	Mitochondrion	Inhibition	Glaser et al. (2010a)
Complex III (cytochrome *b*–*c*1 complex subunit 1)	UQCRC1	0	0	0	12	480	Electron transport	Mitochondrion	Inhibition	Glaser et al. (2010a)
Complex IV (cytochrome *c* oxidase subunit 1)	MT-CO1	0	0	0	1	513	Electron transport	Mitochondrion	Inhibition	Glaser et al. (2010a)

Cytochrome *c*	CYCS	1	0	0	2	105	Apoptosis	Mitochondrion	Release	Shenker et al. (1999, 2002)
Creatine kinase B-type	CKB	0	0	0	5	381	Transferase	Cytoplasm	Inhibition	Glaser et al. (2010b)
Creatine kinase S-type	CKMT2	0	0	0	9	419	Transferase	Mitochondrion	Inhibition	Glaser et al. (2010a)
Cystathionine gamma-lyase	CTH.	2	0	0	10	405	Cysteine synthesis	Cytoplasm	Inhibition	Bridges et al. (2012)
Glutamate–cysteine ligase catalytic subunit	GCLC	0	1	1	14	637	Glutathione synthesis	Cytoplasm	Induction	Thompson et al. (1999)
Glutamine synthetase	GLUL	0	0	0	11	373.	Glutamine and GABA synthesis	Mitochondrion	Inhibition	Kwon and Park (2003)
Glutathione reductase	GSR	0	0	0	10	522	Glutathione reduction	Mitochondrion	Activation	Stringari et al. (2008); Glaser et al. (2010a); Zemolin et al. (2012)
Glutathione-S-transferase	GSTA1	0	0	0	1	222	Conjugation of reduced glutathione	Cytoplasm	Activation	Zemolin et al. (2012)
Phospholipase A2, membrane associated	PLA2G2A	0	2	2	14	144.	Lipid degradation	Plasma membrane	Activation	Mazerik et al. (2007); Verity et al. (1994)

Continued

Table 3 Examples of human thiol-containing proteins involved in the regulation of brain physiology that are targets of $E^{+}Hg$ forms.—cont'd

Protein	Gene	N° CXXC	N° CXC	N° CC	N° C	N° aa	Function	Localization	Hg effects	References
Part B: receptors										
D(1A) dopamine receptor	DRD1	0	0	1	16	446	G-protein coupled receptor	Plasma membrane, endoplasmic reticulum membrane	Dysfunction	Daré et al. (2003)
D(2) dopamine receptor	DRD2	0	1	0	12	443	G-protein coupled receptor	Plasma membrane	Dysfunction	Daré et al. (2003)
D(3) dopamine receptor	DRD3	1	0	1	16	400	G-protein coupled receptor	Plasma membrane	Dysfunction	Daré et al. (2003)
Gamma-aminobutyric acid receptor subunit alpha-1	GABRA1	0	0	0	5	456	Ion channel, receptor	Plasma membrane	Inhibition	Bondy and Agrawal (1980)
Gamma-aminobutyric acid receptor subunit beta-1	GABRB1	0	0	1	5	474	Ion channel, receptor	Plasma membrane	Inhibition	Bondy and Agrawal (1980)
Muscarinic acetylcholine receptor M3	CHRM3	1	1	0	12	590	Inhibition of adenylate cyclase	Plasma membrane	Inhibition	Abd-Elfattah and Shamoo (1981); Abdallah and Shamoo (1984)
Muscarinic acetylcholine receptor M5	CHRM5	2	0	0	17	532	Inhibition of adenylate cyclase	Plasma membrane	Inhibition	Abd-Elfattah and Shamoo (1981); Abdallah and Shamoo (1984)

Part C: transporters										
Calcium-transporting ATPase 1	ATP2B1	0	0	0	17	1258	Calcium transport	Plasma membrane	Inhibition	Berg and Miles (1979)
Cysteine/glutamate transporter	SLC7A11	0	0	0	7	501	Amino acid transport	Plasma membrane	Inhibition	Allen et al. (2001); Fonfría et al. (2005)
Part D: other										
Kelch-like ECH-associated protein 1	KEAP1	0	1	1	27	624	Transcription regulation	Nucleus, cytoplasm	Modulation	Yoshida et al. (2014)
Nuclear factor erythroid 2-related factor 2	NFE2L2	0	0	0	6	605	DNA binding	Nucleus, cytoplasm	Up-regulation	Ni et al. (2010)
Tubulin beta chain	TUBB	0	1	0	8	444	Nucleotide binding, microtubule	Cytoskeleton	Oxidation	Kromidas et al. (1990)

The Hg effects were based on animal models or in vitro studies. The proteins affected in rodents or in cultured cells were selected, and then the human protein information was obtained at http://www.uniprot.org.

As mentioned above, any thiol group found in a protein is a potential target for $E^{+}Hg$. The total number of cysteinyl residues codified in the human genome is about 550,000, found in about 72,000 proteins recovered from the UniProt database (http://www.uniprot.org). Considering only the proteins with experimental evidence of expression, which means about 19,000 proteins, one calculates the existence of 260,000 cysteinyl residues.

The human cysteinyl proteome has been estimated to be around 214,000 (Go et al., 2015; Jones and Go, 2011), corroborating the number we have calculated above for proteins with experimental evidence. The thiol-containing proteins can have from one to several cysteinyl residues. Cysteine is one of the less abundant amino acids in proteins (Miseta and Csutora, 2000; Poole, 2015; Wible and Sutter, 2017), representing only about 2–3% of amino acids comprising the human proteome. This is possibly due to the reactivity and nucleophilic softness of the cysteinyl group (Fomenko et al., 2008; LoPachin et al., 2012). Indeed, the –SH group is the second most powerful soft nucleophile group found in biomolecules, and the first is the –SeH group (Rocha et al., 2017). Their high reactivity likely restrained their abundance.

In contrast to the thiol groups, which are found in thousands of proteins (Go et al., 2015; Jones and Go, 2011), the human genome codifies only about 25 selenoproteins (proteins containing the selenol group found in the selenocysteinyl, U or Sec residues) (Labunskyy et al., 2014), and normally they have 1 to a maximum of about 10 Sec residues in selenoprotein P (Table 4), which is a protein involved in the transport of Se to different tissues in mammals (Labunskyy et al., 2014; Oliveira et al., 2017a). In fact, the majority of selenoproteins have only one Sec residue and, in some cases, it participates in the analog motif of thiol oxidoreductase –C–X–X–C–, i.e., –C–X–X–U– or –U–X–X–C– (Table 4).

In short, there are thousands of targets for $EtHg^{+}$ forms in the case of HMM-SH molecules, and only a few for HMM-SeH molecules. One important aspect that determines the toxicity of $E^{+}Hg$ forms is the accessibility and the intrinsic reactivity of the functional or catalytic role of a given cysteinyl or selenocysteinyl residue found in HMM-SH or HMM-SeH molecules. In the case of thiol groups, the cysteinyl residues in several HMM-SH molecules have no significant functional or catalytic role. Consequently, the binding of $E^{+}Hg$ to their –SH groups will not modify the protein activity. For instance, keratin has several cysteinyl residues that will form disulfide bonds in the secreted protein found in nails and hair. Keratinocytes can accumulate mercury in hair keratin either by incorporating cysteine-$E^{+}Hg$ complexes in the place of cysteine

Table 4 Examples of human selenoproteins involved in the regulation of brain physiology that can be inhibited by $E^{+}Hg$.

Protein	Gene	CXXC	CXC	CC	C	CXXU	CXU	CU	U	aa	Function	Localization	Hg effects	References
Glutathione peroxidase 1	GPX1	0	0	0	5	0	0	0	1	203	Oxidoreductase	Cytoplasm	Inhibition	Franco et al. (2009)
Glutathione peroxidase 4	GPX4	0	0	0	9	0	0	0	1	197	Peroxidase	Mitochondrion	Inhibition	Zemolin et al. (2012)
Iodothyronine deiodinase type I	DIO1	0	0	0	4	0	1	0	1	249	Thyroid hormone synthesis	Endoplasmic reticulum	Inhibition	Pantaleão et al. (2017)
Iodothyronine deiodinase type II	DIO2	0	0	0	6	0	0	0	2	273	Thyroid hormone biosynthesis	Plasma membrane	Inhibition	Mori et al. (2006)
Selenoprotein P	SELENOP	1	1	1	17	1	4	1	10	381	Transport of selenium	Extracellular region	Inhibition	Sasakura and Suzuki (1998)
Selenoprotein W	SELENOW	0	0	0	2	1	0	0	1	87	Antioxidant, redox-related process	Cytoplasm	Down-regulation	Kim et al. (2005)
Thioredoxin reductase 1	TXNRD1	1	0	0	17	0	0	1	1	649	Oxidoreductase	Cytoplasm	Inhibition	Carvalho et al. (2008)

The Hg effects were based on animal or in vitro models. The proteins that affected rodents or in cultured cells were selected and then the human protein information was obtained at http://www.uniprot.org.

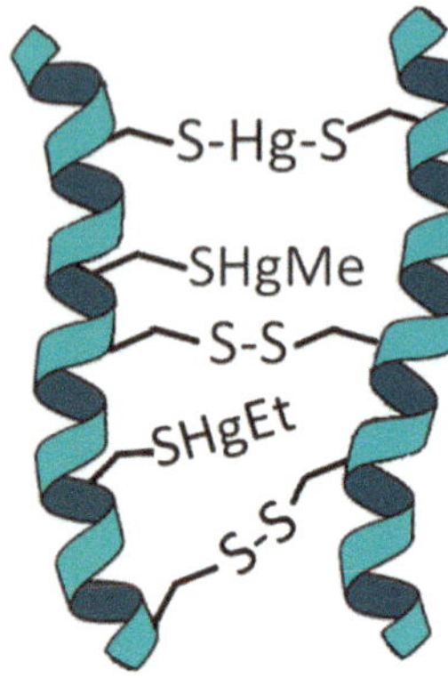

Fig. 7 Schematic representation of accumulation of $E^{+}Hg$ forms in the keratin in hair or nails. The $MeHg^{+}$ and $EtHg^{+}$ coordinate with one $-S^{-}$ group, whereas Hg^{2+} coordinates with two $-S^{-}$ groups. The deposition of mercury in hair or nails usually does not change the biological function of the altered protein, and the quantification of Hg in hair is largely used to evaluate the exposure to Hg (Kusanagi et al., 2018; Marques et al., 2009, 2011, 2012).

(Cernichiari et al., 2007), or $E^{+}Hg$ forms can binding to cysteinyl thiol groups in keratin before the formation of the disulfide bonds in the pre-keratin (Sippel, 1973). Here we have an example where the binding of mercury to a protein will not significantly change the biological function of the protein (Fig. 7). Of possible toxicological significance, the high quantity of mercury incorporated in secreted keratin found in hair and nails can be considered as a route of diversion of $E^{+}Hg$ forms from their target to non-target HMM-SH containing molecules.

The cysteinyl residues (Cys) in HMM-SH molecules can play either a direct or an indirect catalytic role in different classes of oxidoreductases, but an essential role in the catalysis or function of other types of proteins (Fomenko et al., 2008; Miseta and Csutora, 2000; Poole, 2015). Thiol-groups can also participate in the catalysis of non-redox reactions or have a critical structural role in proteins (Fomenko et al., 2008). Here, the binding of $E^{+}Hg$ forms to those cysteinyl residues can inactivate the physiological role of the protein, which can have an important toxicological impact on the exposed organism (Table 3). Some hypothetical types of interaction of $E^{+}Hg$ forms with HMM-SH molecules and the biological consequences of these interactions are depicted in Fig. 8.

Some important motifs containing cysteinyl and/or selenocysteinyl residues are –C–X–X–C– or –C–X–X–U–, which are found in several oxidoreductases (Fomenko et al., 2008). The number of times that the motif –C–X–X–C– appears in human genes are near 19,000 in about 6000 proteins (http://www.uniprot.org), but note that not all of these motifs will

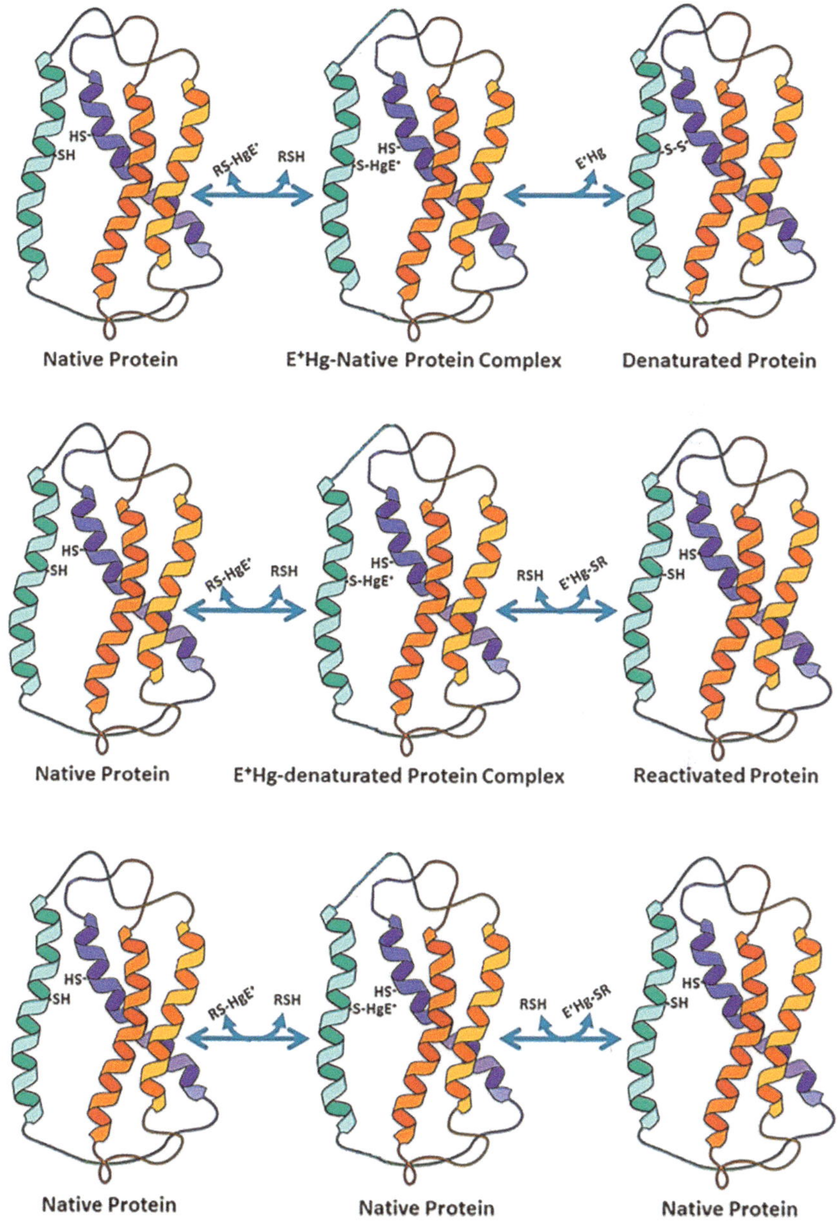

Fig. 8 $MeHg^+$ or $EtHg^+$ can interact with thiol-containing proteins (HMM-SH molecules) and inactivate permanently (top) or transiently (middle) the protein function. In the third type of interaction, the binding of E^+Hg to the HMM-SH did not modify the protein function. This is somewhat similar to that represented in Fig. 7, except that here the E^+Hg will only transiently interact with the HMM-SH molecule.

Table 5 Thiol groups residues found in the proteome of two mammals, human and mouse.

			Thiol groups			
Organism.	**Total protein**	**Protein type**	**CXC**	**CXXC**	**CC**	**C**
Homo sapiens	71,444	sp	8573	19,038	8486	259,182
		tr	9574	17,146	9036	278,316
Mus musculus	51,949	sp	7290	12,629	6277	212,172
		tr	9407	21,372	10,107	290,020

We separated the proteins into two main types: "sp" for UniProtKB/Swiss-Prot and "tr" for UniProtKB/TrEMBL. This is the regular classification used in the UniProt Database.

appear in native proteins, and not all proteins having the motif–C–X–X–C– will be an oxidoreductase (Table 5). For instance, the motif–C–X–X–C– can also be found in metal-coordinating proteins (Fomenko et al., 2008; Miseta and Csutora, 2000), such as in delta-aminolevulinate dehydratase (porphobilinogen synthase; δ-AlaD) (Rocha et al., 2012). This enzyme is an in vitro and in vivo target for E^+Hg (Rocha et al., 2012); however, a causal link between the enzyme inhibition and the triggering of toxic effects is difficult to establish. Broadly speaking, this is a critical caveat in the neurotoxicology of E^+Hg forms, i.e. the observation that E^+Hg forms can inhibit or disrupt a target protein does not establish a causal relationship between the phenomenon and the toxicological outcomes. Some examples of proteins that have been reported to be inhibited after either in vitro or in vivo exposure to $MeHg^+$ are depicted in Table 3 (note that the proteins present in Table 3 are from the human genome, and the experimental studies were usually done in rodents or in vitro). The number of codified cysteinyl residues in these proteins varied from 1 to 27, and some of them have at least one of the motifs –C–X–X–C–, –C–X–C– or –C–C–. Others have only different numbers of "independent" –C– residues (Tables 3 and 5).

An examination of Table 3 provides some examples of the bottlenecks found in the search for targets of E^+Hg. $MeHg^+$ can either inhibit or stimulate HMM-SH molecules (see for instance, glutathione-S-transferase and phospholipase A2, Table 3). One additional problem that can be inferred from Table 3 is the fact that the inhibition of E^+Hg can be indirect. For instance, acetylcholinesterase has been reported to be inhibited after exposure to $MeHg^+$ or Hg^{2+} (Adedara et al., 2015; Franciscato et al., 2009; Moraes-Silva et al., 2014); however, the inhibition cannot be mediated by modifications of its thiol groups because none of them are reactive. In fact, the majority of them participate in –S–S– bridge formation.

Another important factor that we have commented on above is the reactivity of thiol groups. The presence of vicinal thiol groups can increase the potential reactivity of these groups (as found in –C–X–X–C– motifs), but some proteins have "independent," but highly reactive –C– residues, for instance, in human muscle and mitochondrial creatine kinase (CK). CK and mitochondrial CK have one cysteinyl residue with a very low pK_a (Naor and Jensen, 2004; Wang et al., 2001), which explains their strong reactivity with electrophiles (Alvarez and Radi, 2003; Glaser et al., 2010a, 2014; Stachowiak et al., 1998). Thus, it is imperative to develop new in silico tools in the area of predictive toxicology to envisage the structure and the reactivity of thiol groups using the primary structure of a given protein. Another aspect that has to be investigated is the interaction of E^+ Hg forms with nascent proteins, i.e. prior to post-translational processing (this may be the case of pre-keratin, where E^+ Hg forms are incorporated in the protein structure before the assembly of –S–S– bonds).

The physiological role of thiol-containing protein can be modulated by transitory S-glutathionylation, nitrosylation or reversible oxidation of cysteinyl residues (Comini, 2016; Dröse et al., 2014; Pereira et al., 2015; Poole, 2015). The binding of E^+ Hg to this class of cysteinyl residues can drastically modify the normal dynamics of activating–deactivating redox-regulated thiol-containing proteins. As noted above, the stable binding of E^+ Hg to cysteinyl residues can permanently inactivate the HMM-SH molecules involved in the regulation of cell physiology via thiol switches. Of particular toxicological significance, those proteins that are modulated by thiolation or nitrosylation can be preferential targets of E^+ Hg forms, because their cysteinyl residues involved in these reactions are normally highly accessible and reactive toward strong and soft electrophiles, such as E^+ Hg. However, there are no systematic studies about the interaction of E^+ Hg with these types of proteins. Thus, it is imperative to determine the effects of E^+ Hg in proteins regulated by thiolation and other similar mechanisms. Here we posit that E^+ Hg forms will disrupt the cell metabolism via blockage of cysteinyl residues involved in protein regulation via glutathionylation or nitrosylation. One example of such proteins is complex I in respiratory the chain (Dröse et al., 2014), and data from the literature have indicated that Complex I is inhibited by $MeHg^+$ treatment (Table 3; Glaser et al., 2010a, 2013). The detailed in silico and in vitro systematic studies about the interaction of E^+ Hg with thiol groups in such proteins can help in the construction of virtual and molecular predictive tools to facilitate the difficult task of determining the primary molecular targets of E^+ Hg.

In contrast to thiol groups, the selenol group found in selenoproteins is normally involved in the oxidoreductase catalytic activity of selenoenzymes (Fomenko et al., 2007) and possibly in the physiological role of all the selenoproteins. Thus, the binding of E^+Hg to the selenocysteinyl residues in selenoproteins is expected to inactivate them. Most characterized selenoproteins are oxidoreductases and catalyze important redox reactions, such as the degradation of peroxides (glutathione peroxidase or GPx isoforms) and the reduction of thioredoxin (thioredoxin reductase or TrxR isoforms). Thus, the inhibition of these antioxidant enzymes is believed to play a role in causing oxidative stress in $MeHg^+$ poisoning (Farina et al., 2011a,b).

In addition, the binding mode of E^+Hg species in the respective protein target, depends on the ligand (Cys, GSH or H_2O) bound to the Hg atom; i.e., the molecular structure of the Hg molecule will determine which is its target.

For example, the molecular docking simulations demonstrated that $MeHg^+$ and $EtHg^+$ species (H_2O-HgMe, Cys-HgMe, GSH-HgMe, H_2O-HgEt, Cys-HgEt, GSH-HgEt) (Fig. 9) and Hg^{2+} forms ($[H_2O\text{-}Hg\text{-}OH]^+$, $[Cys\text{-}Hg\text{-}H_2O]^+$, $[GSH\text{-}Hg\text{-}H_2O]^+$, Cys-Hg-Cys, Cys-Hg-GSH, GSH-Hg-GSH) (Fig. 10) can access the active site of the δ-AlaD enzyme. These E^+Hg species can establish H-bonds, π–π and cation–π interactions, and the Zn···O and Hg···S coordination that helps to stabilize the Hg species in the active site of δ-AlaD. Specifically, the Hg···S coordination from the C124 and/or C132 residues can lead to the formation of an adduct between the enzyme and the Hg species, inhibiting the enzyme due to the denaturation of the active site, as depicted in Fig. 11.

On the other hand, the docking simulation between the E^+Hg species and the CK enzyme showed that only the small molecules (E^+Hg species that do not present GSH bound to the Hg atom) can interact in the active site of the enzyme. The C283 residue could be a target for the organic mercury molecules H_2O-HgMe, H_2O-HgEt (Fig. 12A), Cys-HgMe and Cys-HgEt (Fig. 12B). In addition, the Hg^{2+} molecule, Cys-Hg-H_2O, also showed interactions with C283 (Fig. 12C). However, the species H_2O-Hg-OH showed S···Hg interactions with the C146 residue (Fig. 12C), indicating that this residue could be a target as well. Due to the proximity of the C283 residue to the active site of CK, the conjugation between C283 and MeHg (Fig. 12D) can lead to the enzyme inhibition. The simulations with the Hg conjugated with GSH do not present S···Hg interactions, due to the large distance between the molecule and the cysteinyl residue from the protein (C283(S)···Hg distance >7 Å) (data not shown).

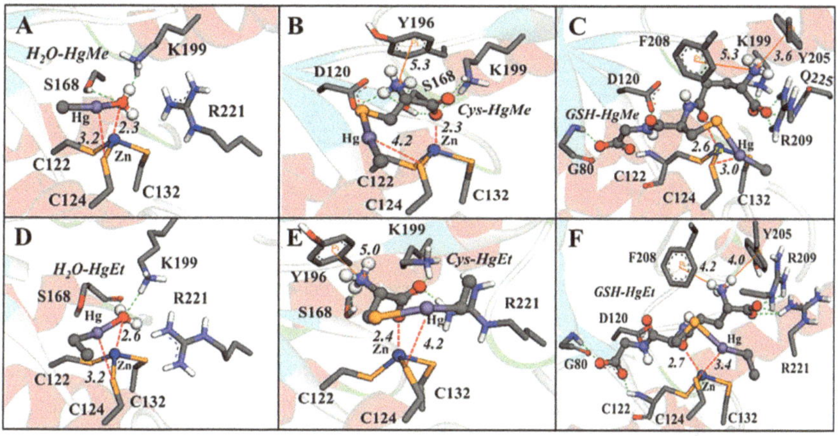

Fig. 9 Interactions between the $MeHg^{+}$ and $EtHg^{+}$ species with δ-AlaD. Putative MeHg complexes [H_2O-HgMe (A), Cys-HgMe (B), and GSH-HgMe (C)] and EtHg complexes [H_2O-HgEt (D), Cys-HgEt (E), and GSH-HgEt (F)] interact in the active site of δ-AlaD. H-bonds are shown in green dotted lines, Zn···O and Hg···S interactions are in red dotted lines, and the cation–π interactions are in orange lines. Only the main residues involved in the interactions are shown. The distances are in Å and the molecular docking simulations were carried out with the Auto Dock Vina 1.1.1 program (Trott and Olson, 2010) using the docking configuration according to Nogara and Rocha (2017).

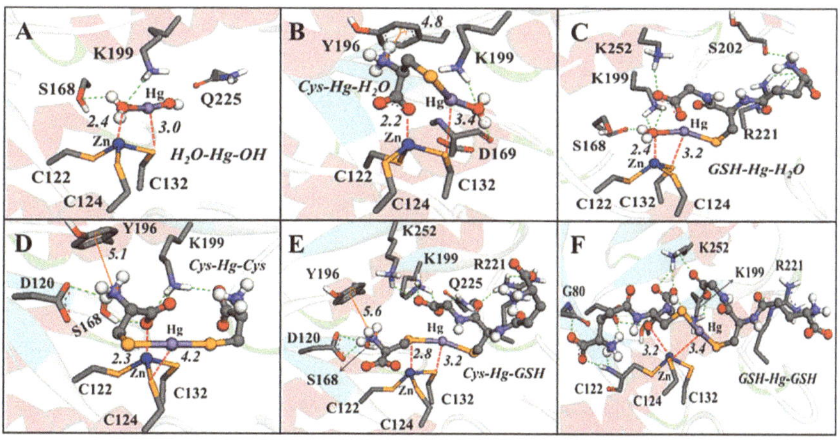

Fig. 10 Interactions between the Hg^{2+} species with δ-AlaD. Putative Hg^{2+} complexes {$[H_2O\text{-}Hg\text{-}OH]^{+}$ (A), $[Cys\text{-}Hg\text{-}H_2O]^{+}$ (B), $[GSH\text{-}Hg\text{-}H_2O]^{+}$ (C), Cys-Hg-Cys (D), Cys-Hg-GSH (E), and GSH-Hg-GSH (F)} interact in the active site of δ-AlaD. H-bonds are shown in green dotted lines, Zn···O and Hg···S interactions are in red dotted lines, and the cation–π interactions are in orange lines. Only the main residues involved in the interactions are shown. The distances are in Å and the molecular docking simulations were carried out with the Auto Dock Vina 1.1.1 program (Trott and Olson, 2010) using the docking configuration according to Nogara and Rocha (2017).

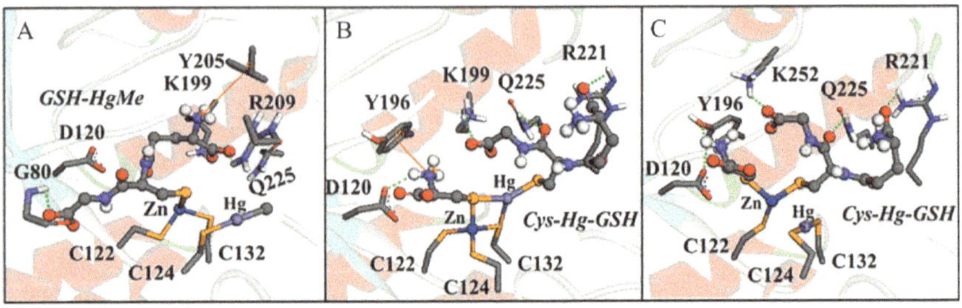

Fig. 11 Simulation of the δ-AlaD denaturation caused by $MeHg^+$ (A) and Hg^{2+} (B–C). H-bonds are showed in green dotted lines and the cation–π interactions are in orange lines. Only the main residues involved in the interactions are shown. The simulation of the enzyme denaturation was done manually in the software Avogadro 1.1.1 (Hanwell et al., 2012) using the amino acid residues at 5 Å of the ligands (with the Cα frozen) and geometrically optimized with the universal force field (UFF) (5000 steps).

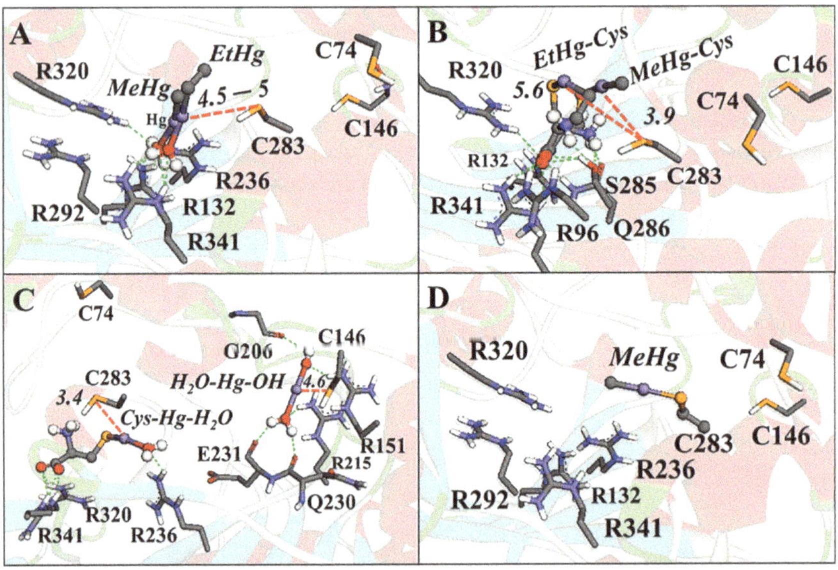

Fig. 12 Interactions between the Hg species with CK. (A) MeHg–H_2O and EtHg–H_2O interactions. (B) Cys-HgMe and Cys-HgEt interactions. (C) Cys-Hg-H_2O and $[H_2O\text{-}Hg\text{-}OH]^+$ interactions. (D) Simulation of the S–Hg bond formation. H-bonds are shown in green dotted lines, Zn···O and Hg···S interactions are in red dotted lines, and the cation–π interactions are in orange lines. Only the main residues involved in the interactions are shown. The distances are in Å. The molecular blind docking simulations were carried out with the Auto Dock Vina 1.1.1 program (Trott and Olson, 2010) using the human brain-type Creatine Kinase (PDB: 3DRB) (Bong et al., 2008), with the coordinates $x = -58.518$, $y = 1.229$, $z = 5.558$, and the grid box size of $64 \times 68 \times 50$ Å, and an exhaustiveness = 100. The simulation of the enzyme S–Hg conjugation was done manually in the software Avogadro 1.1.1 (Hanwell et al., 2012) using the amino acid residues at 5 Å of the ligands (with the Cα frozen) and geometrically optimized with the universal force field (UFF) (5000 steps).

As discussed above, the fate of E^+Hg species in the body will depend greatly on the exchange reactions of the type:

$$R_1\text{-S-HgE}^+ + R_2SH \leftrightarrow R_1SH + R_2\text{-S-HgE}^+$$

Taking into consideration the results of the docking presented in Fig. 12A–D, we can deduce that the neurotoxicity of E^+Hg–S– complexes will vary depending on the size of the complexes and the reactivity/accessibility of the cysteinyl or selenocysteinyl residues involved in the reaction. Consequently, the targets of different types of LMM-S-HgE^+ complexes must be identified for a better understanding of Hg neurotoxicity.

Another crucial aspect that has yet to be studied in detail in animal models is the simultaneous exposure to E^+Hg forms and other potential neurotoxicants (e.g., Pb^{2+}, Al^{3+}, Sn^{2+}, PCBs, etc.). This type of study is needed, because humans and animals are normally exposed to a variety of chemical agents (Grandjean and Landrigan, 2006). Of particular toxicological significance, synergistic, additive or antagonistic interactions among the toxicants can occur after the simultaneous exposure to E^+Hg forms and one or more neurotoxicants. The determination of specific metabolic pathways and/or molecular targets disrupted after the simultaneous exposure to different neurotoxicants is also required, particularly on how the binary, tertiary, quaternary, etc. combinations of toxicants can have additive or synergistic neurotoxic effects. In the next section, some examples of the interaction between mercury and other neurotoxicants in developing humans are exemplified.

6.2 Hg and Other Neurotoxicants: Additive Effect?

As reviewed in Palumbo et al. (2000), hair Hg in pregnancy has been associated with mild neurological effects from 5 to 10 ppm (based on data from the Iraq $MeHg^+$ poisoning), while studies of hair Hg derived from fish-eating (in New Zealand and Canada) have indicated an exposure range between 10 and 20 ppm may provoke adverse effects on the central nervous system (Palumbo et al., 2000).

Nevertheless, important longitudinal studies linking Hg-associated neurological deficits have shown inconsistent outcomes; while adverse associations were shown for Faroese children, no adverse effects have been reported for Seychelles children (Palumbo et al., 2000). Among differences between studies related to neurobehavioral-testing methods, source of exposure (whale meat and fish) and concomitant exposure to neurotoxicants may explain inconsistency between the two cohorts (Palumbo et al., 2000).

Often children-based studies examined neurobehavioral effects of one neurotoxicant at the time, when in real-life scenarios low-dose exposure to Hg occurs simultaneously with multiple co-occurring neurotoxic substances. Association between exposure to mercury and neurobehavioral outcomes requires consideration of co-exposure (to other neurotoxicants) and accompanying confounding factors. Indeed there is a body of literature that addressed combined low-level Hg ($MeHg^+$ and $EtHg^+$) exposures for additive and/or interactions with other environmental toxicants that can contribute to deficits in neurobehavioral test scores. Some of those studies are summarized in Tables 6 and 7.

The effects of multiple exposures during pre- and postnatal life are linked to children's neuropsychological development, but the weight of Hg as the neurotoxic causative factors will depend on the tested parameter. The results revealed a great variability in Hg exposure related to time (prenatal and postnatal) and route of exposure (injected and breastfeeding), marker of exposure (blood, hair, vaccination records), and co-occurring neurotoxicants (Al, Pb, PCBs, and unidentified sources of pollutants related to mining and smoking). In most of the studies, prenatal Hg exposure did not show a predominant neurodevelopment outcome in the first months of postnatal age.

Infants (less than 6 months) have the highest combined exposures of neurotoxicants. Most of the exposures originate during pregnancy and likely continue during breastfeeding. The potential for the highest body burden at this early stage is further increased by the EtHg-Al exposure after a full schedule of TCVs. Nevertheless, the measured outcomes seem to be limited (Table 6). On the basis of prenatal exposure to neurotoxicants, the additional burden of EtHg-Al (in TCVs) is a main cause of concern; the risk of insults that may change neurodevelopmental outcomes has been shown in some studies (Table 6).

Table 7 summarizes the outcomes of neurobehavioral tests in preschoolers and schoolchildren. In preschoolers, most studies did not show significant associations of Hg and deficits in test outcomes; coincidentally, the few studies showing statistically significant Hg interactions were in children living in mining environments (Dórea et al., 2014; Marques et al., 2012), or exposed to a larger suite of neurotoxic metals (Boucher et al., 2014; Freire et al., 2018). It is worth mentioning that in 30 month-old children, internalizing behavior was affected by prenatal levels of PCBs but not by Hg (Tatsuta et al., 2012); neurodevelopment measured at 42 months by the

Table 6 Summary of studies addressing co-occurring exposures to neurotoxicants and their effects on neurobehavioral tests in infants.

Neurotoxic agents	Aim of study	Neurobehavioral tests (age): outcomes	References (country)
MeHg; EtHg	To compare the effect of time hepatitis-B shots (<24 h or >24 h after birth) on neurodevelopment	*Gesell Developmental Schedules (6 months)*: Infants immunized 2–4 days after delivery showed no significant difference in neurodevelopment compared to the group immunized within 24 h	Marques et al. (2007) (Brazil)
MeHg; EtHg	To use principal component analysis as a screening of variables linked to pre- and postnatal Hg exposure	*Gesell Developmental Schedules (6 months)*: Principal component analysis discriminated variability of neurodevelopment outcomes associated with variables that included pre- and postnatal Hg exposure	Marques et al. (2008) (Brazil)
Hg^{2+}; MeHg; EtHg	To evaluate the association between infant Hg and Gesell Developmental Schedules	*Gesell Developmental Schedules (6, 36, 60 months)*: Length of lactation was positively correlated with GDS scores and hair-Hg concentrations were negatively correlated with Gesell Developmental Schedules scores	Marques et al. (2009) (Brazil)
MeHg; EtHg	To study the neurodevelopment of infants exposed to thimerosal in tetanus–diphtheria vaccines during pregnancy	*Gesell Developmental Schedules (6 months)*: No difference was observed	Marques et al. (2010) (Brazil)
PCB; MeHg	To study a birth cohort to clarify the effects of PCB and MeHg on infants' neurodevelopment	*Neonatal Behavioral Assessment Scale (3 days)*: MeHg adversely affects neurobehavioral function	Suzuki et al. (2010) (Japan)
MeHg; EtHg	To compare Hg effects on neurodevelopment of infants from one urban center and two rural villages	*Gesell Developmental Schedules (6 months)*: A higher score of neurodevelopment was negatively associated with exposure to additional TCV-EtHg in urban infants	Dórea et al. (2012) (Brazil)

Continued

Table 6 Summary of studies addressing co-occurring exposures to neurotoxicants and their effects on neurobehavioral tests in infants.—cont'd

Neurotoxic agents	Aim of study	Neurobehavioral tests (age): outcomes	References (country)
PCB; MeHg; Pb	To examine the effects of prenatal exposure to PCBs, MeHg, and Pb on cognitive development infants	*Fagan Test of Infant Intelligence; A-not-B test; Bayley Scales of Infant Development—2nd Edition (6.5 and 11 months)*: Outcomes of impairment in specific domains were different for the tested neurotoxicants; Hg was associated with poorer performance on A-not-B test	Boucher et al. (2014) (Canada)
MeHg; EtHg; Pb; Al	To study the neurodevelopment of children living in the vicinity of tin-ore kilns and smelters and of children living in a fishing village	*Bayley Scales of Infant Development—2nd Edition (6 and 24 months)*: Mental Development Index and Psychomotor Development Index were lower at 24 months of age for children living in the vicinity of tin-ore kilns and smelters	Marques et al. (2014) (Brazil)
MeHg; EtHg	To study Hg exposure and neurodevelopment in children from a tin-ore open-pit mine	*Bayley Scales of Infant Development—2nd Edition (6 and 24 months)*: There was a significant sex difference with boys showing more sensitivity related to Bayley Scales of Infant Development—2nd Edition delays and Thimerosal-containing vaccines	Marques et al. (2015) (Brazil)
MeHg; EtHg	To study neurodevelopment of children in relation to prolonged breastfeeding and Hg exposure	*Bayley Scales of Infant Development—2nd Edition (6 and 24 months)*: MeHg and EtHg exposure did not show a significant dose effect (related to the length of breastfeeding) on neurodevelopment among groups	Marques et al. (2016b) (Brazil)
MeHg; EtHg	To study neurodevelopment of children exposed to MeHg and EtHg	*Bayley Scales of Infant Development—2nd Edition (6 and 24 months)*: There was a significant delay in Bayley Scales of Infant Development related to the combined exposure to Hg	Marques et al. (2016a) (Brazil)

EtHg, ethylmercury; MeHg, methylmercury; PCBs, polychlorinated biphenyls; TCV, thimerosal-containing vaccines.
Adapted and expanded from Dórea (2017a,b).

Table 7 Summary of studies addressing co-occurring exposures of neurotoxicants and their effects on neurobehavioral tests in preschooler and school-aged children.

Neurotoxic agents	Aim of study	Neurobehavioral tests (age): Outcomes	References (country)
Preschoolers			
PCB; MeHg	To verify the possible additive and/or interactive effects between PCB and MeHg	*McCarthy Scales of Children's Abilities (38, 54 months)*: A significant interaction between cord-blood PCBs and maternal hair Hg was found	Stewart et al. (2003) (United States)
PCB; MeHg; Pb	To evaluate the impact of chronic exposure to PCBs and MeHg on visual brain processing	*Visual Evoked Potentials (5–6 years)*: Exposure to MeHg and PCBs was associated with alterations of visual evoked potential responses, especially for the latency of the N75 and of the P100 components	Saint-Amour et al. (2006) (Canada)
MeHg; EtHg	To assess the dependence on fish consumption of families and its impact on nutritional status and neurodevelopment of preschool children	*Gesell Developmental Schedules (1–59 months)*: Fish consumption (as hair Hg) had no impact on Gesell Developmental Schedules	Marques et al. (2011) (Brazil)
PCB; MeHg; Pb	To assess whether PCB, MeHg, and Pb influence infants' maladaptive behavior problems	*Child Behavior Checklist (30 months)*: Internalizing behavior was affected by prenatal exposure to low levels of PCBs. No statistically significant association with Hg or Pb	Tatsuta et al. (2012) (Japan)
EtHg; Pb	To examine the relationship between neonatal exposure to Thimerosal-containing vaccines and child development	*Bayley Scales of Infant Development—2nd Edition (12, 24 months)*: The overall deficit in the psychomotor development index was significantly higher in TCV group	Mrozek-Budzyn et al. (2012) (Poland)

Continued

Table 7 Summary of studies addressing co-occurring exposures of neurotoxicants and their effects on neurobehavioral tests in preschooler and school-aged children.—cont'd

Neurotoxic agents	Aim of study	Neurobehavioral tests (age): Outcomes	References (country)
MeHg; EtHg; Al	To assess the effects of fish consumption and undernutrition on neurodevelopment of preschool children	*Gesell Developmental Schedules (1–59 months)*: Among others, hair Hg was associated with a delay in language and personal-social development	Marques et al. (2012) (Brazil)
Hg, Mn; As; Pb	To explore the relationship between in utero exposure to environmental neurotoxic metals and neurodevelopment	*Comprehensive Developmental Inventory for Infants and Toddlers (24 months)*: Significant adverse association with neurodevelopment only for manganese and lead	Lin et al. (2013) (Taiwan)
MeHg; EtHg	To assess milestone achievements and neurodevelopment in toddlers associated with MeHg and EtHg exposure	*Gesell Developmental Schedules (12, 59 months)*: Milestone achievement was delayed in toddlers from tin-ore mining communities	Dórea et al. (2014) (Brazil)
PCB; MeHg; Pb	To study the relationships between PCB, MeHg and Pb and IQ scores	*Kaufman Assessment Battery for Children (42 months)*: Scores were negatively affected by prenatal exposure to PCBs. No statistically significant association with Hg or Pb	Tatsuta et al. (2014) (Japan)
PCB; Hg; Pb; Cd	To assess the neurocognitive development of children exposed in utero to environmental neurotoxicants	*Bayley Scales of Infant Development—2nd Edition (24 months)*: Only PCB118 had a negative impact on neurocognitive development	Brucker-Davis et al. (2015) (France)
Hg; Pb	To study the associations between prenatal and early life exposure to Hg and Pb with toddler temperament	*Toddler Temperament Scale (24 months):* No significant association between Pb and Hg exposure was observed in the temperament classification	Stroustrup et al. (2016) (Mexico)

Bisphenol A; PCB; Phthalate; Mn; Hg; Pb	To evaluate the association of prenatal exposure to neurotoxicants and neurodevelopment	*Bayley Scales of Infant Development—2nd Edition (12, 24 months)*: Maternal exposure to several xenobiotics was associated with adverse neurodevelopmental performances among the children	Kim et al. (2017) (South Korea)
Hg, Pb; Cd	To investigate cognitive development during the first 5 years of life in relation to heavy metal exposure	*Bayley Scales of Infant Development—2nd Edition and Wechsler Preschool and Primary Scale of Intelligence (6–60 months)*: The stability of cognitive development was unrelated to the umbilical cord level of heavy metals	Lee et al. (2017) (South Korea)
As; Cd; Hg; Mn; Pb	To investigate the effects of prenatal co-exposure to As, Cd, Hg, Mn, and Pb on infants' cognitive and motor function	*McCarthy Scales of Children's Abilities (4–5 years)*: Hg was associated in a dose–response manner with lower scores on the verbal function, and non-linearly related to poorer motor function and gross motor skills	Freire et al. (2018) (Spain)
School-aged children			
PCB; MeHg; Pb	To study performance in delayed reinforcement schedules in children exposed to environmental mixtures of PCBs, MeHg, and Pb	*Differential reinforcement of low rates (9.5 years)*: Children exposed to PCB, MeHg, and Pb demonstrated an impaired performance	Stewart et al. (2006) (United States)
Hg; Pb; Al	Possible contribution of environmental neurotoxicants on attention-deficit hyperactivity disorder	*Test battery for attention performance of children (8–12 years)*: Evidence confirms that elements of attention-deficit hyperactivity disorder are adversely affected by low levels of Pb, but not by other neurotoxic trace metals	Nicolescu et al. (2010) (Romania)

Continued

Table 7 Summary of studies addressing co-occurring exposures of neurotoxicants and their effects on neurobehavioral tests in preschooler and school-aged children.—cont'd

Neurotoxic agents	Aim of study	Neurobehavioral tests (age): Outcomes	References (country)
MeHg; Pb	To evaluate the effects of MeHg and Pb exposure on infants' neurodevelopment	*Bayley Scales of Infant Development—2nd Edition (2, 5, 7 years)*: Tested postnatal MeHg exposure levels had no adverse effects on children's IQ	Cao et al. (2010) (United States)
MeHg; Pb	To assess the effects of prenatal Pb and MeHg exposure on cognitive deficits	*Wechsler Intelligence Scale for Children (7, 14 years)*: Some interaction terms between lead and methylmercury suggested that the combined effect of the exposures was less than additive	Yorifuji et al. (2011) (Faroe Islands)
PCB; MeHg	To determine brain function alterations related to neurobehavior in subjects with high prenatal exposure to MeHg and PCBs	*Functional Magnetic Resonance Imaging (15 years)*: Subjects with high mixed exposure to MeHg and PCBs showed activation in more areas of the brain and different and wider patterns of activation than the low mixed exposure group	White et al. (2011) (Faroe Islands)
PCB; MeHg; Pb	To examine the neurophysiological correlations of response inhibition deficits in children exposed to PCB, MeHg, and Pb	*Go/No-go Performance (11.3 years)*: Hg concentrations were not related to any outcome on this task but showed significant interactions with other contaminants on certain outcomes	Boucher et al. (2012a) (Canada)
PCB; MeHg; Pb	To examine the relationship of developmental exposure to MeHg, PCBs, and Pb to behavioral problems at school age	*Disruptive Behavior Disorders Rating Scale (11.3 years)*: Hg was not related to impairment in response inhibition but showed significant interactions with Pb and PCB.	Boucher et al. (2012b) (Canada)

PCB; Hg; Pb	To assess the impact of exposure to contaminants on visual brain development in school-age children	*Visual Evoked Potentials (10.9 years)*: Cord-blood Hg was associated with a reduction of the N75 amplitude at the highest contrast level and with a delay of the N75 latency at the 12% contrast level	Ethier et al. (2012) (Canada)
Cigarette smoke; Hg; Pb	To assess the interactions of prenatal smoking with Pb and Hg	*Teacher Report Form and the Disruptive Behavior Disorders Rating Scale (11.3 years)*: Co-exposure to Pb and Hg do not appear to exacerbate tobacco effects	Desrosiers et al. (2013) (Canada)
MeHg; Pb	To study the effects of low-level postnatal MeHg and Pb exposure	*Rapid Sequential Movements from the Neurological Examination for Subtle Signs, Full-Scale IQ, Broad Reading, Memory and Learning Slope, Attention/Executive Functions from the Neuropsychological Assessment, Adaptive Skills, and Hit Reaction Time (7 years)*: No detectable adverse effect on neuropsychological development due to the relatively low MeHg exposure	Wang et al. (2014) (United States)
PCB; Hg; organochlorine	To evaluate the effects of prenatal neurotoxicant exposures on memory and learning	*Wide Range Assessment of Memory and Learning (8 years)*: Hg was associated with decrements of Visual Memory, Learning, and Verbal Memory	Orenstein et al. (2014) (United States)
PCB; MeHg; Pb	To verify the association of prenatal exposure to neurotoxicants and fine-motor function	*Santa Ana Form Board, Finger Tapping Test (11.3 years)*: High Hg levels were independently associated with poorer Finger Tapping performance	Boucher et al. (2016) (Canada)

Adapted and expanded from Dórea (2017a,b).

Kaufman Assessment Battery for Children showed similar results (Tatsuta et al., 2014).

Studies in schoolchildren (Table 7) and prenatal Hg exposure used different neurobehavioral tests (assessing sensory processing, speech, hearing, and socialization). Sensitivities of tests regarding motor, cognitive, and mental performance are not discussed here. However, the diversity of constitutional and environmental factors coupled to the different combinations of neurotoxicants makes it challenging to extricate the weight of the Hg toxicity. Nevertheless, the outcomes highlight the importance of the statistical models chosen. Neurophysiological outcomes related to response inhibition deficits were differently affected by cord-blood PCBs and Pb but not by Hg (Boucher et al., 2012a).

Significant association of neurodevelopmental delays with cord-blood Hg was seen in few studies (Boucher et al., 2012a, 2016; Tratnik et al., 2017; Yorifuji et al., 2011). Prenatal exposure showed distinct domain-specific adverse effects for PCBs (impaired visual recognition memory), Hg (work memory), and Pb (slower speed of information processing) measured by the Fagan Test of Infant Intelligence and the A-not-B tests (Boucher et al., 2014). Altered fine-motor function in school children that were prenatally exposed to these neurotoxicants in traditional diets showed distinct neurological outcomes: higher current PCB levels were associated with poor Santa Ana Form Board and Finger Tapping performances; higher current Hg levels were independently associated with poor Finger Tapping (Boucher et al., 2016). Recently, Kalia et al. (2017) reported a beneficial effect on Gesell Development Scores in Chinese children by reducing polycyclic aromatic hydrocarbon exposure but no significant association with Hg. Indeed, several studies showed neurotoxic effects of Hg in parallel with other pollutants: Boucher et al. (2012b) reported no MeHg effect per se, but significant interactions with lead and PCB. White et al. (2011) showed that interaction of MeHg and PCB displays more Functional Magnetic Resonance Imaging in wide areas of the brain.

Children on a traditional diet that were prenatally exposed to MeHg, PCBs, and Pb showed significant associations between attention-deficit hyperactivity disorder and neurotoxicants from children's own dietary sources at the time of tests (Boucher et al., 2012b). Gump et al. (2017) showed distinct domain influences of low-level exposures Pb (children with higher Pb levels "exhibit higher levels of hostile distrust and oppositional defiant behaviors, were more dissatisfied and uncertain about their emotions, and had difficulties with communication") and Hg (positively associated with autism spectrum behaviors). Wang et al. (2014) studied low-level MeHg exposure in school children already exposed to Pb and reported no adverse effects on several neurodevelopment tests (Rapid Sequential Movements from the Neurological Examination for Subtle Signs; Full-Scale IQ, Broad Reading, List A Memory and Learning Slope, Attention/Executive Functions from the Neuropsychological Assessment, Adaptive Skills, and Hit Reaction Time).

In order to evaluate cognitive and psychomotor functions in older children, the results of Table 7 are organized as preschool (1–5 years) and school children (>5 years). In preschoolers, performance tests were assessed by cognitive and psychomotor scales (Gesell Development Scores; Bayley Scales of Infant Development; Comprehensive Developmental Inventory for Infants and Toddlers; McCarthy Scales of Children's abilities; Fagan Test of Infant Intelligence; Kaufman Assessment Battery for Children), visual function (visual evoked potentials; A-not-B). Personality tests were assessed as behavioral outcomes (Child Behavior Checklist; Toddler Temperament Scale). During this intermediary stage, the weight of mercury was not outstanding; most outcomes with a significant weight of mercury happened with an unmeasured mining pollution (Table 7).

As the child develops and matures other tests measure behavioral problems (Differential Reinforcement of Low Rates; attention-deficit hyperactivity disorder; Teacher Report Form; Disruptive Behavior Disorders Rating Scale; Neurological Examination for Subtle Signs), learning ability (Wide Range Assessment of Memory and Learning), motor performance (Santa Ana Form Board; Finger Tapping Test) and intelligence (Intelligence Quotient; Wechsler Intelligence Scale for Children-Revised) in school children. Because of the variety and specificity of tests, it is challenging to draw an outcome profile for the exposure combinations. However, it is possible to conclude that unaccounted variables can interfere positively/negatively with neurodevelopment.

Gascon et al. (2017) summarized the pre- and postnatal exposures to neurotoxicants and neuroprotective components co-occurring during childhood; they showed that depending on the study characteristic, for the individual role of Hg there was no effect or a negative effect. In their Spanish cohort, neurodevelopment was seen to be positively impacted by breastfeeding, fish consumption, and nutritional profiles. This overall positive effect of prenatal fish consumption and breastfeeding were also observed in association with students' achievement scores (Schmiedel et al., 2017).

The variation of a specific time window for exposure and comparable age of testing, along with a defined combination of neurotoxicants, makes neurobehavioral outcomes challenging to interpret. Due to environmental and constitutional confounding factors, the children's neurologic outcomes related to low-level exposures to either fish-MeHg or EtHg per se remain unpredicted. In most cases of multiple low-level Hg exposures with other neurotoxicants, deficits are in the normal range and are not prodromes of neurological diseases. However, the clinically unperceived or "subtle" individual neurodevelopmental deficit has societal costs that are reflected in public spending on special education resources and/or in diminished social achievements or work productivity. Economic models to estimate these costs were made for exposures to MeHg (Bellinger et al., 2016; Gaskin et al., 2015; Trasande et al., 2016) and EtHg (Geier et al., 2016, 2018b).

7. CONCLUDING REMARKS

Mercury is one of the most neurotoxic elements found in the environment, and exposure to E^+Hg forms can have long-lasting behavioral consequences. Therefore, the banning of intentional exposure to E^+Hg forms (e.g., thimerosal) should be extended to all young children (not just in rich countries). In addition, unintentional exposure to MeHg should be avoided by adopting preventive attitudes; for instance, by decreasing or eliminating the consumption of piscivorous or predatory fish during gestation and lactation. A link between exposure to low levels of mercury during human brain development and permanent behavioral disabilities remains debatable. However, a massive set of experimental data supports the deleterious neurobehavioral and neurochemical effects of low levels of exposure to Hg, strongly suggesting that E^+Hg forms are contributing factors in neurodevelopmental diseases. Indeed, the neurochemical indications that E^+Hg forms disrupt normal brain functioning at different subcellular and molecular levels are conclusive. For instance, E^+Hg forms have been shown to impair the synaptic biochemistry and physiology at multiple points. Of particular importance, the deregulation of synaptic neurochemistry and molecular structures can have a causal role in neurodegenerative diseases (Poirel et al., 2018). Nevertheless, although many targets for E^+Hg forms have been described, we know little about which of them are involved in the initiating events that will culminate in synaptic dysfunction. As discussed above, co-exposure to mercurial and other neurotoxicants can exacerbate the neurodevelopmental effects of Hg in developing humans, but our knowledge about the molecular events that synergistically or additively exacerbate the neurotoxicity of E^+Hg forms is still elusive. Thus, in addition to taking preventative measures to avoid Hg exposure, the development of new experimental and virtual approaches to study the molecular targets of E^+Hg forms and other neurotoxicants will be crucial to better understand how neurotoxicants inflict permanent neurodevelopmental disabilities on humans.

ACKNOWLEDGMENTS

M.A. was supported in part by grants from the National Institute of Environmental Health Sciences R01 ES07331, R01 ES020852. J.B.T.R. was supported in part by CAPES, CNPq and FAPERGS. C.S.O. and P.A.N. were supported by CAPES.

REFERENCES

Abdallah, E.A.M., Shamoo, A.E., 1984. Protective effect of dimercaptosuccinic acid on methylmercury and mercuric chloride inhibition of rat brain muscarinic acetylcholine receptors. Pest. Biochem. Physiol. 21, 385–393.

Abd-Elfattah, A.S.A., Shamoo, A.E., 1981. Regeneration of a functionally active rat brain muscarinic receptor by D-penicillamine after inhibition with methylmercury and mercuric chloride. Evidence for essential sulfhydryl groups in muscarinic receptor binding sites. Mol. Pharmacol. 20, 492–497.

Adams, J.B., Romdalvik, J., Ramanujam, V.M.S., Legator, M.S., 2007. Mercury, lead, and zinc in baby teeth of children with autism versus controls. J. Toxicol. Environ. Health 70, 1046–1051.

Adedara, I.A., Rosemberg, D.B., Souza, D.O., Kamdem, J.P., Farombi, E.O., Aschner, M., Rocha, J.B.T., 2015. Biochemical and behavioral deficits in the lobster cockroach *Nauphoeta cinerea* model of methylmercury exposure. Toxicol. Res. 4, 442–451.

Agrawal, S.S., Sharma, P., 2017. Current status of mercury level in skin whitening creams. Curr. Med. Res. Pract. 30, 47–50.

Allen, J.W., Mutkus, L.A., Aschner, M., 2001. Methylmercury-mediated inhibition of 3H-D-aspartate transport in cultured astrocytes is reversed by the antioxidant catalase. Brain Res. 902, 92–100.

Alvarez, B., Radi, R., 2003. Peroxynitrite reactivity with amino acids and proteins. Amino Acids 25, 295–311.

Alves, A.C., Monteiro, M.S., Machado, A.L., Oliveira, M., Bóia, A., Correia, A., Oliveira, N., Soares, A.M., Loureiro, S., 2017. Mercury levels in parturient and newborns from Aveiro region, Portugal. J. Toxicol. Environ. Health A 9, 1–3.

AMAP/UNEP, 2013. Technical background report for the global mercury assessment 2013. Arctic Monitoring and Assessment Programme, Oslo, Norway/UNEP Chemicals Branch, Geneva, Switzerland.

American Psychiatric Association (Ed.), 2013. Diagnostic and Statistical Manual of Mental Disorders, fifth ed. American Psychiatric Association, Arlington.

Anderegg, G., Arnaud-Neu, F., Delgado, R., Felcman, J., Popov, K., 2003. Critical evaluation of stability constants of metal complexes of complexones for biomedical and environmental applications. Pure Appl. Chem. 77, 1445–1495.

Andersen, S.L., 2003. Trajectories of brain development: point of vulnerability or window of opportunity? Neurosci. Biobehav. Rev. 27, 3–18.

Arnold, A.P., Canty, A.J., 1983. Methylmercury(II) sulfhydryl interactions. Potentiometric determination of the formation constants for complexation of methylmercury(II) by sulfhydryl containing amino acids and related molecules, including glutathione. Can. J. Chem. 61, 1428–1434.

Arnold, A.P., Tan, K.-S., Rabenstein, D.L., 1986. Nuclear magnetic resonance studies of the solution chemistry of metal complexes. 23. Complexation of methylmercury by selenohydryl-containing amino acids and related molecules. Inorg. Chem. 25, 2433–2437.

Bakir, F., Damluji, S.F., Amin-Zaki, L., Murtadha, M., Khalidi, A., Al-Rawi, N.Y., Tikriti, S., Dhahir, H.I., Clarkson, T.W., Smith, J.C., Doherty, R.A., 1973. Methylmercury poisoning in Iraq. Science 181, 230–241.

Bellinger, D.C., O'Leary, K., Rainis, H., Gibb, H.J., 2016. Country-specific estimates of the incidence of intellectual disability associated with prenatal exposure to methylmercury. Environ. Res. 147, 159–163.

Berg, G.G., Miles, E.F., 1979. Mechanisms of inhibition of active transport ATPases by mercurial. Chem. Biol. Interact. 27, 199–219.

Bernard, A.M., Collette, C., Lauwerys, R., 1992. Renal effects of in utero exposure to mercuric chloride in rats. Arch. Toxicol. 66, 508–513.

Berthon, G., 1995. Critical evaluation of the stability constants of metal complexes of amino acids with polar side chains. Pure Appl. Chem. 67, 1117–1240.

Bilbo, S.D., Block, C.L., Bolton, J.L., Hanamsagar, R., Tran, P.K., 2018. Beyond infection—maternal immune activation by environmental factors, microglial development, and relevance for autism spectrum disorders. Exp. Neurol. 299, 241–251.

Bjørklund, G., Dadar, M., Mutter, J., Aaseth, J., 2017a. The toxicology of mercury: current research and emerging trends. Environ. Res. 159, 545–554.

Bjørklund, G., Mutter, J., Aaseth, J., 2017b. Metal chelators and neurotoxicity: lead, mercury, and arsenic. Arch. Toxicol. 91, 3787–3797.

Bjørklund, G., Aaseth, J., Ajsuvakova, O.P., Nikonorov, A.A., Skalny, A.V., Skalnaya, M.G., Tinkov, A.A., 2017c. Molecular interaction between mercury and selenium in neurotoxicity. Coord. Chem. Rev. 332, 30–37.

Bondy, S.C., Agrawal, A.K., 1980. Methylmercury modulates $GABA_A$ receptor complex differentially in rat cortical and cerebellar membranes in vitro. Arch. Toxicol. 46, 249–256.

Bong, S.M., Moon, J.H., Nam, K.H., Lee, K.S., Chi, Y.M., Hwang, K.Y., 2008. Structural studies of human brain-type creatine kinase complexed with the ADP-Mg2+-NO3-creatine transition-state analogue complex. FEBS Lett. 582, 3959–3965.

Bosch, A.C., O'Neill, B., Sigge, G.O., Kerwath, S.E., Hoffman, L.C., 2016. Mercury accumulation in Yellowfin tuna (*Thunnus albacares*) with regards to muscle type, muscle position and fish size. Food Chem. 190, 351–356.

Boucher, O., Burden, M.J., Muckle, G., Saint-Amour, D., Ayotte, P., Dewailly, É., Nelson, C.A., Jacobson, S.W., Jacobson, J.L., 2012a. Response inhibition and error monitoring during a visual go/no-go task in Inuit children exposed to lead, polychlorinated biphenyls, and methylmercury. Environ. Health Perspect. 120, 608–615.

Boucher, O., Jacobson, S.W., Plusquellec, P., Dewailly, E., Ayotte, P., Forget-Dubois, N., Jacobson, J.L., Muckle, G., 2012b. Prenatal methylmercury, postnatal lead exposure, and evidence of attention deficit/hyperactivity disorder among Inuit children in Arctic Québec. Environ. Health Perspect. 120, 1456–1461.

Boucher, O., Muckle, G., Jacobson, J.L., Carter, R.C., Kaplan-Estrin, M., Ayotte, P., Dewailly, É., Jacobson, S.W., 2014. Domain-specific effects of prenatal exposure to PCBs, mercury, and lead on infant cognition: results from the Environmental Contaminants and Child Development Study in Nunavik. Environ. Health Perspect. 122, 310–316.

Boucher, O., Muckle, G., Ayotte, P., Dewailly, E., Jacobson, S.W., Jacobson, J.L., 2016. Altered fine motor function at school age in Inuit children exposed to PCBs, methylmercury, and lead. Environ. Int. 95, 144–151.

Branco, V., Canário, J., Holmgren, A., Carvalho, C., 2011. Inhibition of the thioredoxin system in the brain and liver of zebra-seabreams exposed to waterborne methylmercury. Toxicol. Appl. Pharmacol. 251, 95–103.

Branco, V., Ramos, P., Canário, J., Lu, J., Holmgren, A., Carvalho, C., 2012. Biomarkers of adverse response to mercury: histopathology versus thioredoxin reductase activity. Biomed. Res. Int. 19, https://doi.org/10.1155/2012/359879.

Branco, V., Godinho-Santos, A., Gonçalves, J., Lu, J., Holmgren, A., Carvalho, C., 2014. Mitochondrial thioredoxin reductase inhibition, selenium status, and Nrf-2 activation are determinant factors modulating the toxicity of mercury compounds. Free Radic. Biol. Med. 73, 95–105.

Branco, V., Caito, S., Farina, M., Rocha, J.B.T., Aschner, M., Carvalho, C., 2017. Biomarkers of mercury toxicity: past, present, and future trends. J. Toxicol. Environ. Health B 20, 119–154.

Brenden, N., Rabbani, H., Abedi-Valugerdi, M., 2001. Analysis of mercury-induced immune activation in nonobese diabetic (NOD) mice. Clin. Exp. Immunol. 125, 202–210.

Bridges, C.C., Zalups, R.K., 2005. Molecular and ionic mimicry and the transport of toxic metals. Toxicol. Appl. Pharmacol. 204, 274–308.

Bridges, C.C., Zalups, R.K., 2010. Transport of inorganic mercury and methylmercury in target tissues and organs. J. Toxicol. Environ. Health B 13, 385–410.

Bridges, C.C., Zalups, R.K., 2017. Mechanisms involved in the transport of mercuric ions in target tissues. Arch. Toxicol. 91, 63–81.

Bridges, C.C., Bauch, C., Verrey, F., Zalups, R.K., 2004. Mercuric conjugates of cysteine are transported by the amino acid transporter System $b^{0,+}$: implications of molecular mimicry. J. Am. Soc. Nephrol. 15, 663–673.

Bridges, C.C., Krasnikov, B.F., Joshee, L., Pinto, J.T., Hallen, A., Li, J., Zalups, R.K., Cooper, A.J.L., 2012. New insights into the metabolism of organomercury compounds: mercury-containing cysteine S-conjugates are substrates of human glutamine transaminase K and potent inactivators of cystathionine γ-lyase. Arch. Biochem. Biophys. 517, 20–29.

Brombach, C.-C., Manorut, P., Kolambage-Dona, P.P.P., Ezzeldin, M.F., Chen, B., Corns, W.T., Feldmann, J., Krupp, E.M., 2017. Methylmercury varies more than one order of magnitude in commercial European rice. Food Chem. 214, 360–365.

Brucker-Davis, F., Ganier-Chauliac, F., Gal, J., Panaïa-Ferrari, P., Pacini, P., Fénichel, P., Hiéronimus, S., 2015. Neurotoxicant exposure during pregnancy is a confounder for assessment of iodine supplementation on neurodevelopment outcome. Neurotoxicol. Teratol. 51, 45–51.

Burbacher, T.M., Shen, D.D., Liberato, N., Grant, K.S., Cernichiari, E., Clarkson, T., 2005. Comparison of blood and brain mercury levels in infant monkeys exposed to methylmercury or vaccines containing thimerosal. Environ. Health Perspect. 113, 1015–1021.

Burger, J., Stern, A.H., Gochfeld, M., 2005. Mercury in commercial fish: optimizing individual choices to reduce risk. Environ. Health Perspect. 113, 266–271.

Cao, Y., Chen, A., Jones, R.L., Radcliffe, J., Caldwell, K.L., Dietrich, K.N., Rogan, W.J., 2010. Does background postnatal methyl mercury exposure in toddlers affect cognition and behavior? Neurotoxicology 31, 1–9.

Carvalho, C.M.L., Chew, E.-H., Hashemy, S.I., Lu, J., Holmgren, A., 2008. Inhibition of the human thioredoxin system: a molecular mechanism of mercury toxicity. J. Biol. Chem. 283, 11913–11923.

Carvalho, C.M.L., Lu, J., Zhang, X., Arnér, E.S.J., Holmgren, A., 2011. Effects of selenite and chelating agents on mammalian thioredoxin reductase inhibited by mercury: implications for treatment of mercury poisoning. FASEB J. 25, 370–381.

Ceccatelli, S., Daré, E., Moors, M., 2010. Methylmercury-induced neurotoxicity and apoptosis. Chem. Biol. Interact. 188, 301–308.

Cernichiari, E., Myers, G.J., Ballatori, N., Zareba, G., Vyas, J., Clarkson, T., 2007. The biological monitoring of prenatal exposure to methylmercury. Neurotoxicology 28, 1015–1022.

Chakraborty, P., 2017. Mercury exposure and Alzheimer's disease in India—an imminent threat? Sci. Total Environ. 589, 232–235.

Chan, T.Y., 2011. Inorganic mercury poisoning associated with skin-lightening cosmetic products. Clin. Toxicol. 49, 886–891.

Chang, Y., Lee, W.Y., Lin, Y.J., Hsu, T., 2017. Mercury (II) impairs nucleotide excision repair (NER) in zebrafish (*Danio rerio*) embryos by targeting primarily at the stage of DNA incision. Aquat. Toxicol. 92, 97–104.

Chauhan, V., Srikumar, S., Aamer, S., Pandareesh, M.D., Chauhan, A., 2017. Methylmercury exposure induces sexual dysfunction in male and female *Drosophila Melanogaster*. Int. J. Environ. Res. Public Health 14, 1108.

Choi, J., Bae, S., Lim, H., Lim, J.A., Lee, Y.H., Ha, M., Kwon, H.J., 2017. Mercury exposure in association with decrease of liver function in adults: a longitudinal study. J. Prev. Med. Public Health 50, 377–385.

Clarkson, T.W., 1972. The pharmacology of mercury compounds. Annu. Rev. Pharmacol. 12, 375–406.

Clarkson, T.W., Magos, L., 2006. The toxicology of mercury and its chemical compounds. Crit. Rev. Toxicol. 36, 609–662.

Comini, M.A., 2016. Measurement and meaning of cellular thiol: disufhide redox status. Free Radic. Res. 50, 246–271.

Copan, L., Fowles, J., Barreau, T., McGee, N., 2015. Mercury toxicity and contamination of households from the use of skin creams adulterated with mercurous chloride (calomel). Int. J. Environ. Res. Public Health 12, 10943–10954.

Cui, W., Liu, G., Bezerra, M., Lagos, D.A., Li, Y., Cai, Y., 2017. Occurrence of methylmercury in rice-based infant cereals and estimation of daily dietary intake of methylmercury for infants. J. Agric. Food Chem. 65, 9569–9578.

Dalla Corte, C.L., Wagner, C., Sudati, J.H., Comparsi, B., Leite, G.O., Busanello, A., Soares, F.A.A., Aschner, M., Rocha, J.B.T., 2013. Effects of diphenyl diselenide on methylmercury toxicity in rats. Biomed. Res. Int. https://doi.org/10.1155/2013/983821

Danielsson, B.R.G., Dencker, L., Orsen, I., Khayat, A., 1984. Fetotoxicity of inorganic mercury in the mouse: distribution and effects on nutrient uptake by placenta and fetus. Biol. Res. Pregnancy Perinatol. 5, 102–109.

Danielsson, B.R.G., Fredriksson, A., Dahlgren, L., Gardlund, A.T., Olsson, L., Dencker, L., Archer, T., 1993. Behavioural effects of prenatal metallic mercury inhalation exposure in rats. Neurotoxicol. Teratol. 15, 391–396.

Daré, E., Fetissov, S., Hökfelt, T., Hall, H., Ore Ögren, S., Ceccatelli, S., 2003. Effects of prenatal exposure to methylmercury on dopamine-mediated locomotor activity and dopamine D2 receptor binding. Arch. Pharm. 367, 500–508.

Davidson, P.W., Myers, G.J., Cox, C., Axtell, C., Shamlaye, C., Sloane-Reeves, J., Cernichiari, E., Needham, L., Choi, A., Wang, Y., Berlin, M., Clarkson, T.W., 1998. Effects of prenatal and postnatal methylmercury exposure from fish consumption on neurodevelopment: outcomes at 66 months of age in the Seychelles Child Development Study. JAMA 280, 701–707.

Davidson, P.W., Myers, G.J., Cox, C., Wilding, G.E., Shamlaye, C.F., Huang, L.S., Cernichiari, E., Sloane-Reeves, J., Palumbo, D., Clarkson, T.W., 2006. Methylmercury and neurodevelopment: longitudinal analysis of the Seychelles child development cohort. Neurotoxicol. Teratol. 28, 529–535.

Davidson, P.W., Strain, J.J., Myers, G.J., Thurston, S.W., Bonham, M.P., Shamlaye, C.F., Stokes-Riner, A., Wallace, J.M., Robson, P.J., Duffy, E.M., Georger, L.A., Sloane-Reeves, J., Cernichiari, E., Canfield, R.L., Cox, C., Huang, L.S., Janciuras, J.,

Clarkson, T.W., 2008a. Neurodevelopmental effects of maternal nutritional status and exposure to methylmercury from eating fish during pregnancy. Neurotoxicology 29, 767–775.

Davidson, P.W., Jean Sloane, R., Myers, G.J., Hansen, O.N., Huang, L.S., Georger, L.A., Cox, C., Thurston, S.W., Shamlaye, C.F., Clarkson, T.W., 2008b. Association between prenatal exposure to methylmercury and visuospatial ability at 10.7 years in the Seychelles child development study. Neurotoxicology 29, 453–459.

Davidson, P.W., Leste, A., Benstrong, E., Burns, C.M., Valentin, J., Sloane-Reeves, J., Huang, L.S., Miller, W.A., Gunzler, D., van Wijngaarden, E., Watson, G.E., Zareba, G., Shamlaye, C.F., Myers, G.J., 2010. Fish consumption, mercury exposure, and their associations with scholastic achievement in the Seychelles Child Development Study. Neurotoxicology 31, 439–447.

De Simone, F., Gencarelli, C.N., Hedgecock, I.M., Pirrone, N., 2016. A modeling comparison of mercury deposition from current anthropogenic mercury emission inventories. Environ. Sci. Technol. 50, 5154–5162.

Debes, F., Budtz-Jorgensen, E., Weihe, P., White, R.F., Grandjean, P., 2006. Impact of prenatal methylmercury exposure on neurobehavioral function at age 14 years. Neurotoxicol. Teratol. 28, 363–375.

Debes, F., Weihe, P., Grandjean, P., 2016. Cognitive deficits at age 22 years associated with prenatal exposure to methylmercury. Cortex 74, 358–369.

Desrosiers, C., Boucher, O., Forget-Dubois, N., Dewailly, E., Ayotte, P., Jacobson, S.W., Jacobson, J.L., Muckle, G., 2013. Associations between prenatal cigarette smoke exposure and externalized behaviors at school age among Inuit children exposed to environmental contaminants. Neurotoxicol. Teratol. 39, 84–90.

Do, S.Y., Lee, C.G., Kim, J.Y., Moon, Y.H., Kim, M.S., Bae, I.H., Song, H.S., 2017. Cases of acute mercury poisoning by mercury vapor exposure during the demolition of a fluorescent lamp factory. Ann. Occup. Environ. Med. 29, 19.

Dórea, J.G., 2017a. Low-dose thimerosal in pediatric vaccines: adverse effects in perspective. Environ. Res. 152, 280–293.

Dórea, J.G., 2017b. Abating mercury exposure in young children should include thimerosal-free vaccines. Neurochem. Res. 42, 2673–2685.

Dórea, J.G., Marques, R.C., Isejima, C., 2012. Neurodevelopment of Amazonian infants: antenatal and postnatal exposure to methyl- and ethylmercury. J. Biomed. Biotechnol. 2012, 132876.

Dórea, J.G., Farina, M., Rocha, J.B.T., 2013. Toxicity of ethylmercury (and Thimerosal): a comparison with methylmercury. J. Appl. Toxicol. 33, 700–711.

Dórea, J.G., Marques, R.C., Abreu, L., 2014. Milestone achievement and neurodevelopment of rural Amazonian toddlers (12 to 24 months) with different methylmercury and ethylmercury exposure. J. Toxicol. Environ. Health A 77, 1–13.

Dröse, S., Brandt, U., Wittig, I., 2014. Mitochondrial respiratory chain complexes as sources and targets of thiol-based redox-regulation. Biochim. Biophys. Acta 1844, 1344–1354.

Drum, D.A., 2009. Are toxic biometals destroying your children's future? Biometals 22, 697–700.

Eide, I., Syversen, T.L.M., 1983. Relationship between catalase activity and uptake of elemental mercury by rat brain. Acta Pharmacol. Toxicol. 52, 217–223.

Ekino, S., Susa, M., Ninomiya, T., Imamura, K., Kitamura, T., 2007. Minamata disease revisited: an update on the acute and chronic manifestations of methyl mercury poisoning. J. Neurol. Sci. 262, 131–144.

Ekstrand, J., Nielsen, J.B., Havarinasab, S., Zalups, R.K., Söderkvist, P., Hultman, P., 2010. Mercury toxicokinetics—dependency on strain and gender. Toxicol. Appl. Pharmacol. 243, 283–291.

El-Ansary, A., Bjorklund, G., Tinkov, A.A., Skalny, A., Dera, H.A., 2017. Relationship between selenium, lead, and mercury in red blood cells of Saudi autistic children. Metab. Brain Dis. 32, 1073–1080.

El-baz, F., Elhossiny, R.M., Elsayed, A.B., Gaber, G.M., 2010. Hair mercury measurement in Egyptian autistic children. Egyptian J. Med. Human Gen. 11, 135–141.

Estes, M.L., McAllister, A.K., 2016. Maternal immune activation: implications for neuropsychiatric disorders. Science 353, 772–777.

Ethier, A.A., Muckle, G., Bastien, C., Dewailly, É., Ayotte, P., Arfken, C., Jacobson, S.W., Jacobson, J.L., Saint-Amour, D., 2012. Effects of environmental contaminant exposure on visual brain development: a prospective electrophysiological study in school-aged children. Neurotoxicology 33, 1075–1085.

Fakhoury, M., 2018. Imaging genetics in autism spectrum disorders: linking genetics and brain imaging in the pursuit of the underlying neurobiological mechanisms. Prog. Neuropsychopharmacol. Biol. Psychiatry 80, 101–114.

Farina, M., Aschner, M., Rocha, J.B.T., 2011a. Oxidative stress in MeHg-induced neurotoxicity. Toxicol. Appl. Pharmacol. 256, 405–417.

Farina, M., Rocha, J.B.T., Aschner, M., 2011b. Mechanisms of methylmercury-induced neurotoxicity: evidence from experimental studies. Life Sci. 89, 555–563.

Farina, M., Aschner, M., Rocha, J.B.T., 2017. The catecholaminergic neurotransmitter system in methylmercury-induced neurotoxicity. In: Aschner, M., Costa, L.G. (Eds.), Advances in Neurotoxicology. vol. 1. Academic Press, pp. 47–81.

Fernández de Cossío, L., Guzmán, A., van der Veldt, S., Luheshi, G.N., 2017. Prenatal infection leads to ASD-like behavior and altered synaptic pruning in the mouse offspring. Brain Behav. Immun. 63, 88–98.

Ferrer-Sueta, G., Manta, B., Botti, H., Radi, R., Trujillo, M., Denicola, A., 2011. Factors affecting protein thiol reactivity and specificity in peroxide reduction. Chem. Res. Toxicol. 24, 434–450.

Fido, A., Al-Saad, S., 2005. Toxic trace elements in the hair of children with autism. Autism 9, 290–298.

Florentine, M.J., Sanfilippo, D.J., 1991. Elemental mercury poisoning. Clin. Pharmacokinet. 10, 213–221.

Fomenko, D.E., Xing, W., Adair, B.M., Thomas, D.J., Gladyshev, V.N., 2007. High-throughput identification of catalytic redox-active cysteine residues. Science, 387–389.

Fomenko, D.E., Marino, S.M., Gladyshev, V.N., 2008. Functional diversity of cysteine residues in proteins and unique features of catalytic redox-active cysteines in thiol oxidoreductases. Mol. Cells 26, 228–235.

Fonfría, E., Vilaró, M.T., Babot, Z., Rodríguez-Farré, E., Suñol, C., 2005. Mercury compounds disrupt neuronal glutamate transport in cultured mouse cerebellar granule cells. J. Neurosci. Res. 79, 545–553.

Franciscato, C., Goulart, F.R., Lovatto, N.M., Duarte, F.A., Flores, E.M.M., Dressler, V.L., Peixoto, N.C., Pereira, M.E., 2009. $ZnCl_2$ exposure protects against behavioral and acetylcholinesterase changes induced by $HgCl_2$. Int. J. Dev. Neurosci. 27, 459–468.

Franco, J.L., Posser, T., Dunkley, P.R., Dickson, P.W., Mattos, J.J., Martins, R., Bainy, A.C.D., Marques, M.R., Dafre, A.L., Farina, M., 2009. Methylmercury neurotoxicity is associated with inhibition of the antioxidant enzyme glutathione peroxidase. Free Radic. Biol. Med. 47, 449–457.

Freire, C., Amaya, E., Gil, F., Fernández, M.F., Murcia, M., Llop, S., Andiarena, A., Aurrekoetxea, J., Bustamante, M., Guxens, M., Ezama, E., Fernández-Tardón, G., Olea, N., 2018. Prenatal co-exposure to neurotoxic metals and neurodevelopment in preschool children: the Environment and Childhood (INMA) Project. Sci. Total Environ. 621, 340–351.

Fuhr, B.J., Rabenstein, D.L., 1973. Nuclear magnetic resonance studies of the solution chemistry of metal complexes. IX. The binding of cadmium, zinc, lead, and mercury by glutathione. J. Amer. Chem. Soc. 95, 6944–6950.

Fujimura, M., Usuki, F., 2017. In situ different antioxidative systems contribute to the site-specific methylmercury neurotoxicity in mice. Toxicology 392, 55–63.

Gao, Z., Ying, X., Yan, J., Wang, J., Cai, S., Yan, C., 2017. Acute mercury vapor poisoning in a 3-month-old infant: a case report. Clin. Chim. Acta 465, 119–122.

Garrecht, M., Austin, D.W., 2011. The plausibility of a role for mercury in the etiology of autism: a cellular perspective. Toxicol. Environ. Chem. 93, 1251–1273.

Gascon, M., Guxens, M., Vrijheid, M., Torrent, M., Ibarluzea, J., Fano, E., Llop, S., Ballester, F., Fernández, M.F., Tardón, A., Fernández-Somoano, A., Sunyer, J., 2017. The INMA-INfancia y Medio Ambiente-(Environment and Childhood) project: more than 10 years contributing to environmental and neuropsychological research. Int. J. Hyg. Environ. Health 220, 647–658.

Gaskin, J., Rennie, C., Coyle, D., 2015. Reducing periconceptional methylmercury exposure: cost-utility analysis for a proposed screening program for women planning a pregnancy in Ontario, Canada. Environ. Health Perspect. 123, 1337–1344.

Geier, D.A., Geier, M.R., 2006. A meta-analysis epidemiological assessment of neurodevelopmental disorders following vaccines administered from 1994 through 2000 in the United States. Neuroendocrinol. Lett. 27, 401–413.

Geier, D.A., Kern, J.K., Homme, K.G., Sykes, L.K., Geier, M.R., 2016. Thimerosal-containing hepatitis B vaccine exposure is highly associated with childhood obesity: a case-control study using the Vaccine Safety Datalink. N. Am. J. Med. Sci. (7), 297–306.

Geier, D.A., Kerna, J.K., Homm, K.G., Geier, M.R., 2018a. A cross-sectional study of the relationship between infant Thimerosal containing hepatitis B vaccine exposure and attention-deficit/hyperactivity disorder. J. Trace Elem. Med. Biol. 46, 1–9.

Geier, D.A., Kern, J.K., Homme, K.G., Geier, M.R., 2018b. A cross-sectional study of the association between infant hepatitis B vaccine exposure in boys and the risk of adverse effects as measured by receipt of special education services. Int. J. Environ. Res. Public Health 15, 123.

Gimenez-Llort, L., Ahlbom, E., Daré, E., Vahter, M., Ogren, S., Ceccatelli, S., 2001. Prenatal exposure to methylmercury changes dopamine-modulated motor activity during early ontogeny: age and gender-dependent effects. Environ. Toxicol. Pharmacol. 9, 61–70.

Glaser, V., Nazari, E.M., Müller, Y.M.R., Feksa, L., Wannmacher, C.M.D., Rocha, J.B.T., de Bem, A.F., Farina, M., Latini, A., 2010a. Effects of inorganic selenium administration in methylmercury-induced neurotoxicity in mouse cerebral cortex. Int. J. Dev. Neurosci. 28, 631–637.

Glaser, V., Leipnitz, G., Straliotto, M.R., Oliveira, J., dos Santos, V.V., Wannmacher, C.M.D., de Bem, A.F., Rocha, J.B.T., Farina, M., Latini, A., 2010b. Oxidative stress-mediated inhibition of brain creatine kinase activity by methylmercury. Neurotoxicology 31, 454–460.

Glaser, V., Moritz, B., Schmitz, A., Dafré, A.L., Nazari, E.M., Müller, Y.M., Feksa, L., Straliottoa, M.R., de Bem, A.F., Farina, M., Rocha, J.B.T., 2013. Protective effects of diphenyl diselenide in a mouse model of brain toxicity. Chem. Biol. Interact. 206, 18–26.

Glaser, V., de Paula Martins, R., Vieira, A.J., de Medeiros Oliveira, E., Straliotto, M.R., Mukdsi, J.H., Torres, A.I., de Bem, A.F., Farina, M., Rocha, J.B.T., De Paul, A.L., 2014. Diphenyl diselenide administration enhances cortical mitochondrial number and activity by increasing hemeoxygenase type 1 content in a methylmercury-induced neurotoxicity mouse model. Mol. Cell. Biochem. 390, 1–8.

Go, Y.M., Chandler, J.D., Jones, D.P., 2015. The cysteine proteome. Free Radic. Biol. Med. 84, 227–245.

Gochfeld, M., Burger, J., 2005. Good fish/bad fish: a composite benefit–risk by dose curve. Neurotoxicology 26, 511–520.

Gonzalez-Raymat, H., Liu, G., Liriano, C., Li, Y., Yin, Y., Shi, J., Jiang, G., Cai, Y., 2017. Elemental mercury: its unique properties affect its behavior and fate in the environment. Environ. Pollut. 229, 69–86.

Grandjean, P., Landrigan, P.J., 2006. Developmental neurotoxicity of industrial chemicals. Lancet 368, 2167–2178.

Grandjean, P., Weihe, P., White, R.F., Debes, F., Araki, S., Yokoyama, K., Murata, K., Sorensen, N., Dahl, R., Jorgensen, P.J., 1997. Cognitive deficit in 7-year-old children with prenatal exposure to methylmercury. Neurotoxicol. Teratol. 19, 417–428.

Grandjean, P., Budtz-Jorgensen, E., Steuerwald, U., Heinzow, B., Needham, L.L., Jorgensen, P.J., Weihe, P., 2003. Attenuated growth of breast-fed children exposed to increased concentrations of methylmercury and polychlorinated biphenyls. FASEB J. 17, 699–701.

Grandjean, P., Budtz-Jorgensen, E., Jorgensen, P.J., Weihe, P., 2005. Umbilical cord mercury concentration as biomarker of prenatal exposure to methylmercury. Environ. Health Perspect. 113, 905–908.

Gu, B., Bian, Y., Miller, C.L., Dong, W., Jiang, X., Liang, L., 2011. Mercury reduction and complexation by natural organic matter in anoxic environments. Proc. Natl. Acad. Sci. U.S.A. 108, 1479–1483.

Gump, B.B., Dykas, M.J., MacKenzie, J.A., Dumas, A.K., Hruska, B., Ewart, C.K., Parsons, P.J., Palmer, C.D., Bendinskas, K., 2017. Background lead and mercury exposures: psychological and behavioral problems in children. Environ. Res. 158, 576–582.

Hachiya, N., 2012. Epidemiological update of methylmercury and Minamata disease. In: Ceccatelli, S., Aschner, M. (Eds.), In: Methylmercury and Neurotoxicity. Current Topics in Neurotoxicity, vol. 2. Springer, Boston, MA.

Halbach, S., Clarkson, T.W., 1978. Enzymatic oxidation of mercury vapor by erythrocytes. Biochim. Biophys. Acta 523, 522–531.

Hancock, R.D., Martell, A.E., 1996. Hard and soft acid-base behavior in aqueous solution: steric effects make some metal ions hard: a quantitative scale of hardness-softness for acids and bases. J. Chem. Educ. 73, 654–661.

Hanwell, M.D., Curtis, D.E., Lonie, D.C., Vandermeersch, T., Zurek, E., Hutchison, G.R., 2012. Avogadro: an advanced semantic chemical editor, visualization, and analysis platform. J. Chem. 4 (1), 17.

Harada, M., 1995. Minamata disease: methylmercury poisoning in Japan caused by environmental pollution. Crit. Rev. Toxicol. 25, 1–24.

Heppner, D.E., Hristova, M., Ida, T., Mijuskovic, A., Dustin, C.M., Bogdándi, V., Fukuto, J.M., Dick, T.P., Nagy, P., Li, J., Akaike, T., van der Vliet, A., 2018. Cysteine perthiosulfenic acid (Cys-SSOH): a novel intermediate in thiol-based redox signaling? Redox Biol. 14, 379–385.

Hodgson, N.W., Waly, M.I., Al-Farsi, Y.M., Al-Sharbati, M.M., Al-Farsi, O., Ali, A., Ouhtit, A., Zang, T., Zhou, Z.S., Deth, R.C., 2014. Decreased glutathione and elevated hair mercury levels are associated with nutritional deficiency-based autism in Oman. Exp. Biol. Med. 239, 697–706.

Horowitz, H.M., Jacob, D.J., Amos, H.M., Streets, D.G., Sunderland, E.M., 2014. Historical mercury releases from commercial products: global environmental implications. Environ. Sci. Technol. 48, 10242–10250.

Hu, Y., Cheng, H., 2012. Mercury risk from fluorescent lamps in China: current status and future perspective. Environ. Int. 44, 141–150.

Huang, X., Wang, C., Li, S., Liu, Y., Zhang, Z., 2014. Highly sensitive visualization of inorganic mercury in mouse neurons using a fluorescent probe. J. Fluoresc. 24, 1313–1317.

Huber, M.L., Laesecke, A., Friend, D.G., 2006. Correlation for the vapor pressure of mercury. Ind. Eng. Chem. Res. 45, 7351–7361.
Hultman, P., Lindh, U., Horsted-Bindslev, P., 1998. Activation of the immune system and systemic immune-complex deposits in Brown Norway rats with dental amalgam restorations. J. Dental Res. 77, 1415–1425.
Ishitobi, H., Stern, S., Thurston, S.W., Zareba, G., Langdon, M., Gelein, R., Weiss, B., 2010. Organic and inorganic mercury in neonatal rat brain after prenatal exposure to methylmercury and mercury vapor. Environ. Health Perspect. 118, 242–248.
Jawaid, M., Ingman, F., 1981. Potentiometric studies on the complex formation between methylmercury(II) and some keto- and amino-carboxylic acids. Talanta 28, 137–143.
Jeong, K.S., Park, H., Ha, E., Shin, J., Hong, Y.-C., Ha, M., Park, H., Kim, B.-N., Lee, B., Lee, S.-J., Lee, K.Y., Kim, J.H., Kim, Y., 2017. High maternal blood mercury level is associated with low verbal IQ in children. J. Korean Med. Sci. 32, 1097–1104.
Johansson, B.B., 1990. The physiology of the blood-brain barrier. In: Porter, J.C., Ježová, D. (Eds.), Circulating Regulatory Factors and Neuroendocrine Function. In: Advances in Experimental Medicine and Biology, vol. 274. Springer, Boston, MA.
Johansson, C., Castoldi, A.F., Onishchenko, N., Manzo, L., Vahter, M., Ceccatelli, S., 2007. Neurobehavioural and molecular changes induced by methylmercury exposure during development. Neurotox. Res. 11, 241–260.
Johnson, F.O., Atchison, W.D., 2009. The role of environmental mercury, lead and pesticide exposure in development of amyotrophic lateral sclerosis. Neurotoxicology 30, 761–765.
Jones, D.P., Go, Y.M., 2011. Mapping the cysteine proteome: analysis of redox-sensing thiols. Curr. Opin. Chem. Biol. 15, 103–112.
Junior, J.M.F.C., Lima, A.A.D.S., Junior, D.R., Khoury, E.D.T., Souza, G.D.S., Silveira, L.C.L., Pinheiro, M.D.C.N., 2017. Emotional and motor symptoms in riverside dwellers exposed to mercury in the Amazon. Rev. Bras. Epidemiol. 20, 212–224.
Kalia, V., Perera, F., Tang, D., 2017. Environmental pollutants and neurodevelopment: review of benefits from closure of a coal-burning power plant in Tongliang, China. Glob. Pediatr. Health 4. 2333794X17721609, https://doi.org/10.1177/2333794X17721609.
Kern, J.K., Geier, D.A., Sykes, L.K., Haley, B.E., Geier, M.R., 2016. The relationship between mercury and autism: a comprehensive review and discussion. J. Trace Elem. Med. Biol. 37, 8–24.
Khaled, E.M., Meguid, N.A., Bjorklund, G., Gouda, A., Bahary, M.H., Hashish, A., Sallam, N.M., Chirumbolo, S., El-Bana, M.A., 2016. Altered urinary porphyrins and mercury exposure as biomarkers for autism severity in Egyptian children with autism spectrum disorder. Metab. Brain Dis. 31, 1419–1426.
Khan, M.A.K., Wang, F., 2009. Mercury-selenium compounds and their toxicological significance: toward a molecular understanding of the mercury-selenium antagonism. Environ. Toxicol. Chem. 28, 1567–1577.
Kim, Y.J., Chai, Y.G., Ryu, J.C., 2005. Selenoprotein W as molecular target of methylmercury in human neuronal cells is down-regulated by GSH depletion. Biochem. Biophys. Res. Commun. 20, 1095–10200.
Kim, S., Eom, S., Kim, H.J., Lee, J.J., Choi, G., Choi, S., Kim, S., Kim, S.Y., Cho, G., Kim, Y.D., Suh, E., Kim, S.K., Kim, S., Kim, G.H., Moon, H.B., Park, J., Kim, S., Choi, K., Eun, S.H., 2017. Association between maternal exposure to major phthalates, heavy metals, and persistent organic pollutants, and the neurodevelopmental performances of their children at 1 to 2years of age- CHECK cohort study. Sci. Total Environ. 624, 377–384.
Kristensen, A.K.B., Thomsen, J.F., Mikkelsen, S., 2014. A review of mercury exposure among artisanal small-scale gold miners in developing countries. Int. Arch. Occup. Environ. Health 87, 579–590.

Kromidas, L., Trombetta, L.D., Jamall, I.S., 1990. The protective effects of glutathione against methylmercury cytotoxicity. Toxicol. Lett. 51, 67–80.

Kudo, A., Turner, R.R., 1999. Mercury contamination of Minamata Bay: historical overview and progress towards recovery. In: Ebinghaus, R., Turner, R.R., Lacerda, L.D., Vasiliev, O., Salomons, W. (Eds.), Mercury Contaminated Sites. Springer, Berlin, Heidelberg, pp. 143–158.

Kusanagi, E., Takamura, H., Chen, S.J., Adachi, M., Hoshi, N., 2018. Children's hair mercury concentrations and seafood consumption in five regions of Japan. Arch. Environ. Contam. Toxicol. 74, 259–272. https://doi.org/10.1007/s00244-017-0502-x.

Kuwabara, J.S., Arai, Y., Topping, B.R., Pickering, I.J., George, G.N., 2007. Mercury speciation in piscivorous fish from mining-impacted reservoirs. Environ. Sci. Technol. 41, 2745–2749.

Kwon, O.-S., Park, Y.-J., 2003. In vitro and in vivo dose-dependent inhibition of methylmercury on glutamine synthetase in the brain of different species. Environ. Toxicol. Pharmacol. 14, 17–24.

Labunskyy, V.M., Hatfield, D.L., Gladyshev, V.N., 2014. Selenoproteins: molecular pathways and physiological roles. Physiol. Rev. 94, 739–777.

Lee, R., Middleton, D., Caldwell, K., Dearwent, S., Jones, S., Lewis, B., Monteilh, C., Mortensen, M.E., Nickle, R., Orloff, K., Reger, M., Risher, J., Rogers, H.S., Watters, M., 2009. A review of events that expose children to elemental mercury in the United States. Cien. Saude Colet. 15, 585–598.

Lee, H., Park, H., Ha, E., Hong, Y.C., Ha, M., Park, H., Kim, B.N., Lee, S.J., Lee, K.Y., Kim, J.H., Jeong, K.S., Kim, Y., 2017. Stability of cognitive development during the first five years of life in relation to heavy metal concentrations in umbilical cord blood: Mothers' and Children's Environmental Health (MOCEH) birth cohort study. Sci. Total Environ. 609, 153–159.

Leterme, B., Blanc, P., Jacques, D., 2014. A reactive transport model for mercury fate in soil—application to different anthropogenic pollution sources. Environ. Sci. Pollut. Res. 21, 12279–12293.

Li, R., Xu, X., Xu, Z., Ao, M., Shang, L., Qiu, G., Tang, S., 2016a. Total mercury and methylmercury concentrations and risk assessments in rice plants collected from a river watershed in a typical mercury mining area of Guizhou. Res. Environ. Sci. 29, 1829–1839.

Li, Y., Zhao, J., Li, Y.-F., Xu, X., Zhang, B., Liu, Y., Cui, L., Li, B., Gao, Y., Chai, Z., 2016b. Comparative metalloproteomic approaches for the investigation proteins involved in the toxicity of inorganic and organic forms of mercury in rice (*Oryza sativa* L.) roots. Metallomics 8, 663–671.

Li, H., Li, H., Li, Y., Liu, Y., Zhao, Z., 2018. Blood mercury, arsenic, cadmium, and lead in children with autism spectrum disorder. Biol. Trace Elem. Res. 181, 31–37.

Liem-Nguyen, V., Skyllberg, U., Nam, K., Björn, E., 2017. Thermodynamic stability of mercury(II) complexes formed with environmentally relevant low-molecular-mass thiols studied by competing ligand exchange and density functional theory. Environ. Chem. 14, 243–253.

Lin, C.C., Chen, Y.C., Su, F.C., Lin, C.M., Liao, H.F., Hwang, Y.H., Hsieh, W.S., Jeng, S.F., Su, Y.N., Chen, P.C., 2013. In utero exposure to environmental lead and manganese and neurodevelopment at 2 years of age. Environ. Res. 123, 52–57.

Loomes, R., Hull, L., Mandy, W.P.L., 2017. What is the male-to-female ratio in autism spectrum disorder? A systematic review and meta-analysis. J. Am. Acad. Child Adolesc. Psychiatry 56, 466–474.

LoPachin, R.M., Gavin, T., DeCaprio, A., Baarber, D.S., 2012. Application of the hard and soft, acids and bases (HSAB) theory to toxicants-target interactions. Chem. Res. Toxicol. 25, 239–251.

Loux, N.T., 1998. An assessment of mercury-species-dependent binding with natural organic carbon. Chem. Spec. Bioavailab. 10, 127–136.
Malagutti, K.S., Silva, A.P., Braga, H.C., Mitozo, P.A., Santos, A.R.S., Dafre, A.L., de Bem, A.F., Farina, M., 2009. 17β-estradiol decreases methylmercury-induced neurotoxicity in male mice. Environ. Toxicol. Pharmacol. 27, 293–297.
Marques, R.C., Dórea, J.G., Manzatto, A.G., Bastos, W.R., Bernardi, J.V., Malm, O., 2007. Time of perinatal immunization, thimerosal exposure and neurodevelopment at 6 months in breastfed infants. Acta Paediatr. 96, 864–868.
Marques, R.C., Bernardi, J.V., Dórea, J.G., Bastos, W.R., Malm, O., 2008. Principal component analysis and discrimination of variables associated with pre- and post-natal exposure to mercury. Int. J. Hyg. Environ. Health 211, 606–614.
Marques, R.C., Dórea, J.G., Bernardi, J.V.E., Bastos, W.R., Malm, O., 2009. Prenatal and postnatal mercury exposure, breastfeeding and neurodevelopment during the first 5 years. Cogn. Behav. Neurol. 22, 134–141.
Marques, R.C., Dórea, J.G., Bernardi, J.V.E., 2010. Thimerosal exposure (from tetanus-diphtheria vaccine) during pregnancy and neurodevelopment of breastfed infants at six months. Acta Paediatr. 99, 1–6.
Marques, R.C., Dórea, J.G., McManus, C., Leão, R.S., Brandão, K.G., Marques, R.C., Vieira, I.H., Guimarães, J.R., Malm, O., 2011. Hydroelectric reservoir inundation (Rio Madeira Basin, Amazon) and changes in traditional lifestyle: impact on growth and neurodevelopment of pre-school children. Public Health Nutr. 14, 661–669.
Marques, R.C., Dórea, J.G., Leão, R.S., Dos Santos, V.G., Bueno, L., Marques, R.C., Brandão, K.G., Palermo, E.F., Guimarães, J.R., 2012. Role of methylmercury exposure (from fish consumption) on growth and neurodevelopment of children under 5 years of age living in a transitioning (tin-mining) area of the western Amazon, Brazil. Arch. Environ. Contam. Toxicol. 62, 341–350.
Marques, R.C., Bernardi, J.V., Dórea, J.G., de Fatima, R., Moreira, M., Malm, O., 2014. Perinatal multiple exposure to neurotoxic (lead, methylmercury, ethylmercury, and aluminum) substances and neurodevelopment at six and 24 months of age. Environ. Pollut. 187, 130–135.
Marques, R.C., Bernardi, J.V., Abreu, L., Dórea, J.G., 2015. Neurodevelopment outcomes in children exposed to organic mercury from multiple sources in a tin-ore mine environment in Brazil. Arch. Environ. Contam. Toxicol. 68, 432–441.
Marques, R.C., Abreu, L., Bernardi, J.V., Dórea, J.G., 2016a. Neurodevelopment of Amazonian children exposed to ethylmercury (from thimerosal in vaccines) and methylmercury (from fish). Environ. Res. 149, 259–265.
Marques, R.C., Abreu, L., Bernardi, J.V., Dórea, J.G., 2016b. Traditional living in the Amazon: extended breastfeeding, fish consumption, mercury exposure and neurodevelopment. Ann. Hum. Biol. 43, 360–370.
Martin, R.B., 2002. Practical hardness scales for metal ion complexes. Inorg. Chim. Acta 339, 27–33.
Martinez-Finley, E., Caito, S., Fretham, S., Chen, P., Aschner, M., 2015. Metal toxicology. In: Mohamed, A. (Ed.), Mammalian Toxicology. John Wiley & Sons, Ltd. https://doi.org/10.1002/9781118683484.
Masur, L.C., 2011. A review of the use of mercury in historic and current ritualistic and spiritual practices. Altern. Med. Rev. 16, 314–320.
Mazerik, J.N., Hagele, T., Sherwani, S., Ciapala, V., Butler, S., Kuppusamy, M.L., Hunter, M., Kuppusamy, P., Marsh, C.B., Parinandi, N.L., 2007. Phospholipase A2 activation regulates cytotoxicity of methylmercury in vascular endothelial cells. Int. J. Toxicol. 26, 553–569.
McCarthy, M.M., Wright, C.L., 2017. Convergence of sex differences and the neuroimmune system in autism spectrum disorder. Biol. Psychiatry 81, 402–410.

Mckeown-Eyssen, G.E., Ruedy, J., Neims, A., 1983. Methyl mercury exposure in northern Quebec. II. Neurologic findings in children. Am. J. Epidemiol. 118, 470–479.
Meltzer, A., Van De Water, J., 2017. The role of the immune system in autism spectrum disorder. Neuropsychopharmacology 42, 284–298.
Miseta, A., Csutora, P., 2000. Relationship between the occurrence of cysteine in proteins and the complexity of organisms. Mol. Biol. Evol. 17, 1232–1239.
Modgil, S., Lahiri, D.K., Sharma, V.L., Anand, A., 2014. Role of early life exposure and environment on neurodegeneration: implications on brain disorders. Transl. Neurodegener. 3, 9.
Mohamed, F.E.B., Zaky, E.A., El-Sayed, A.B., Elhossieny, R.M., Zahra, S.S., Eldin, W.S., Youssef, W.Y., Khaled, R.A., Youssef, A.M., 2015. Assessment of hair aluminum, lead, and mercury in a sample of autistic Egyptian children: environmental risk factors of heavy metals in autism. Behav. Neurol. 2015, 545674.
Montero-Alvarez, A., Fernández de la Campa, M.D., Sanz-Medel, A., 2014. Mercury speciation in Cuban commercial edible fish by HPLC-ICP-MS using the double spike isotope dilution analysis strategy. Int. J. Environ. Anal. Chem. 2, 36–47.
Moraes-Silva, L., Siqueira, L.F., Oliveira, V.A., Oliveira, C.S., Ineu, R.P., Pedroso, T.F., Fonseca, M.M., Pereira, M.E., 2014. Preventive effect of $CuCl_2$ on behavioral alterations and mercury accumulation in central nervous system induced by $HgCl_2$ in newborn rats. J. Biochem. Mol. Toxicol. 28, 328–335.
Mora-Zamorano, F.X., Klingler, R., Basu, N., Head, J., Murphy, C.A., Binkowski, F.P., Larson, J.K., Carvan, M.J., 2017. Developmental methylmercury exposure affects swimming behavior and foraging efficiency of yellow perch (*Perca flavescens*) Larvae. ACS Omega 31, 4870–4877.
Morgan, D.L., Chanda, S.M., Price, H.C., Fernando, R., Liu, J., Brambila, E., O'Connor, R.W., Beliles, R.P., Barone Jr., S., 2002. Disposition of inhaled mercury vapor in pregnant rats: maternal toxicity and effects on developmental outcome. Toxicol. Sci. 66, 261–273.
Mori, K., Yoshida, K., Tani, J., Hoshikawa, S., Ito, S., Watanabe, C., 2006. Ethylmercury inhibits type II 5′-deiodinase activity in NB41A3 neuroblastoma cells. Toxicol. Lett. 161, 96–101.
Mostafazadeh, B., Kiani, A., Mohamadi, E., Shaki, F., Shirazi, F.H., 2013. Mercury exposure of gold mining workers in the northwest of Iran. Pak. J. Pharm. Sci. 26, 1267–1270.
Mousavi, A., 2011. Predicting mercury(II) binding by organic ligands: a chemical model of therapeutic and environmental interests. Environ. Forensics 12, 327–332.
Mrozek-Budzyn, D., Majewska, R., Kieltyka, A., Augustyniak, M., 2012. Neonatal exposure to thimerosal from vaccines and child development in the first 3 years of life. Neurotoxicol. Teratol. 34, 592–597.
Mutter, J., Naumann, J., Sadaghiani, C., Schneider, R., Walach, H., 2004. Alzheimer disease: mercury as pathogenetic factor and apolipoprotein E as a moderator. Neuroendocrinol. Lett. 25, 331–339.
Myers, G.J., Thurston, S.W., Pearson, A.T., Davidson, P.W., Cox, C., Shamlaye, C.F., Cernichiari, E., Clarkson, T.W., 2009. Postnatal exposure to methyl mercury from fish consumption: a review and new data from the Seychelles Child Development Study. Neurotoxicology 30, 338–349.
Naor, M.M., Jensen, J.H., 2004. Determinants of cysteine pKa values in creatine kinase and α1-antitrypsin. Proteins 57, 799–803.
Ni, M., Li, X., Yin, Z., Jiang, H., Sidoryk-Wegrzynowicz, M., Milatovic, D., Cai, J., Aschner, M., 2010. Methylmercury induces acute oxidative stress, altering Nrf2 protein level in primary microglial cells. Toxicol. Sci. 116, 590–603.
Nicolescu, R., Petcu, C., Cordeanu, A., Fabritius, K., Schlumpf, M., Krebs, R., Krämer, U., Winneke, G., 2010. Environmental exposure to lead, but not other neurotoxic metals,

relates to core elements of ADHD in Romanian children: performance and questionnaire data. Environ. Res. 110, 476–483.

Noble, M.J., Decker, S.L., Horowitz, B.Z., 2016. Inhalational mercury toxicity from artisanal gold extraction reported to the Oregon poison center, 2002–2015. Clin. Toxicol. 54, 847–851.

Nogara, P.A., Rocha, J.B.T., 2017. In silico studies of mammalian d-ALAD interactions with selenides and selenoxides. Mol. Inform. 36, 1700091.

Oliveira, C.S., Oliveira, V.A., Ineu, R.P., Moraes-Silva, L., Pereira, M.E., 2012. Biochemical parameters of pregnant rats and their offspring exposed to different doses of inorganic mercury in drinking water. Food Chem. Toxicol. 50, 2382–2387.

Oliveira, C.S., Joshee, L., Zalups, R.K., Pereira, M.E., Bridges, C.C., 2015. Disposition of inorganic mercury in pregnant rats and their offspring. Toxicology 335, 62–71.

Oliveira, C.S., Piccoli, B.C., Aschner, M., Rocha, J.B.T., 2017a. Chemical speciation of selenium and mercury as determinant of their neurotoxicity. In: Aschner, M., Costa, L.G. (Eds.), Neurotoxicity of Metals, Advances in Neurobiology. Springer, Cham. https://doi.org/10.1007/978-3-319-60189-2_4.

Oliveira, C.S., Joshee, L., Bridges, C.C., 2017b. MRP2 and the transport kinetics of cysteine conjugates of inorganic mercury. Biol. Trace Elem. Res. https://doi.org/10.1007/s12011-017-1163-3.

Orenstein, S.T., Thurston, S.W., Bellinger, D.C., Schwartz, J.D., Amarasiriwardena, C.J., Altshul, L.M., Korrick, S.A., 2014. Prenatal organochlorine and methylmercury exposure and memory and learning in school-age children in communities near the New Bedford Harbor Superfund site, Massachusetts. Environ. Health Perspect. 122, 1253–1259.

Oz, S.G., Tozlu, M., Yalcin, S.S., Sozen, T., Sain Guven, G., 2012. Mercury vapor inhalation and poisoning of a family. Inhal. Toxicol. 24, 652–658.

Palmer, N., Beam, A., Agniel, D., Eran, A., Manrai, A., Spettell, C., Steinberg, G., Mandl, K., Fox, K., Nelson, S.F., Kohane, I., 2017. Association of sex with recurrence of autism spectrum disorder among siblings. JAMA Pediatr. 171, 1107–1112.

Palumbo, D.R., Cox, C., Davidson, P.W., Myers, G.J., Choi, A., Shamlaye, C., Sloane-Reeves, J., Cernichiari, E., Clarkson, T.W., 2000. Association between prenatal exposure to methylmercury and cognitive functioning in Seychellois children: a reanalysis of the McCarthy Scales of Children's Ability from the main cohort study. Environ. Res. 84, 81–88.

Pamphlett, R., Jew, S.K., 2013. Heavy metals in locus ceruleus and motor neurons in motor neuron disease. Acta Neuropathol. Commun. 1, 81.

Pamphlett, R., Jew, S.K., 2014. Uptake of inorganic mercury by human locus ceruleus and corticomotor neurons: implications for amyotrophic lateral sclerosis. Acta Neuropathol. Commun. 1, 13.

Pantaleão, T.U., Ferreira, A.C.F., Santos, M.C.S., Figueiredo, A.S.P., Louzada, R.A.N., Rosenthal, D., Carvalho, D.P., Costa, V.M.C., 2017. Effect of thimerosal on thyroid hormones metabolism in rats. Endocr. Connect. 6, 741–747.

Park, J.-D., Zheng, W., 2012. Human exposure and health effects of inorganic and elemental mercury. J. Prev. Med. Public Health 45, 344–352.

Peixoto, N.C., Roza, T., Morsch, V.M., Pereira, M.E., 2007. Behavioral alterations induced by $HgCl_2$ depend on the postnatal period of exposure. Int. J. Dev. Neurosci. 25, 39–46.

Pereira, C.D., Minamino, N., Takao, T., 2015. Free thiol of transthyretin in human plasma most accessible to modification/oxidation. Anal. Chem. 87, 10785–10791.

Poirel, O., Mella, S., Videau, C., Ramet, L., Davoli, M.A., Herzog, E., Katsel, P., Mechawar, N., Haroutunian, V., Epelbaum, J., Daumas, S., Mestikawy, S.E., 2018. Moderate decline in select synaptic markers in the prefrontal cortex (BA9) of patients with Alzheimer's disease at various cognitive stages. Sci. Rep. 8 (938), 2018.

Poole, L.B., 2015. The basics of thiols and cysteines in redox biology and chemistry. Free Radic. Biol. Med. 80, 148–157.

Priya, M.D.L., Geetha, A., 2011. Level of trace elements (copper, zinc, magnesium and selenium) and toxic elements (lead and mercury) in the hair and nail of children with autism. Biol. Trace Elem. Res. 142, 148–158.

Rabenstein, D.L., 1978. The chemistry of methylmercury toxicology. J. Chem. Educ. 54, 292–296.

Rabenstein, D.L., Fairhurst, M.T., 1975. Nuclear magnetic resonance studies of the solution chemistry of metal complexes. XI. The binding of methylmercury by sulfhydryl-containing amino acids and by glutathione. J. Am. Chem. Soc. 97, 2086–2092.

Rabenstein, D.L., Isab, A.A., 1982. A proton nuclear magnetic resonance study of the interaction of mercury with intact human erythrocytes. Biochim. Biophys. Acta 721, 374–384.

Rabenstein, D.L., Reid, R.S., 1984. Nuclear magnetic resonance studies of the solution chemistry of metal complexes. 20. Ligand-exchange kinetics of methylmercury(II)-thiol complexes. Inorg. Chem. 23, 1246–1250.

Rabenstein, D.L., Ozubko, R., Libich, S., Evans, C.A., Fairhurst, M.T., Suvanprakorn, C., 1974. Nuclear magnetic resonance studies of the solution chemistry of metal complexes. X. Determination of formation constants of the methylmercury complexes of selected amines and aminocarboxylic acids. J. Coord. Chem. 3, 263–271.

Rabenstein, D.L., Reid, R.S., Yamashita, G., Tan, K.S., Arnold, A.P., 1986. Determination of thiols and selenols by titration with methylmercury with end point detection by nuclear magnetic resonance spectrometry. Anal. Chem. 58, 1266–1269.

Reinhardt, J.W., 1992. Side-effects: mercury contribution to body burden from dental amalgam. Adv. Dent. Res. 6, 110–113.

Reisz, J.A., Bechtold, E., King, S.B., Poole, L.B., Furdui, C.M., 2013. Thiol-blocking electrophiles interfere with labeling and detection of protein sulfenic acids. FEBS J. 280, 6150–6161.

Riemer, J., Schwarzländer, M., Conrad, M., Herrmann, J.M., 2015. Thiol switches in mitochondria: operation and physiological relevance. Biol. Chem. 396, 465–482.

Risher, J.F., Nickle, R.A., Amler, S.N., 2003. Elemental mercury poisoning in occupational and residential settings. Int. J. Hyg. Environ. Health 206, 371–379.

Rocha, J.B.T., Saraiva, R.A., Garcia, S.C., Gravina, F.S., Nogueira, C.W., 2012. Aminolevulinate dehydratase (δ-ALA-D) as marker protein of intoxication with metals and other pro-oxidant situations. Toxicol. Res. 1, 85–102.

Rocha, J.B.T., Piccoli, B.C., Oliveira, C.S., 2017. Biological and chemical interest in selenium: a brief historical account. Arkivoc Part ii, 457–491. https://doi.org/10.3998/ark.5550190.p009.784.

Rodrigues, J., Branco, V., Lu, J., Holmgren, A., Carvalho, C., 2015. Toxicological effects of thiomersal and ethylmercury: inhibition of the thioredoxin system and NADP+-dependent dehydrogenases of the pentose phosphate pathway. Toxicol. Appl. Pharmacol. 286, 216–223.

Rooney, J.P.K., 2014. The retention time of inorganic mercury in the brain—a systematic review of the evidence. Toxicol. Appl. Pharmacol. 274, 425–435.

Rossi, A.D., Ahlbom, E., Ogren, S.O., Nicotera, P., Ceccatelli, S., 1997. Prenatal exposure to methylmercury alters locomotor activity of male but not female rats. Exp. Brain Res. 117, 428–436.

Rothenberg, S.E., Yu, X., Liu, J., Biasini, F.J., Hong, C., Jiang, X., Nong, Y., Cheng, Y., Korrick, S.A., 2016. Maternal methylmercury exposure through rice ingestion and offspring neurodevelopment: a prospective cohort study. Int. J. Hyg. Environ. Health 219, 832–842.

Rudyk, O., Eaton, P., 2014. Biochemical methods for monitoring protein thiol redox states in biological systems. Redox Biol. 2, 803–813.

Ruggieri, F., Majorani, C., Domanico, F., Alimonti, A., 2017. Mercury in children: current state on exposure through human biomonitoring studies. Int. J. Environ. Res. Public Health 14, 519.

Sahin, D., Erdolu, C.O., Karadenizli, S., Kara, A., Bayrak, G., Beyaz, S., Demir, B., Ateş, N., 2017. Effects of gestational and lactational exposure to low dose mercury chloride ($HgCl_2$) on behaviour, learning and hearing thresholds in WAG/Rij rats. EXCLI J. 15, 391–402.

Saint-Amour, D., Roy, M.S., Bastien, C., Ayotte, P., Dewailly, E., Després, C., Gingras, S., Muckle, G., 2006. Alterations of visual evoked potentials in preschool Inuit children exposed to methylmercury and polychlorinated biphenyls from a marine diet. Neurotoxicology 27, 567–578.

Sasakura, C., Suzuki, K.T., 1998. Biological interaction between transition metals (Ag, Cd and Hg), selenide/sulfide and selenoprotein P. J. Inorg. Biochem. 71, 159–162.

Schmidt, L., Bizzi, C.A., Duarte, F.A., Dressler, V.L., Flores, E.M., 2013. Evaluation of drying conditions of fish tissues for inorganic mercury and methylmercury speciation analysis. Microchem. J. 108, 53–59.

Schmiedel, V., Vogt, H., Walach, H., 2017. Are pupils' 'Programme for International Student Assessment (PISA)' scores associated with a nation's fish consumption? Scand. J. Public Health. https://doi.org/10.1177/1403494817717834.

Schuurs, A.H.B., 1999. Reproductive toxicity of occupational mercury. A review of the literature. J. Dent. 27, 249–256.

Seby, F., Potin-Gautier, M., Giffaut, E., Borge, G., Donard, O.F.X., 2001. A critical review of thermodynamic data for selenium species at 25 °C. Chem. Geol. 171, 173–194.

Shen, J.C., Huang, Z.Y., Zhuang, Z.X., Wang, X.R., Lee, F.S.C., 2005. Investigation of mercury metallothionein complexes in tissues of rat after oral intake of $HgCl_2$. Appl. Organomet. Chem. 19, 140–146.

Shenker, B.J., Guo, T.L., Insug, O., Shapiro, I.M., 1999. Induction of apoptosis in human T-cells by methyl mercury: temporal relationship between mitochondrial dysfunction and loss of reductive reserve. Toxicol. Appl. Pharmacol. 157, 23–35.

Shenker, B.J., Pankoski, L., Zekavat, A., Shapiro, I.M., 2002. Mercury-induced apoptosis in human lymphocytes: caspase activation is linked to redox status. Antioxid. Redox Signal. 4, 379–389.

Shikara, I.A., Al Dabbagh, T.Q., 1973. Neurological manifestations in organic mercury poisoning. Ann. Coll. Med. Mosul. 4, 33–42.

Short, J.D., Downs, K., Tavakoli, S., Asmis, R., 2016. Protein thiol redox signaling in monocytes and macrophages. Antioxid. Redox Signal. 25, 816–835.

Sippel, T.O., 1973. The histochemistry of thiols and disulphides. I. The use of N-(4-aminophenyl) maleimide for demonstrating thiol groups. Histochem. J. 5, 413–423.

Skyllberg, U., 2008. Competition among thiols and inorganic sulfides and polysulfides for Hg and MeHg in wetland soils and sediments under suboxic conditions: illumination of controversies and implications for MeHg net production. J. Geophys. Res. 113. https://doi.org/10.1029/2008jg000745.

Skyllberg, U., Bloom, P.R., Qian, J., Lin, C.M., Bleam, W.F., 2006. Complexation of mercury(II) in soil organic matter: EXAFS evidence for linear two-coordination with reduced sulfur groups. Environ. Sci. Technol. 40, 4174–4180.

Sood, P.P., Bapu, C., Vijayalakshmi, K., 1995. Vitamins and monothiols efficacy in the restoration of adenosine nucleotide degradation enzymes altered during methylmercury intoxication. J. Environ. Pathol. Toxicol. Oncol. 14, 101–105.

Stachowiak, O., Dolder, M., Wallimann, T., Richter, C., 1998. Mitochondrial creatine kinase is a prime target of peroxynitrite-induced modification and inactivation. J. Biol. Chem. 273, 16694–16699.

Stewart, P.W., Reihman, J., Lonky, E.I., Darvill, T.J., Pagano, J., 2003. Cognitive development in preschool children prenatally exposed to PCBs and MeHg. Neurotoxicol. Teratol. 25, 11–22.

Stewart, P.W., Sargent, D.M., Reihman, J., Gump, B.B., Lonky, E., Darvill, T., Hicks, H., Pagano, J., 2006. Response inhibition during differential reinforcement of low rates (DRL) schedules may be sensitive to low-level polychlorinated biphenyl, methylmercury, and lead exposure in children. Environ. Health Perspect. 114, 1923–1929.

Stricks, W., Kolthoff, I.M., 1953. Reactions between mercuric mercury and cysteine and glutathione. Apparent dissociation constants, heats and entropies of formation of various forms of mercuric mercaptocysteine and -glutathione. J. Am. Chem. Soc. 75, 5673–5681.

Stringari, J., Nunes, A.K.C., Franco, J.L., Bohrer, D., Garcia, S.C., Dafre, A.L., Milatovic, D., Souza, D.O., Rocha, J.B.T., Aschner, M., Farina, M., 2008. Prenatal methylmercury exposure hampers glutathione antioxidant system ontogenesis and causes long-lasting oxidative stress in the mouse brain. Toxicol. Appl. Pharmacol. 227, 147–154.

Stroustrup, A., Hsu, H.H., Svensson, K., Schnaas, L., Cantoral, A., Solano González, M., Torres-Calapiz, M., Amarasiriwardena, C., Bellinger, D.C., Coull, B.A., Téllez-Rojo, M.M., Wright, R.O., Wright, R.J., 2016. Toddler temperament and prenatal exposure to lead and maternal depression. Environ. Health 15, 71.

Sugiura, Y., Hojo, Y., Tamai, Y., Tanaka, H., 1976. Selenium protection against mercury toxicity. Binding of methylmercury by the selenohydryl-containing ligand. J. Am. Chem. Soc. 98, 2339–2341.

Sun, G.F., Hu, W.T., Yuan, Z.H., Zhang, B.A., Lu, H., 2017. Characteristics of mercury intoxication induced by skin-lightening products. Chin. Med. J. (Engl.) 1, 130.

Suzuki, K., Nakai, K., Sugawara, T., Nakamura, T., Ohba, T., Shimada, M., Hosokawa, T., Okamura, K., Sakai, T., Kurokawa, N., Murata, K., Satoh, C., Satoh, H., 2010. Neurobehavioral effects of prenatal exposure to methylmercury and PCBs, and seafood intake: neonatal behavioral assessment scale results of Tohoku study of child development. Environ. Res. 110, 699–704.

Tatsuta, N., Nakai, K., Murata, K., Suzuki, K., Iwai-Shimada, M., Yaginuma-Sakurai, K., Nakamura, T., Hosokawa, T., Satoh, H., 2012. Prenatal exposures to environmental chemicals and birth order as risk factors for child behavior problems. Environ. Res. 114, 47–52.

Tatsuta, N., Nakai, K., Murata, K., Suzuki, K., Iwai-Shimada, M., Kurokawa, N., Hosokawa, T., Satoh, H., 2014. Impacts of prenatal exposures to polychlorinated biphenyls, methylmercury, and lead on intellectual ability of 42-month-old children in Japan. Environ. Res. 133, 321–326.

Thompson, S.A., White, C.C., Krejsa, C.M., Diaz, D., Woods, J.S., Eaton, D.L., Kavanagh, T.J., 1999. Induction of glutamate-cysteine ligase (γ-glutamylcysteine synthetase) in the brains of adult female mice subchronically exposed to methylmercury. Toxicol. Lett. 110, 1–9.

Trasande, L., DiGangi, J., Evers, D.C., Petrlik, J., Buck, D.G., Šamánek, J., Beeler, B., Turnquist, M.A., Regan, K., 2016. Economic implications of mercury exposure in the context of the global mercury treaty: hair mercury levels and estimated lost economic productivity in selected developing countries. J. Environ. Manage. 183, 229–235.

Tratnik, J.S., Falnoga, I., Trdin, A., Mazej, D., Fajon, V., Miklavčič, A., Kobal, A.B., Osredkar, J., Sešek Briški, A., Krsnik, M., Neubauer, D., Kodrič, J., Stropnik, S., Gosar, D., Lešnik Musek, P., Marc, J., Jurkovič Mlakar, S., Petrović, O., Vlašić-Cicvarić, I., Prpić, I., Milardović, A., Radić Nišević, J., Vuković, D., Fišić, E., Špirić, Z., Horvat, M., 2017. Prenatal mercury exposure, neurodevelopment and apolipoprotein E genetic polymorphism. Environ. Res. 152, 375–385.

Trott, O., Olson, A.J., 2010. AutoDock Vina: improving the speed and accuracy of docking with a new scoring function, efficient optimization and multithreading. J. Comput. Chem. 31, 455–461.

Verity, M.A., Sarafian, T., Pacifici, E.H.K., Sevanian, A., 1994. Phospholipase A2 stimulation by methyl mercury in neuron culture. J. Neurochem. 62, 705–714.

Vimy, M.J., Takahashi, Y., Lorscheider, F.L., 1990. Maternal-fetal distribution of mercury (^{203}Hg) released from dental amalgam fillings. Am. J. Physiol. 258, 939–945.

Wagner, C., Sudati, J.H., Nogueira, C.W., Rocha, J.B.T., 2010. In vivo and in vitro inhibition of mice thioredoxin reductase by methylmercury. Biometals 23, 1171–1177.

Walach, H., Mutter, J., Deth, R., 2014. Inorganic mercury and Alzheimer's disease—results of a review and a molecular mechanism. In: Martin, C.R., Preedy, V. (Eds.), Diet and Nutrition in Dementia and Cognitive Decline. Elsevier Inc., pp. 593–601. (Chapter 55).

Wang, P.F., McLeish, M.J., Kneen, M.M., Lee, G., Kenyon, G.L., 2001. An unusually low pKa for Cys282 in the active site of human muscle creatine kinase. Biochemistry 40, 11698–11705.

Wang, Y., Chen, A., Dietrich, K.N., Radcliffe, J., Caldwell, K.L., Rogan, W.J., 2014. Postnatal exposure to methyl mercury and neuropsychological development in 7-year-old urban inner-city children exposed to lead in the United States. Child Neuropsychol. 20, 527–538.

Watson, G.E., Evans, K., Thurston, S.W., van Wijngaarden, E., Wallace, J.M.W., McSorley, E.M., Bonham, M.P., Mulhern, M.S., McAfee, A.J., Davidson, P.W., Shamlaye, C.F., Strain, J.J., Love, T., Zareba, G., Myers, G.J., 2012. Prenatal exposure to dental amalgam in the Seychelles Child Development Nutrition Study: associations with neurodevelopmental outcomes at 9 and 30 months. Neurotoxicology 33, 1511–1517.

Watson, G.E., van Wijngaarden, E., Love, T.M.T., McSorley, E.M., Bonham, M.P., Mulhern, M.S., Yeates, A.J., Davidson, P.W., Shamlaye, C.F., Strain, J.J., Thurston, S.W., Harrington, D., Zareba, G., Wallace, J.M.W., Myers, G.J., 2013. Neurodevelopmental outcomes at 5 years in children exposed prenatally to maternal dental amalgam: The Seychelles Child Development Nutrition Study. Neurotoxicol. Teratol. 39, 57–62.

Weeb, M., Cain, K., 1982. Functions of metallothionein. Biochem. Pharmacol. 31, 137–142.

Weerapana, E., Wang, C., Simon, G.M., Richter, F., Khare, S., Dillon, M.B.D., Bachovchi, D.A., Mowen, K., Baker, D., Cravatt, B.F., 2010. Quantitative reactivity profiling predicts functional cysteines in proteomes. Nature 468, 790–797.

White, R.F., Palumbo, C.L., Yurgelun-Todd, D.A., Heaton, K.J., Weihe, P., Debes, F., Grandjean, P., 2011. Functional MRI approach to developmental methylmercury and polychlorinated biphenyl neurotoxicity. Neurotoxicology 32, 975–980.

WHO (World Health Organization), 2007. Exposure to Mercury: A Major Public Health Concern. Environmental Health Criteria. World Health Organization, Geneva.

Wible, R.S., Sutter, T.R., 2017. Soft cysteine signaling network: the functional significance of cysteine in protein function and the soft acids/bases thiol chemistry that facilitates cysteine modification. Chem. Res. Toxicol. 30, 729–762.

Wu, C., Wu, G., Wang, Z., Zhang, Z., Qian, Y., Ju, L., 2018. Soil mercury speciation and accumulation in rice (*Oryza sativa* L.) grown in wastewater-irrigated farms. Appl. Geochem. 89, 202–209.

Yamamoto, R., Suzuki, T., Satoh, H., Kawais, K., 1986. Generation and dose as modifying factors of inorganic mercury accumulation in brain, liver, and kidneys of rats fed methylmercury. Environ. Res. 41, 309–318.

Yorifuji, T., Debes, F., Weihe, P., Grandjean, P., 2011. Prenatal exposure to lead and cognitive deficit in 7- and 14-year-old children in the presence of concomitant exposure to similar molar concentration of methylmercury. Neurotoxicol. Teratol. 33, 205–211.

Yoshida, M., Satoh, M., Shimada, A., Yamamoto, E., Yasutake, A., Tohyama, C., 2002. Maternal-to-fetus transfer of mercury in metallothionein-null pregnant mice after exposure to mercury vapor. Toxicology 175, 215–222.

Yoshida, M., Watanabe, C., Horie, K., Satoh, M., Sawada, M., Shimada, A., 2005. Neurobehavioral changes in metallothionein-null mice prenatally exposed to mercury vapor. Toxicol. Lett. 155, 361–368.

Yoshida, E., Abiko, Y., Kumagai, Y., 2014. Glutathione adduct of methylmercury activates the Keap1–Nrf2 pathway in SH-SY5Y cells. Chem. Res. Toxicol. 27, 1780–1786.
Zalups, R.K., 1993. Early aspects of the intrarenal distribution of mercury after the intravenous administration of mercuric chloride. Toxicology 79, 215–228.
Zalups, R.K., 2000. Molecular interactions with mercury in the kidney. Pharmacol. Rev. 52, 113–143.
Zalups, R.K., Ahmad, S., 2004. Homocysteine and the renal epithelial transport and toxicity of inorganic mercury: role of basolateral transporter organic anion transporter 1. J. Am. Soc. Nephrol. 15, 2023–2031.
Zalups, R.K., Barfuss, D.W., 1995. Pretreatment with p-aminohippurate inhibits the renal uptake and accumulation of injected inorganic mercury in the rat. Toxicology 103, 23–35.
Zemolin, A.P.P., Meinerz, D.F., de Paula, M.T., Mariano, D.O.C., Rocha, J.B.T., Pereira, A.B., Posser, T., Franco, J.L., 2012. Evidences for a role of glutathione peroxidase 4 (GPx4) in methylmercury induced neurotoxicity in vivo. Toxicology 302, 60–67.

FURTHER READING

Go, Y.M., Jones, D.P., 2011. Cysteine/cystine redox signaling in cardiovascular disease. Free Radic. Biol. Med. 50, 495–509.

CHAPTER THREE

Connections Between Manganese Neurotoxicity and Neurological Disease

Rachana Nitin[*,†,‡], **Aaron B. Bowman**[*,†,‡,1]
[*]Vanderbilt University, Vanderbilt University Medical Center, Department of Pediatrics, 1161 21st Avenue S, D-4105 MCN, Nashville, TN, United States
[†]Vanderbilt University, Vanderbilt University Medical Center, Department of Neurology, 1161 21st Avenue S, D-4105 MCN, Nashville, TN, United States
[‡]Vanderbilt University, Vanderbilt University Medical Center, Department of Biochemistry, 1161 21st Avenue S, D-4105 MCN, Nashville, TN, United States
[1]Corresponding author: e-mail address: aaron.bowman@vanderbilt.edu

Contents

1. INTRODUCTION

The normal, physiological concentration of manganese in the human brain is about 5.32–14.03 ng Mn/mg protein, while 15.96–42.09 ng Mn/mg protein is the pathophysiological threshold (Aschner and Dorman, 2006; Chen et al., 2015). Mn is an essential divalent trace metal, required as a cofactor for a diverse set of enzymes like arginase, Mn-superoxide dismutase, and glutamine synthetase. As a result, Mn like other essential metals is highly regulated to ensure proper health (Horning et al., 2015). The highest levels of Mn within the body are found in the liver, pancreas, kidney, bone and

Advances in Neurotoxicology, Volume 2
ISSN 2468-7480
https://doi.org/10.1016/bs.ant.2018.03.001

brain (Chen et al., 2018). Although the brain does not have the highest comparative Mn concentrations normally, it is the primary target of Mn accumulation and neurotoxicity. The naturally elevated levels of Mn that occur normally in brain, as compared to other peripheral organs, present a role for Mn in brain physiology and function (Kumar et al., 2014; O'Neal et al., 2014). Within the brain, Mn is differentially localized, with the highest distribution in the caudate, putamen and the globus pallidus and the lowest in the pons and medulla (Aschner et al., 2007; Chen et al., 2016). The primary route of Mn intake is dietary sources, and it is absorbed in the intestinal lumen especially in the duodenum and jejunum and excreted through the biliary system (Horning et al., 2015). With a multitude of dietary sources, Mn deficiency is thought to occur rarely; Mn accumulation on the other hand, especially through repeated environmental exposure (mines, metal factories, pesticides, water/food contamination), via the digestive, pulmonary, and olfactory bulb (inhalation of Mn-laden particulates) poses a hazard to the nervous system.

2. ESTABLISHED ROLES OF Mn IN NEUROLOGICAL DYSFUNCTION

2.1 Role of Mn in Neurodegenerative Disease

Mn accumulation upon overexposure occurs primarily in nuclei of the basal ganglia (e.g., globus pallidus, striatum, substantia nigra), but also in other brain regions including thalamus and cortex. Neurotoxic exposure is associated with a disorder known as manganism, an irreversible, progressive, neurodegenerative disorder classified as a parkinsonian disorder due to its similarity to Parkinson's disease (PD). It is not a common disorder, since it is dependent on repeated high dose exposures (Chakraborty and Aschner, 2012; Chen et al., 2015; Dobson et al., 2004; Earley et al., 2011). The physical symptoms of manganism are distinct from PD and are dystonia, rigid posture and gait instability. Patients with manganism also show cognitive loss and emotional instability, once the neuronal degeneration spreads to the substantia nigra, thereby affecting DA neurons (Bouabid et al., 2014; Lao et al., 2017; Marshall et al., 1997). Some of the differences between PD and manganism are that dopaminergic degeneration is not as severe as in PD, and levodopa treatment is not as effective (Chen et al., 2018). Huntington's disease (HD) is another neurodegenerative disease that shows Mn imbalance. In HD there are a loss of striatal neurons accompanied by chorea and other motor and non-motor symptoms. There is a growing body of evidence that a deficiency in bioavailable Mn in these neurons is due to impaired Mn handling (Horning et al., 2015; Madison et al., 2012; Williams et al., 2010). HD, which also affects the basal ganglia, has motor symptoms and a lower enzymatic activity of arginase II, and other Mn-dependent enzymes (Bichell et al., 2017; Horning et al., 2015; O'Neal and Zheng, 2015; Tidball et al., 2015). Since the imbalance of Mn in neural disorders mostly occurs in the substantia nigra, globus pallidus and striatum—regions which either produce DA or receive abundant dopaminergic inputs, it supports a link between Mn and DA metabolism or function in Mn associated neurological disease (Sistrunk et al., 2007; Stredrick et al., 2004).

2.2 Connections Between Mn Neurotoxicity to DA Neural Dysfunction

Mn is an essential metal, but excessive exposure in occupational and environmental settings may be accompanied by neurotoxicity and its accumulation in subcortical structures of the basal ganglia namely the substantia nigra, globus pallidus and the caudate/putamen of the striatum all of which are highly sensitive to oxidative injury (Cersosimo and Koller, 2006; Guilarte, 2010; Newland et al., 1989; Olanow, 2004). Chronic childhood exposure to Mn via drinking water preferentially impacts basal ganglia structures (Lao et al., 2017). Excessive exposure to Mn also occurs in subjects administered parenteral nutrition (Hsieh et al., 2007; Iwase et al., 2002), and other patients including those with liver dysfunction or Mn associated drug-abuse (Aschner et al., 2015; Henn et al., 2011; Ohtake et al., 2005; Pratesi et al., 2008; Rivera-Mancía et al., 2011; Sikk et al., 2011; Stepens et al., 2008). Finally, iron (Fe) deficiency anemia, a risk factor for increased Mn accumulation, affects ~1.6 billion people worldwide (McLean et al., 2008).

Excessive brain Mn deposition leads to an extrapyramidal syndrome referred to as manganism with parkinsonian clinical features including rigidity, tremor, dystonic movements and bradykinesia (Calne et al., 1994; Cersosimo and Koller, 2006; Guilarte, 2010; Olanow, 2004). Idiopathic PD is a neurodegenerative disorder with a strong environmental etiology resulting in impaired speech and facial expressions, bradykinesia and motor rigidity associated with the dysfunction and degeneration of dopaminergic neurons in the substantia nigra (Sulzer, 2007). Exposure to excessive Mn is a putative risk factor for PD (Gorell et al., 2004). Further, a dose-dependent progression of parkinsonism features has been observed with exposure to Mn-containing welding fumes (Racette et al., 2017). Parkinsonism in welders (vs. non-welders) is clinically distinguishable only by age of onset

(46 vs. 63 years, respectively), with higher prevalence of PD among welders compared with age-standardized individuals in the general population (Criswell et al., 2011; Racette et al., 2005, 2012). However, the relationship between Mn, DA neurobiology and PD/parkinsonism is not straightforward. Mn toxicity has been associated with alpha-synuclein accumulation (a major component of Lewy bodies, and a pathological hallmark of PD), but this effect may be neuroprotective (Harischandra et al., 2015; Peres et al., 2016; Tong et al., 2009). In contrast, at least some patients with Mn-induced parkinsonism do not produce Lewy bodies and are unresponsive to levodopa (i.e., DA replacement therapy) (Cersosimo and Koller, 2006; Guilarte and Gonzales, 2015; Kwakye et al., 2015). Some early reports suggest levodopa treatment and/or DA may even enhance Mn neurotoxicity (Benedetto et al., 2010; Parenti et al., 1986, 1988). However, Mn toxicity has been linked to (a) reduced tyrosine hydroxylase (TH—the rate limiting enzyme in DA production), (b) alterations in DA and its metabolites (either up or down, potentially related to the Mn exposure timing or other exposure differences), (c) reduced DA transporter (DAT) cell surface expression, and (d) inconsistent reports of impaired DA neurotransmission and reduced dopaminergic neuron viability (Aschner and Erikson, 2009; Bowman et al., 2011; Guilarte, 2010; Horning et al., 2015; Kwakye et al., 2015; Madison et al., 2012; Zhang et al., 2011). Thus, while excess Mn exposure is clearly neurotoxic, eliciting cognitive deficits and parkinsonian features, the mechanism of this neurotoxicity remains unclear, but likely involves dopaminergic neuronal dysfunction or DA-related toxicity.

2.3 Role of Mn Neurotoxicity in Neurodevelopment Disorders

Although Mn exposure during adulthood can lead to neural dysfunction, excessive accumulation during fetal and childhood stages has profound consequences on neural development. Studies conducted in North Carolina and Bangladesh have shown that there is an increase in the infant mortality rate associated with an increase in Mn contamination of water, and lower IQ and cognitive scores, respectively (Chen et al., 2016; Spangler and Spangler, 2009; Wasserman et al., 2011). Further, children ingesting Mn-laden well water show inattention, hyperactivity and impulsivity (Brooks et al., 2017; Dion et al., 2018; Grandjean and Landrigan, 2014; Lao et al., 2017). Early post-natal Mn exposure in rats showed a persistent impairment of motor coordination and cognitive function (object recognition) (Beaudin et al., 2013, 2017a,b). Similarly, rats exposed to Mn during the fetal stages through Mn-fed pregnant dams demonstrated impaired learning and spatial memory (Amos-Kroohs et al., 2017; Peres et al., 2015). Other chronic Mn-based developmental studies in early post-natal mice found that there is an aberration in adult neurogenesis in the hippocampus (Kikuchihara et al., 2015), while Mn-exposed pregnant rats produced offspring with increased microglial activation in the dentate gyrus, which negatively affected the late-stage differentiation of granular cells in that region (Abe et al., 2015; Wang et al., 2013). These studies point to a disruptive effect of excessive Mn levels on the developing brain which may adversely affect memory, learning and attention.

3. EVIDENCE TO SUPPORT A LINK BETWEEN Mn AND OTHER NEUROLOGICAL DISORDERS

3.1 Restless Legs Syndrome

Primary restless legs syndrome (RLS) or Willis–Ekbom disease is a neurological sensorimotor disorder associated with an intense need to move one's legs to ease discomfort (Klingelhoefer et al., 2016; Koo et al., 2016). Primary RLS has both familial and idiopathic forms, with ~60% of patients having a family history of RLS in a first degree relative (Allen et al., 2017a; Mitterling et al., 2015; Picchietti et al., 2017). RLS has a preponderance in people of European descent, who show an occurrence rate up to ~14.5%. It is one of the most common sleep-related disorders in the European and North American general population, and normally diagnosed between 45 and 55 years of age, with a marked prevalence in females over males (Allen et al., 2005; Koo 2015; Ohayon et al., 2012; Trenkwalder et al., 2009; Yeh et al., 2012). The RLS phenotype includes a feeling of discomfort (creeping, tingling) or pain especially in the lower limbs, an urge to move to relieve unease (akathisia), increased wakefulness and reduced non-REM stage 2 sleep (Allen et al., 2017a; Koo et al., 2016; Ohayon et al., 2012; Yeh et al., 2012). Dysregulation of Fe metabolism is another characteristic commonly seen in patients, indicated by brain and peripheral iron deficiency and accompanying changes in iron storage and transport proteins (Connor, 2008; Connor et al., 2017; Zecca et al., 2004). Studies have also shown that dopamine agonists alleviate motor and sensory symptoms, suggesting an involvement of the dopamine system (Allen, 2015). The sensory features of RLS are defined by three characteristic features—(1) symptoms worsen with rest or inactivity; (2) discomfort is alleviated by activity; and (3) symptoms have a circadian cycle and are worse in the evening/night (worse during night-time rest than day-time rest).

The relief from symptoms felt on activity is immediate, but is most often temporary and lasts only for the duration of the activity (Allen and Earley, 2001; Schapira, 2004; Yeh et al., 2012). Patients with RLS may also show periodic limb movements in their sleep (PLMS) in the form of leg motion, elbow flexes, and twitches, though this is not a specific and definitive symptom (Koo et al., 2016). Apart from the sensorimotor phenotype, RLS also has a psychological aspect, with patients being more susceptible to depression, impulsivity, and have issues with arousal and decision making (Allen et al., 2017a; Koo et al., 2016).

3.1.1 Genetic Contribution to RLS

RLS and narcolepsy are the two sleep disorders with the most definitive genetic causes (Allen et al., 2017a; Winkelmann et al., 2017). Genome wide association studies (GWASs) have shown an increased RLS risk associated with single nucleotide polymorphisms (SNPs) in *MEIS1* (*rs2300478*), *BTBD9* (*rs9357271*, *rs3923809*), *MAP2K5/SKOR1* (*rs12593813*), and *PTPRD* (*rs1975197*) (Schulte and Winkelmann, 2015; Spieler et al., 2014; Winkelmann et al., 2007, 2011). In non-familial forms of RLS SNPs for *MEIS1* and *BTBD9* show the highest odds ratio for genetic association, with presence of at-risk alleles increasing the risk for RLS anywhere between 30% and 70%, though the mechanisms by which they alter Fe metabolism or DA biology are unknown (Schormair et al., 2011; Stefansson et al., 2007; Yang et al., 2011).

MEIS1 is a highly conserved homeobox gene which has been studied extensively in various animal models. It is responsible for limb patterning in *Drosophila* and for endothelial determination and vascularization in mice (Allen et al., 2017a; Thireau et al., 2017; Yang et al., 2011). In *Caenorhabditis elegans* studies, a loss of *unc-62* (*MEIS1* homologue) by RNAi shows increased ferritin expression and increased life span (Schormair et al., 2011). In mice, a homozygous knockout of *MEIS1* shows embryonic lethality, developmental defects and hemorrhaging, while in heterozygote knockouts mice show Fe deficiency and locomotor hyperactivity, similar perhaps to RLS (Spieler et al., 2014). Further, carriers of *MEIS1* risk haplotype in humans show reduced *MEIS1* mRNA and protein expression as well as increased ferritin and divalent metal transporter (DMT-1) transporter in brain tissue samples (Catoire et al., 2011; Spieler et al., 2014; Stefansson et al., 2007; Xiong et al., 2009). Homozygous occurrence of the *MEIS1* risk SNP has also been shown to increase ferritin and DMT-1 protein in the thalamus of patients (Li et al., 2016; Sørensen et al., 2012). Further, when human HeLa cells and

SKN-SHF cells were treated with deferoxamine mesylate, an Fe chelator, it resulted in a twofold lower expression of MEIS1 protein (Connor et al., 2011b; Li et al., 2016).

BTBD9 is a BTB/POZ domain containing protein, primarily involved in protein–protein interactions, and widely expressed in the central nervous system. In RLS *Drosophila* models with the functional knockdown of the *BTBD9* homologue *CG1826* in dopaminergic neurons with RNAi, studies report fragmented sleep patterns and increased locomotor activity which can be rescued by neural reexpression of *CG1826* (Allen et al., 2017b; Freeman et al., 2012; Winkelmann et al., 2017). Similar phenotypes are seen in heterozygous and homozygous *Btbd9* knockout mice that show increased locomotor activity, sleep disturbances and elevated serum Fe levels all of which can be treated with dopamine agonists indicating a mechanistic involvement of the dopamine pathway (Stefansson et al., 2007; Trenkwalder et al., 2009). In humans, the *BTBD9* risk haplotype is implicated in Fe metabolism as well—with an Icelandic proband population with RLS showing a 13% decrease in serum ferritin with each risk allele (Stefansson et al., 2007).

3.1.2 RLS and Role of Iron

The association between Fe-deficiency anemias and the prevalence of RLS was made in the early 1950s by Norlander (1954). Neural pathophysiological studies in primary RLS have shown that low serum and brain Fe are a common symptom of the disease, and oral Fe supplementation has been shown to alleviate symptoms for several patients (Connor, 2008; Deng et al., 2017). Imbalances in Fe homeostasis in the central nervous system have been indicated to play a larger role in exacerbating RLS symptoms as compared to peripheral Fe (Khan et al., 2017; Moos et al., 2007; Zecca et al., 2004). Mice sustained on an Fe deficient diet show sleep disturbance, increased tactile sensitivity, and increased locomotor activity (Connor, 2008; Daubian-Nosé et al., 2014). The role of Fe in disease pathology is reinforced, since populations with iron-deficient anemia have a 31% prevalence of RLS—almost triple the average prevalence in the US population (Jellen et al., 2009; Khan et al., 2017). Serum Fe levels, however, do not reflect deficiency; in fact, the loss of Fe appears limited to the central nervous system. MRI of RLS patients has shown a reduction in white matter tracts, another effect of reduced Fe which results in hypomyelination in anemic animals and humans (Connor et al., 2011a, 2017; Ganz, 2013; Zecca et al., 2004). Most studies have focused on the substantia nigra, thalamus and the striatum because extrapyramidal regions tend to have higher Fe stores than non-motor regions of the brain. MRI studies have shown sustained and significant (almost 2.5-fold) Fe deficiency in the substantia nigra and the thalamus (Daubian-Nosé et al., 2014; Li et al., 2016; Sørensen et al., 2012). Protein quantification of Fe transport elements in brain slices as well as cerebrospinal fluid (CSF) has shown reduced ferritin and increased circulating transferrin (Tf), DMT-1, and ferroportin (FPN) in choroid plexus, all consistent with altered brain Fe status (Connor et al., 2004; DeAndrade et al., 2012; Moos and Morgan, 2000). Transferrin receptor (TfR), however, does not show increased expression as expected, but shows an opposing effect, with lower membrane TfR in patients than in controls (Connor et al., 2004; Moos and Morgan, 2000; Moos et al., 2007). Histochemistry of substantia nigra slices shows lowered enzymatic activity and protein concentration of iron response protein 1 (IRP-1) along with lower CSF prohepcidin in patients, all in keeping with the lack of Fe in the brain (Connor, 2008; Connor et al., 2017; Earley et al., 2014). Further, when looking at the substantia nigra, western blots of micro-dissected dopaminergic neurons showed that there is a decrease in intracellular ferritin, Fe, Tf and, as seen before, a paradoxical decrease in TfR (Connor et al., 2004; Patton et al., 2011). The same study has shown that though there is a paucity of Fe in the cytosol of the DA neurons, there is a significant increase in the mitochondrial ferritin and cytochrome oxidase C indicating an increase in the number of mitochondria within the neurons (Farina et al., 2013; Roth and Garrick, 2003). Taken together with the CSF observations, it suggests that there is an insufficiency of Fe availably to the neurons due to low Fe stores in the brain, which results in the above protein and heme profiles.

This discrepancy between serum Fe sufficiency and brain Fe deficiency is a unique feature of RLS. In control samples, both serum and brain Fe show a positive correlation, which is abolished in RLS. Experimental evidence has demonstrated that patients with RLS may have a fourfold decrease in the CSF ferritin–serum ferritin ratio as compared to unaffected controls (Connor et al., 2017; Ganz, 2013; Li et al., 2016). This led to the hypothesis that there is an imbalance in the influx/efflux system of the brain Fe system. Majority of the Fe in the brain is transported via the endothelial cells of the blood–brain barrier (BBB), through routes that have not been clearly defined. When the BBB, choroid plexus and the microvasculature of RLS patients were studied, the results were inconclusive. Fe and ferritin profiles of the choroid plexus and the BBB imitate those of the CSF and the neurons (Moos et al., 2007; Skjørringe et al., 2015). However, in the microvasculature there was a significant decrease in the Tf and TfR presence—unlike the choroid plexus that has increased Tf, but low TfR (much like the substantia nigra) (Skjørringe et al., 2015; Wijemanne and Jankovic, 2015). The translation of Tf is influenced by the activity of the IRP1, since Tf mRNA has an iron response element (IRE). Reduced activity of IRP-1 in the brain, and in the neurons, which is seen in RLS, would result in lowered Tf expression (Allen, 2015; Earley et al., 2014; Patton et al., 2011). These observations imply that there is a global mismanagement of Fe homeostasis in the brain which is preventing the correct sensing, uptake and transport of Fe from the blood and CSF across the BBB and choroid plexus, into the cells.

3.1.3 Co-regulation of Fe and Mn Brain Transport

Fe and Mn share several key cell transporters. Brain Mn status has been found to be inversely linked to brain Fe status, such that anemia can trigger increased Mn influx and accumulation (Aschner and Aschner, 1990; Au et al., 2008; Gunter et al., 2013; Tai et al., 2016). With several shared influx/efflux proteins for Fe and Mn, it is a logical extrapolation that changes in the homeostasis of either essential element would affect the cellular status of the other. Key shared transporters are described next and their role in the transport of both Fe and Mn under normal homeostatic and dyshomeostatic conditions is elaborated upon.

3.1.3.1 Divalent Metal Transporter 1

DMT-1 (formerly divalent cation transporter) is a promiscuous metal transporter. It has been shown to modulate the transport of zinc, cadmium, copper, nickel, lead, cobalt, as well as Fe and Mn, all in their +2 oxidative states (Au et al., 2008; Earley et al., 2008). Metal transport by DMT-1 is either Tf independent or Tf dependent (discussed in the following section). DMT-1 is a highly conserved, transmembrane protein (with 11 or 12 transmembrane domains) which acts as a proton–metal symporter; that is, at neutral pH, it symports one proton for every divalent metal cation, in a Tf independent fashion (Moos and Morgan, 2000; Roth and Garrick, 2003; Zhang et al., 2005). Its importing capabilities can thus be enhanced by increasing the proton concentration in the extracellular space. Since DMT-1 has such a non-selective and ubiquitous presence, it is unclear if it is the main transporter of Mn; though it has been positively shown to transport Mn, the transport depends on the relative density of DMT-1 and concentration of other metals in the environment. DMT-1 translation is influenced by Fe stores, since it has two mRNA isoforms—one which is +IRE and the other without (Au et al., 2008; Chakraborty and Aschner, 2012; Connor et al., 2011b; Skjørringe et al., 2015; Xiong et al., 2007). With change in the cellular Fe stores, it is possible that the translation rates and subsequently the presence and density of DMT-1 on cell surfaces are modulated, which could in turn impact the transport of other metals (Earley et al., 2008; Madejczyk and Ballatori, 2012).

3.1.3.2 Transferrin and Transferrin Receptor

Tf–TfR system is an iron scavenging system, where transferrin is a freely circulating Fe-binding protein and the TfR is a membrane bound protein, functionally associated with DMT-1 in endosomes for divalent metal transport. Tf is a single chain glycoprotein that has two Fe binding domains, which can also bind other metals such as aluminum, cadmium, manganese and copper, but only in their +3 oxidative states. Tf exists as apo-Tf (without bound cation), monovalent and divalent Tf with one or two cations bound to the metal-binding domain, respectively (Connor et al., 2011b; Li et al., 2016). Tf binds Fe^{3+} or Mn^{3+} and this metal–Tf complex binds to TfR on the cellular membrane. Tf–TfR binding triggers the formation of a clathrin coated pit and finally an endocytic vesicle with the TfR–Tf–metal complex and the juxtaposed DMT-1 within it. Once the endosome is internalized, vacuolar-ATPase (V-ATPase) is recruited to the endosomal membrane. V-ATPase pumps H^+ ions into the endosome and makes the internal environment acidic, which allows the Fe^{3+} or Mn^{3+} to be released from the Tf–TfR and converted to Fe/Mn^{2+}. The divalent cations traverse the membrane of the endosome via DMT-1, and the apo-Tf–TfR complex is recycled back to the cell surface and the Tf is released from the TfR (Arosio et al., 2017; Dobson et al., 2004). In the brain, Tf mRNA is present in oligodendrocytes (maximal production), astrocytes and neurons, and Tf protein is present in brain capillary and endothelial cells which form a part of the BBB (Connor et al., 2004; Smith et al., 2012; Zhang et al., 2005). TfR distribution in the brain is heterogeneous but overlaps as expected with Tf localization. TfR is found in endothelial capillary cells of the BBB, neurons, astrocytes, microglia, and choroid plexus epithelial cells. This provides evidence that the Tf–TfR system is active at the BBB and capable of transporting Fe and under different circumstances Mn, from the blood and CSF into the brain (Earley et al., 2014; Farina et al., 2013; Madejczyk and Ballatori, 2012).

3.1.3.3 Ferroportin as an Exporter

A third Fe-binding protein FPN is present on the basolateral surface of cells and acts as an Fe and Mn exporter. It binds to divalent cation and allows it to exit cells. Expression of FPN increases in a dose-dependent manner on exposure to Mn and thus acts as a neuroprotective measure to prevent intracellular accumulation in human embryonic kidney cells (Chen et al., 2015; Connor et al., 2017; Ganz, 2013; Madejczyk and Ballatori, 2012). At the BBB, it acts to shuttle Mn and Fe across the endothelial cells into the brain. FPN is ubiquitously expressed by all neurons, oligodendrocytes, astrocytes, the endothelial cells of the BBB, and cells of the choroid plexus, for which Mn has been shown to be a substrate (Au et al., 2008; Chakraborty and Aschner, 2012), leading to the conclusion that FPN does help in Mn export out of cells, and across the BBB.

3.1.4 Dopaminergic Alterations in RLS

The first evidence of involvement of the dopaminergic system in RLS was the observation that there was symptomatic relief when the patients were treated with DA agonists, and symptomatic worsening when DA antagonists were delivered (Carlos et al., 2017; Ferré et al., 2017). DA is the major neurotransmitter in the substantia nigra and acts on two major families of receptors—D1-like family and the D2-like family (further abbreviated as D1R and D2R). The substantia nigra plays a significant role in control of movements and is the most severely affected region in PD, other parkinsonisms and movement related disorders (Peeraully and Tan, 2012; Pringsheim et al., 2014; Wang et al., 2017). Since RLS involves PLMS and an urge to move relentlessly, it seemed but natural to look at the basal ganglia—especially the substantia nigra and the striatum of patients. Functional MRI (fMRI) studies have shown activation of the thalamus, cerebellum, striatum, and inferior olive, preceding involuntary movements in RLS, but no activation of the cortex (Koo et al., 2016; Trenkwalder et al., 2009). This observation bolsters the evidence supporting the involvement of the subcortical dopaminergic system in RLS, rather than voluntary cortical involvement (Allen et al., 2005; Koo et al., 2016; Yeh et al., 2012). Further, Fe is very closely linked to dopaminergic biology, it is an important cofactor for TH, which is the rate limiting step in DA production (DeAndrade et al., 2012; Li et al., 2016), and since there is robust evidence for anemia in RLS, there is a high chance that the dopaminergic system is also involved. Experimental evidence in support of dopamine dysregulation is seen in immunohistochemical (IHC) staining, positron emission tomography (PET) and single-photon emission computed tomography (SPECT) analysis of patient brains. IHC of substantia nigra brain slices has shown a reduction in the D2 receptors in the nigra and putamen, and an increase in TH expression in the same regions (Connor et al., 2009; Earley et al., 2014). PET analysis of subjects has shown that there is a decrease in the DAT binding potential at the presynaptic regions in the striatum. This reduction in DAT binding potential is not seen with SPECT analysis, implying that there is no loss in total DAT production, but a downregulation of membrane bound DAT (Cubo et al., 2011; Earley et al., 2011). Patients also exhibit increased free DA and homovanillic acid (DA metabolite) in CSF, and an increased concentration of DA in extracellular space of the striatum and substantia nigra when studied using microdialysis (Kawada, 2017; Ruppert et al., 2017; Weinstock et al., 2012). Most of these observations indicate a hyperactivated dopaminergic system in RLS.

Hyperactivated DA signaling in RLS seems contradictory to the clinical observations where restlessness is reduced by DA agonists. DA agonists are a common medication for PD which is characterized by a loss of DA neurons and thus an insufficiency of nervous DA in the basal ganglia. Also, Fe is a cofactor of TH—thus anemia should result in lowered DA production (Connor et al., 2017; Khan et al., 2017). A plausible reason for the paradoxical increase in TH production is hypothesized to be because of the role of hypoxia-inducible factor 1 alpha (HIF-1α) (Earley et al., 2014; Patton et al., 2011). The hypoxia system is highly sensitive to Fe status, and protein evaluation of autopsied brains of patients shows an increase in HIF-1α, which in turn activates TH production. Increased TH production results in increased DA synthesis and hyperactivity of the dopaminergic neurons at the synapse (Arosio et al., 2017; Koo et al., 2016; Patton et al., 2011; Weinstock et al., 2012). This increase in the DA can cause a subsequent downregulation of the postsynaptic D2R, to maintain the DA stasis. However, the increased DA has no effect on the postsynaptic D1R. This preferential D2R downregulation is currently unexplained and requires further investigation. mRNA expression studies have also shown that there is no change in the D1R or D2R transcriptional profiles, indicating that it is solely the trafficking of the D2R which is affected (Allen, 2015; Earley et al., 2011). As seen in SPECT and PET studies it is only the membrane bound DAT that is reduced, while total DAT remains constant. This increased DAT recycling and decreased DAT binding potential at the synapse is a result of increased localized DA competition for DAT binding (Khan et al., 2017). An overall increase in DA production and activity also explains the increased hyperactivity and wakefulness during night. DA production follows a circadian cycle, and levels in normal subjects drop during the night. A relative lowering of the synaptic DA coupled with low D2R presence might be sufficient to reduce the inhibition of the thalamocortical neurons, resulting in heightened thalamic activity, wakefulness and locomotor activity at night. These experimental results and clinical observations add credence to the theory that there is dopaminergic imbalance in RLS, which plays a role in its pathophysiology (Allen, 2015; Allen et al., 2017a; Koo 2015; Koo et al., 2016; Picchietti et al., 2017).

3.1.5 Summary of the Putative Role of Mn Neurotoxicity in RLS

In RLS, there is compelling evidence for Fe dysregulation being a part of the pathophysiology of the disease. Since the biochemistry of metals is interrelated, changes in Fe can result in dysregulation of other divalent metal as well. Mn like Fe is an important cofactor, has similar neural distribution, and competes for transport proteins such as DMT-1, Tf, and FPN (Aschner and Dorman, 2006; Aschner et al., 2007; Chen et al., 2015; Farina et al., 2013). As a result, Fe and Mn biology can exhibit inverse relationships—such that deficiency in Fe is associated with Mn accumulation (Aschner and Aschner, 1990; Fitsanakis et al., 2008; Williams et al., 2010). Developmental studies in rats have shown that an Fe deficient diet in pregnant dams and subsequent nutritional manganese overexposure can cause neurological changes in the pups like increased head dips and increased open field locomotion (Amos-Kroohs et al., 2015; O'Neal et al., 2014; Sistrunk et al., 2007; Stredrick et al., 2004). Clinical studies in humans have also shown that Fe deficient anemia results in the increased whole blood Mn levels (Farina et al., 2013). This makes a compelling argument for the involvement of Mn in Fe related disorders.

RLS also shows neuronal deficiency of Fe, and a subsequent increase in iron-related proteins like DMT-1 and Tf and a reduction in ferritin. It has been demonstrated in a clinical study in women that reduced ferritin, which is indicative of Fe stores, affects intestinal Mn absorption and retention (Arosio et al., 2017; Dobson et al., 2004; Finley, 1999; Madejczyk and Ballatori, 2012; Smith et al., 2012). In response to anemia there is a compensatory increase in DMT-1, Tf and FPN within the brain, the endothelium of the BBB and the choroid plexus. In such a physiological situation, where loss of ferritin increases Mn absorption, it follows that the heightened expression of the Fe transporters would act on Mn (Li et al., 2016; Moos et al., 2007; Smith et al., 2012; Tai et al., 2016; Zhang et al., 2005). A result of this would be greater Mn transport into the brain, across the

BBB, and into the substantia nigra, striatum and thalamus—which are the places within the brain that show increased expression of Tf, DMT-1 and FPN. Effects of Mn accumulation within the brain can result in neuronal loss and malfunction (Chen et al., 2015; Sistrunk et al., 2007; Stredrick et al., 2004). In RLS, since the substantia nigra and striatum would be most susceptible to Mn, the dopaminergic system would likely be affected. Studies in Sprague-Dawley rats have shown that Mn toxicity changes the burst rates of neurons in the globus pallidus and subthalamic nucleus and increases DA production in the striatum (Molina et al., 2007; Tai et al., 2016). DAT transfected HEK cells show reduced DA uptake and increased rate of internalization of the DAT on exposure to Mn (Roth et al., 2013). DA producing mouse catecholamine cells have shown increased cytotoxicity due to Mn-induced oxidative stress and irregularities in DA production. Ex vivo whole brain tissue exposures to Mn have shown decreases in DA and DOPAC release. In DMT-1 overexpressing SH-SY5Y cells, Mn import is greatly increased, finally leading to autophagy via the activation of the JNK–ASK pathway (Carmona et al., 2014; Sistrunk et al., 2007; Stredrick et al., 2004). This autophagy can be rescued through the supplementation of Fe which increases ferritin stores in the cell, reduces Mn uptake, and prevents the activation of the autophagic pathway. These results further bolster the possible role that Mn might play in mediating the symptoms and etiology of RLS.

3.2 Attention Deficit Hyperactive Disorder

ADHD is a common, heritable and chronic neurobehavioral disorder affecting school-aged children, and ~6% of the affected population may persist into adulthood (Azeredo et al., 2017). The exact etiology of ADHD remains a mystery, but there is evidence that indicates that genetic and environmental factors may both be involved (Azeredo et al., 2017; Stubbe, 2000). Classical symptoms of ADHD include inattention, impulsivity, hyperactivity and behavioral issues. Clinical observations of patients have shown reduced volume and slower activity of the prefrontal cortex, caudate and the cerebellum (Guilarte, 2013; Lao et al., 2017; Sharma and Couture, 2014). Several other studies have demonstrated lower density of the dopamine receptor in these same brain regions (Fusar-Poli et al., 2012; Tripp and Wickens, 2009). Genetic studies indicate that polymorphisms coding for those variants of D4, D5 and DAT-1 which decrease DA function are correlated with ADHD occurrence (Azeredo et al., 2017). Further studies have shown that there is a bias toward neurodevelopmental genes being affected by rare alterations in copy numbers in patients with ADHD (Elia et al., 2010). A recent study in DAT-KO rats shows behavioral hyperactivity and cognitive dysfunction which can be remedied by methylphenidate and methamphetamines, both highly effective ADHD treatments (Bono et al., 2018; Leo et al., 2018). Overall, these observations point toward a reduced activity of the dopaminergic system in the brain.

Mn has been implicated as an environmental risk factor for ADHD. Studies indicate a correlation between ADHD diagnosis and positive occurrence of associated symptoms with an increase in serum levels of Mn (Lucchini et al., 2017). Exposure during the sensitive prenatal and early post-natal ages, where neural development is active, is perhaps the most detrimental (Bhang et al., 2013; Hong et al., 2014). Studies in Long-Evans rats demonstrate that early post-natal (D1–25) exposure to Mn through water results in fine motor skill deficits, lowered attention, increased susceptibility to distraction and lack of arousal control, which lasts into adulthood. Long term Mn exposures also show deficits in attention and fine motor skills, along with lowered DA release in the prefrontal cortex (Beaudin et al., 2013, 2015, 2017a). Administration of methylphenidate to the Mn-exposed rats alleviated the loss of motor coordination and impulsivity control (Beaudin et al., 2015, 2017b). Similar Mn associated neural deficits have been observed in children as well. A study of students in Quebec, Canada, demonstrated that tap water Mn exposure caused low memory and lowered attention spans, though hyperactivity was not strongly correlated (Oulhote et al., 2014). In Brazil, a study conducted in Salvador, where children were exposed to air-borne Mn, found lower verbal memory scores and a higher tendency to be distracted—indicating a role for the frontostriatal circuitry (Carvalho et al., 2014; Carvalho et al., 2017). Further studies in the same region indicate sex-biased observations of behavioral issues, such that girls in the study cohort showed increased externalizing behaviors like aggression and inattention with increasing Mn exposure as compared to boys (Carvalho et al., 2017; Menezes-Filho et al., 2014). Treatment of children diagnosed with Ritalin (methylphenidate) also showed a decrease in serum levels of Mn (Farias et al., 2010). Evidence also shows that excessive Mn exposure by inhalation in rats decreased brain DAT levels (Saputra et al., 2016). This level of exposure was not sufficient to trigger DA neuron degeneration and motor symptoms though. Other studies in rats and honeybees show that subacute levels of Mn injection (rats) or environmental exposure (honey bees and fruit flies) show a contrary increase in monoamines like dopamine and a subsequent increase in its metabolites

(DOPAC and HVA) (Dion et al., 2018; Søvik et al., 2015). These data indicate that Mn might play a role in maintaining DA function—and disruption of Mn homeostatic levels results in an imbalance of the dopaminergic system. Since the dopaminergic system is closely linked to motor and mood regulation, it seems intuitive that dysfunction could impinge upon impulse and motor control. Overall, it is unlikely that Mn will play a singular role in ADHD, especially since genetic factors are also implicated, but given the cumulative evidence a strong premise exists for a potential link between Mn neurotoxicity and ADHD risk.

ADHD has also been correlated to the prevalence of RLS in patients. Studies show that there is an increase in the rate of sleep disturbances and periodic limb movements during sleep in children diagnosed with ADHD (Montoya et al., 2018). At the same time, there is evidence which reveals that children diagnosed with pediatric forms of RLS record inferior performance on cognitive tests and manifest other behavioral symptoms similar to ADHD (Fliers et al., 2011). Although a GWAS study looking at the major RLS associated SNPs in MEIS1, BTBD9 and MAP2K5 did not find a significant allelic association with ADHD, there is evidence linking a polymorphism in the neuronal nitric oxide synthase (NOS1) gene to both RLS and ADHD. This is relevant, as NOS1 uses arginine as a resource, and neuronal arginine levels are under direct control of the Mn-dependent enzyme Arg2 (Horning et al., 2015; Jiménez-Jiménez et al., 2018; Reif, 2010; Schimmelmann et al., 2009). Furthermore, published results show that adults who have children diagnosed with ADHD have a significantly increased risk of being affected by RLS (Gao et al., 2011). This would suggest a genetic heritability which is common between both these disorders. Since the etiology of RLS and ADHD both involves issues with impulse control, increased motor disturbances and dopaminergic dysfunction, a common pathophysiology might be at play.

3.3 Tourette Syndrome

Tourette syndrome (TS) is a neurodevelopmental disorder which affects up to 0.8% of the school going population (Scharf et al., 2012). It is characterized by repeated vocal tics and involuntary movements which are exacerbated by stress and anxiety. The most common comorbidities are ADHD, obsessive–compulsive disorder (OCD) and depression (Du et al., 2010; Eddy et al., 2009; Robertson, 2000, 2005). Like RLS, TS shows an involvement of the frontostriatal circuitry and the basal ganglia and responds to pharmacological modifications of the dopamine system. A GWAS study found that those SNPs associated with RLS in the BTBD9 gene were over-represented in TS. Further, other non-RLS BTBD9 SNPs were positively associated with TS without OCD (Qi et al., 2017; Rivière et al., 2009). The common recruitment of the DA system and the TS–ADHD comorbidity makes it another candidate neural disorder which may include Mn dysregulation. A study looking at changes in placental DNA methylation alterations due to fetal Mn exposure found that several of the differentially methylated genes are associated with neural development and various neural diseases including, but not limited to Schizophrenia, Alzheimer's, ADHD, autism and TS (Maccani et al., 2015). Further research regarding the role of Mn and DA in TS and its pathophysiological links to RLS need to be elucidated.

4. CONCLUSIONS

RLS, ADHD and TS are neurological disorders with common phenotyping manifestations which include alterations in motor function, mood and impulsivity (Hallett, 2015; Koo et al., 2016; Stubbe, 2000). The combination of motor, mood and impulsivity symptoms indicates an involvement of the basal ganglia and dopaminergic circuitry. This hypothesis is bolstered by clinical observations like—alleviating restlessness with DA replacement in RLS, volumetric decrease of caudate nucleus, putamen and globus pallidus in children and adults with TS, and similar volumetric asymmetry (reduced or increased density) in the caudate of ADHD patients (Koo et al., 2016; Riva et al., 2018; Trotti, 2017). Brain imaging in all three disorders shows activation of the thalamus and malfunctioning of the frontostriatal circuits (Koo 2015; Riva et al., 2018). Moreover, evidence points to the involvement of Mn in RLS and ADHD—where anemia in RLS is hypothesized to cause an increase in Mn influx through shared transporters and serum, and Mn blood levels in ADHD patients are elevated and have been shown to reduce with Ritalin treatment (Bhang et al., 2013; Farias et al., 2010). An important point to note is that in the brain the basal ganglia is the region of highest Mn concentration, making neurons within that region susceptible to changes in Mn homeostasis (Guilarte, 2013; Racette et al., 2012). Mn accumulation has been shown to affect dopaminergic function and cause DA neuron toxicity (Benedetto et al., 2010; Kwakye et al., 2015; Lao et al., 2017; Lucchini et al., 2017; Pfalzer and Bowman, 2017; Rivera-Mancía et al., 2011; Roth et al., 2013;

Stredrick et al., 2004). Thus, if brain Mn is affected in RLS, ADHD and Tourette, it could explain the dopaminergic effects in these diseases. Future directions would need to investigate the role of Mn transport across the BBB and the choroid plexus, and the changes in its distribution within the basal ganglia and how it affects dopaminergic function. Being able to parse this out would help us better understand the etiology and pathophysiology, which would undoubtedly help better treatment and diagnosis.

REFERENCES

Abe, H., Ohishi, T., Nakane, F., Shiraki, A., Tanaka, T., Yoshida, T., Shibutani, M., 2015. Exposure to MnCl2 · 4H2O during development induces activation of microglial and perivascular macrophage populations in the hippocampal dentate gyrus of rats. J. Appl. Toxicol. 35, 529–535.

Allen, R., 2015. Restless leg syndrome/Willis-Ekbom disease pathophysiology. Sleep Med. Clin. 10, 207–214.

Allen, R.P., Earley, C.J., 2001. Restless legs syndrome: a review of clinical and pathophysiologic features. J. Clin. Neurophysiol. 18, 128–147.

Allen, R.P., Walters, A.S., Montplaisir, J., Hening, W., Myers, A., Bell, T.J., Ferini-Strambi, L., 2005. Restless legs syndrome prevalence and impact: REST general population study. Arch. Intern. Med. 165, 1286–1292.

Allen, R.P., Montplaisir, J., Walters, A.S., Ferini-Strambi, L., Hogl, B., 2017a. Restless legs syndrome and periodic limb movements during sleep. In: Kryger, M., Roth, T., Dement, W.C. (Eds.), Principles and Practice of Sleep Medicine. Elsevier Inc., Pennsylvania, pp. 923–935.

Allen, R., Donelson, N., Jones, B., Li, Y., Manconi, M., Rye, D., Sanyal, S., Winkelmann, J., 2017b. Animal models of RLS phenotypes. Sleep Med. 31, 23–28.

Amos-Kroohs, R., Bloor, C., Qureshi, M., Vorhees, C., Williams, M., 2015. Effects of developmental exposure to manganese and/or low iron diet: changes to metal transporters, sucrose preference, elevated zero-maze, open-field, and locomotion in response to fenfluramine, amphetamine, and MK-801. Toxicol. Rep. 2, 1046–1056.

Amos-Kroohs, R.M., Davenport, L.L., Atanasova, N., Abdulla, Z.I., Skelton, M.R., Vorhees, C.V., Williams, M.T., 2017. Developmental manganese neurotoxicity in rats: cognitive deficits in allocentric and egocentric learning and memory. Neurotoxicol. Teratol. 59, 16–26.

Arosio, P., Elia, L., Poli, M., 2017. Ferritin, cellular iron storage and regulation. IUBMB Life.

Aschner, M., Aschner, J.L., 1990. Manganese transport across the blood-brain barrier: relationship to iron homeostasis. Brain Res. Bull. 24, 857–860.

Aschner, M., Dorman, D.C., 2006. Manganese. Toxicol. Rev.

Aschner, M., Erikson, K., Ronald, 2009. Manganese and its role in Parkinson's disease: from transport to neuropathology. Neuromolecular Med.

Aschner, M., Guilarte, T.R.R., Schneider, J.S., Zheng, W., 2007. Manganese: recent advances in understanding its transport and neurotoxicity. Toxicol. Appl. Pharmacol. 221, 131–147.

Aschner, J., Anderson, A., Slaughter, J., Aschner, M., Steele, S., Beller, A., Mouvery, A., Furlong, H., Maitre, N., 2015. Neuroimaging identifies increased manganese deposition in infants receiving parenteral nutrition. Am. J. Clin. Nutr. 102, 1482–1489.

Au, C., Benedetto, A., Aschner, M., 2008. Manganese transport in eukaryotes: the role of DMT1. Neurotoxicology 29, 569–576.

Azeredo, A., Moreira, D., Barbosa, F., 2017. ADHD, CD, and ODD: systematic review of genetic and environmental risk factors. Res. Dev. Disabil. https://doi.org/10.1016/j.ridd.2017.12.010

Beaudin, S., Nisam, S., Smith, D., 2013. Early life versus lifelong oral manganese exposure differently impairs skilled forelimb performance in adult rats. Neurotoxicol. Teratol. 38, 36–45.

Beaudin, S., Strupp, B., Lasley, S., Fornal, C., Mandal, S., Smith, D., 2015. Oral methylphenidate alleviates the fine motor dysfunction caused by chronic postnatal manganese exposure in adult rats. Toxicol. Sci. 144, 318–327.

Beaudin, S., Strupp, B., Strawderman, M., Smith, D., 2017a. Early postnatal manganese exposure causes lasting impairment of selective and focused attention and arousal regulation in adult rats. Environ. Health Perspect. 125 (2), 230–237.

Beaudin, S., Strupp, B., Uribe, W., Ysais, L., Strawderman, M., Smith, D., 2017b. Methylphenidate alleviates manganese-induced impulsivity but not distractibility. Neurotoxicol. Teratol. 61, 17–28.

Benedetto, A., Au, C., Avila, D.S., Milatovic, D., Aschner, M., 2010. Extracellular dopamine potentiates Mn-induced oxidative stress, lifespan reduction, and dopaminergic neurodegeneration in a BLI-3-dependent manner in C. elegans. PLoS Genet. 6 (8), e1001084.

Bhang, S.-Y., Cho, S.-C., Kim, J.-W., Hong, Y.-C., Shin, M.-S., Yoo, H., Cho, I., Kim, Y., Kim, B.-N., 2013. Relationship between blood manganese levels and children's attention, cognition, behavior, and academic performance—a nationwide cross-sectional study. Environ. Res. 126, 9–16.

Bichell, T.J.V.J., Wegrzynowicz, M., Tipps, K.G., Bradley, E.M., Uhouse, M.A., Bryan, M., Horning, K., Fisher, N., Dudek, K., Halbesma, T., Umashanker, P., Stubbs, A.D., Holt, H.K., Kwakye, G.F., Tidball, A.M., Colbran, R.J., Aschner, M., Neely, M.D., Di Pardo, A., Maglione, V., Osmand, A., Bowman, A.B., 2017. Reduced bioavailable manganese causes striatal urea cycle pathology in Huntington's disease mouse model. Biochim. Biophys. Acta 1863, 1596–1604.

Bono, F., Savoia, P., Guglielmi, A., Gennarelli, M., Piovani, G., Sigala, S., Leo, D., Espinoza, S., Gainetdinov, R., Devoto, P., Spano, P., Missale, C., Fiorentini, C., 2018. Role of dopamine D2/D3 receptors in development, plasticity, and neuroprotection in human iPSC-derived midbrain dopaminergic neurons. Mol. Neurobiol. 55 (2), 1054–1067.

Bouabid, S., Delaville, C., Deurwaerdère, P., Lakhdar-Ghazal, N., Benazzouz, A., 2014. Manganese-induced atypical parkinsonism is associated with altered basal ganglia activity and changes in tissue levels of monoamines in the rat. PLoS One 9, e98952.

Bowman, A., Kwakye, G., Hernández, E.H., Aschner, M., 2011. Role of manganese in neurodegenerative diseases. J. Trace Elem. Med. Biol. 25 (4), 191–203.

Brooks, E.M., Gonzalez, M., Eden, A.R., O'Neal, J., Sabo, R.T., Etz, R.S., 2017. What family physicians really think of maintenance of certification part II activities. J. Contin. Educ. Heal. Prof. 37, 223–229.

Calne, D.B., Chu, N.S., Huang, C.C., Lu, C.S., Olanow, W., 1994. Manganism and idiopathic parkinsonism: similarities and differences. Neurology 44, 1583–1586.

Carlos, K., Prado, G., Teixeira, C., Conti, C., de Oliveira, M., Prado, L., Carvalho, L., 2017. Benzodiazepines for restless legs syndrome. Cochrane Database Syst. Rev. 3, CD006939.

Carmona, A., Roudeau, S., Perrin, L., Veronesi, G., Ortega, R., 2014. Environmental manganese compounds accumulate as Mn(II) within the Golgi apparatus of dopamine cells: relationship between speciation, subcellular distribution, and cytotoxicity. Metallomics 6, 822–832.

Carvalho, C., Menezes-Filho, J., Matos, V., Bessa, J., Coelho-Santos, J., Viana, G., Argollo, N., Abreu, N., 2014. Elevated airborne manganese and low executive function in school-aged children in Brazil. Neurotoxicology 45, 301–308.

de Carvalho, C., Oulhote, Y., Martorelli, M., de Carvalho, C., Menezes-Filho, J., Argollo, N., Abreu, N., 2017. Environmental manganese exposure and associations with memory, executive functions, and hyperactivity in Brazilian children. Neurotoxicology.

Catoire, H., Dion, P.A., Xiong, L., Amari, M., 2011. Restless legs syndrome-associated MEIS1 risk variant influences iron homeostasis. Ann. Neurol. 70, 170–175.

Cersosimo, M., Koller, W., 2006. The diagnosis of manganese-induced parkinsonism. Neurotoxicology 27, 340–346.

Chakraborty, S., Aschner, M., 2012. Altered manganese homeostasis: implications for BLI-3-dependent dopaminergic neurodegeneration and SKN-1 protection in C. elegans. J. Trace Elem. Med. Biol. 26, 183–187.

Chen, P., Chakraborty, S., Mukhopadhyay, S., Lee, E., Paoliello, M., Bowman, A., Aschner, M., 2015. Manganese homeostasis in the nervous system. J. Neurochem. 134, 601–610.

Chen, P., Culbreth, M., Aschner, M., 2016. Exposure, epidemiology, and mechanism of the environmental toxicant manganese. Environ. Sci. Pollut. Res. Int. 23, 13802–13810.

Chen, P., Bornhorst, J., Aschner, M., 2018. Manganese metabolism in humans. Front. Biosci. 23, 1655–1679.

Connor, J., 2008. Pathophysiology of restless legs syndrome: evidence for iron involvement. Curr. Neurol. Neurosci. 8, 162–166.

Connor, J.R., Wang, X.S., Patton, S.M., Menzies, S.L., Troncoso, J.C., Earley, C.J., Allen, R.P., 2004. Decreased transferrin receptor expression by neuromelanin cells in restless legs syndrome. Neurology 62, 1563–1567.

Connor, J.R., Wang, X.-S., Allen, R.P., Beard, J.L., Wiesinger, J.A., Felt, B.T., Earley, C.J., 2009. Altered dopaminergic profile in the putamen and substantia nigra in restless leg syndrome. Brain 132, 2403–2412.

Connor, J., Ponnuru, P., Lee, B.-Y., Podskalny, G., Alam, S., Allen, R., Earley, C., Yang, Q., 2011a. Postmortem and imaging based analyses reveal CNS decreased myelination in restless legs syndrome. Sleep Med. 12, 614–619.

Connor, J.R., Ponnuru, P., Wang, X.-S., Patton, S.M., Allen, R.P., Earley, C.J., 2011b. Profile of altered brain iron acquisition in restless legs syndrome. Brain 134 (4), 959–968.

Connor, J., Patton, S., Oexle, K., Allen, R., 2017. Iron and restless legs syndrome: treatment, genetics and pathophysiology. Sleep Med. 31, 61–70.

Criswell, S.R., Perlmutter, J.S., Videen, T.O., Moerlein, S.M., Flores, H.P., Birke, A.M., Racette, B.A., 2011. Reduced uptake of [^{18}F]FDOPA PET in asymptomatic welders with occupational manganese exposure. Neurology 76 (15), 1296–1301.

Cubo, E., López, M., Ceberio, J., Alfonso, I., Martinez, B., Berciano, J., Iglesias, F., Obeso, J., 2011. Striatal dopamine function in a family with multiple SCA-3 phenotypes. J. Neurol. 258, 308–310.

Daubian-Nosé, P., Frank, M., Esteves, A., 2014. Sleep disorders: a review of the interface between restless legs syndrome and iron metabolism. Sleep Sci 7, 234–237.

DeAndrade, M., Johnson, R., Unger, E., Zhang, L., van Groen, T., Gamble, K., Li, Y., 2012. Motor restlessness, sleep disturbances, thermal sensory alterations and elevated serum iron levels in Btbd9 mutant mice. Hum. Mol. Genet. 21, 3984–3992.

Deng, Y., Wu, J., Jia, Q., 2017. Efficacy of intravenous iron sucrose in hemodialysis patients with restless legs syndrome (RLS): a randomized, placebo-controlled study. Med. Sci. Monit. 23, 1254–1260.

Dion, L.-A.A., Saint-Amour, D., Sauvé, S., Barbeau, B., Mergler, D., Bouchard, M.F., 2018. Changes in water manganese levels and longitudinal assessment of intellectual function in children exposed through drinking water. Neurotoxicology 64, 118–125.

Dobson, A.W., Erikson, K.M., Aschner, M., 2004. Manganese neurotoxicity. Ann. N. Y. Acad. Sci. 1012, 115–128.

Du, J.-C., Chiu, T.-F., Lee, K.-M., Wu, H.-L., Yang, Y.-C., Hsu, S.-Y., Sun, C.-S., Hwang, B., Leckman, J., 2010. Tourette syndrome in children: an updated review. Pediatr. Neonatol. 51, 255–264.

Earley, C.J., Ponnuru, P., Wang, X., Patton, S.M., Conner, J.R., Beard, J.L., Taub, D.D., Allen, R.P., 2008. Altered iron metabolism in lymphocytes from subjects with restless legs syndrome. Sleep 31, 847–852.

Earley, C.J., Kuwabara, H., Wong, D.F., Gamaldo, C., Salas, R., Brasic, J., Ravert, H.T., Dannals, R.F., Allen, R.P., 2011. The dopamine transporter is decreased in the striatum of subjects with restless legs syndrome. Sleep 34, 341–347.

Earley, C., Connor, J., Garcia-Borreguero, D., Jenner, P., Winkelman, J., Zee, P.C., Allen, R., 2014. Altered brain iron homeostasis and dopaminergic function in restless legs syndrome (Willis–Ekbom disease). Sleep Med. 15 (11), 1288–1301.

Eddy, C., Rizzo, R., Cavanna, A., 2009. Neuropsychological aspects of Tourette syndrome: a review. J. Psychosom. Res. 67, 503–513.

Elia, J., Gai, X., Xie, H., Perin, J., Geiger, E., Glessner, J., D'arcy, M., deBerardinis, R., Frackelton, E., Kim, C., Lantieri, F., Muganga, B., Wang, L., Takeda, T., Rappaport, E., Grant, S., Berrettini, W., Devoto, M., Shaikh, T., Hakonarson, H., White, P., 2010. Rare structural variants found in attention-deficit hyperactivity disorder are preferentially associated with neurodevelopmental genes. Mol. Psychiatry 15, 637.

Farias, A., Cunha, A., Benko, C., McCracken, J., Costa, M., Farias, L., Cordeiro, M., 2010. Manganese in children with attention-deficit/hyperactivity disorder: relationship with methylphenidate exposure. J. Child Adol. Psychop. 20, 113–118.

Farina, M., Avila, D.S., da Rocha, J.B.B., Aschner, M., 2013. Metals, oxidative stress and neurodegeneration: a focus on iron, manganese and mercury. Neurochem. Int. 62, 575–594.

Ferré, S., Earley, C., Gulyani, S., Garcia-Borreguero, D., 2017. In search of alternatives to dopaminergic ligands for the treatment of restless legs syndrome: iron, glutamate, and adenosine. Sleep Med. 31, 86–92.

Finley, J.W., 1999. Manganese absorption and retention by young women is associated with serum ferritin concentration. Am. J. Clin. Nutr. 70, 37–43.

Fitsanakis, V.A., Zhang, N., Anderson, J.G., Erikson, K.M., Avison, M.J., Gore, J.C., Aschner, M., 2008. Measuring brain manganese and iron accumulation in rats following 14 weeks of low-dose manganese treatment using atomic absorption spectroscopy and magnetic resonance imaging. Toxicol. Sci. 103, 116–124.

Fliers, E., Vasquez, A., Poelmans, G., Rommelse, N., Altink, M., Buschgens, C., Asherson, P., Banaschewski, T., Ebstein, R., Gill, M., Miranda, A., Mulas, F., Oades, R., Roeyers, H., Rothenberger, A., Sergeant, J., Sonuga-Barke, E., Steinhausen, H.-C., Faraone, S., Buitelaar, J., Franke, B., 2011. Genome-wide association study of motor coordination problems in ADHD identifies genes for brain and muscle function. World J. Biol. Psychiatry 13, 211–222.

Freeman, A., Pranski, E., Miller, D., Radmard, S., Bernhard, D., Jinnah, H.A., Betarbet, R., Rye, D., Sanyal, S., 2012. Sleep fragmentation and motor restlessness in a Drosophila model of restless legs syndrome. Curr. Biol. 22, 1142–1148.

Fusar-Poli, P., Rubia, K., Rossi, G., Sartori, G., Balottin, U., 2012. Striatal dopamine transporter alterations in ADHD: pathophysiology or adaptation to psychostimulants? A meta-analysis. Am. J. Psychiatry 169 (3), 264–272.

Ganz, T., 2013. Systemic iron homeostasis. Physiol. Rev. 93, 1721–1741.

Gao, X., Lyall, K., Palacios, N., Walters, A., Ascherio, A., 2011. RLS in middle aged women and attention deficit/hyperactivity disorder in their offspring. Sleep Med. 12, 89–91.

Gorell, J., Peterson, E., Rybicki, B., Johnson, C., 2004. Multiple risk factors for Parkinson's disease. J. Neurol. Sci. 217, 169–174.

Grandjean, P., Landrigan, P.J., 2014. Neurobehavioural effects of developmental toxicity. Lancet Neurol. 13, 330–338.
Guilarte, T., 2010. Manganese and Parkinson's disease: a critical review and new findings. Environ. Health Perspect. 118, 1071–1080.
Guilarte, T., 2013. Manganese neurotoxicity: new perspectives from behavioral, neuroimaging, and neuropathological studies in humans and non-human primates. Front. Aging Neurosci. 5, 23.
Guilarte, T., Gonzales, K., 2015. Manganese-induced parkinsonism is not idiopathic Parkinson's disease: environmental and genetic evidence. Toxicol. Sci. 146, 204–212.
Gunter, T.E., Gerstner, B., Gunter, K.K., Malecki, J., Gelein, R., Valentine, W.M., Aschner, M., Yule, D.I., 2013. Manganese transport via the transferrin mechanism. Neurotoxicology 34, 118–127.
Hallett, M., 2015. Tourette syndrome: update. Brain Dev. 37, 651–655.
Harischandra, D., Jin, H., Anantharam, V., Kanthasamy, A., Kanthasamy, A., 2015. α-Synuclein protects against manganese neurotoxic insult during the early stages of exposure in a dopaminergic cell model of Parkinson's disease. Toxicol. Sci. 143, 454–468.
Henn, B., Kim, J., Wessling-Resnick, M., Téllez-Rojo, M., Jayawardene, I., Ettinger, A., Hernández-Avila, M., Schwartz, J., Christiani, D., Hu, H., Wright, R., 2011. Associations of iron metabolism genes with blood manganese levels: a population-based study with validation data from animal models. Environ. Health 10, 1–11.
Hong, S.-B., Kim, J.-W., Choi, B.-S., Hong, Y.-C., Park, E.-J., Shin, M.-S., Kim, B.-N., Yoo, H.-J., Cho, I.-H., Bhang, S.-Y., Cho, S.-C., 2014. Blood manganese levels in relation to comorbid behavioral and emotional problems in children with attention-deficit/hyperactivity disorder. Psychiatry Res. 220, 418–425.
Horning, K.J., Caito, S.W., Tipps, G.K., Bowman, A.B., Aschner, M., 2015. Manganese is essential for neuronal health. Annu. Rev. Nutr. 35, 1–38.
Hsieh, C.-T., Liang, J.-S., Peng, S., Lee, W.-T., 2007. Seizure associated with total parenteral nutrition-related hypermanganesemia. Pediatr. Neurol. 36, 181–183.
Iwase, K., Higaki, J., Mikata, S., Tanaka, Y., Kondoh, H., Yoshikawa, M., Hori, S., Kamiike, W., 2002. Manganese deposition in basal ganglia due to perioperative parenteral nutrition following gastrointestinal surgeries. Dig. Surg. 19, 174–183.
Jellen, L., Beard, J., Jones, B., 2009. Systems genetics analysis of iron regulation in the brain. Biochimie 91, 1255–1259.
Jiménez-Jiménez, F., Alonso-Navarro, H., García-Martín, E., Agúndez, J., 2018. Genetics of restless legs syndrome: an update. Sleep Med. Rev. 39, 108–121.
Kawada, T., 2017. Restless leg syndrome and periodic limb movements in sleep: risk for cardiovascular disease. Sleep Med. 36, 180.
Khan, F., Ahlberg, C., Chow, C., Shah, D., Koo, B., 2017. Iron, dopamine, genetics, and hormones in the pathophysiology of restless legs syndrome. J. Neurol. 1–8.
Kikuchihara, Y., Abe, H., Tanaka, T., Kato, M., Wang, L., Ikarashi, Y., Yoshida, T., Shibutani, M., 2015. Relationship between brain accumulation of manganese and aberration of hippocampal adult neurogenesis after oral exposure to manganese chloride in mice. Toxicology 331, 24–34.
Klingelhoefer, L., Bhattacharya, K., Reichmann, H., 2016. Restless legs syndrome. Clin. Med. 16, 379–382.
Koo, B., 2015. Restless leg syndrome across the globe: epidemiology of restless legs syndrome/Willis-Ekbom disease. Sleep Med. Clin. 10 (3), 189–205.
Koo, B., Bagai, K., Walters, A., 2016. Restless legs syndrome: current concepts about disease pathophysiology. Tremor Other Hyperkinet. Mov. 6, 401.
Kumar, K., Lowe Jr., E.W., Aboud, A.A., Neely, M.D., Redha, R., Bauer, J.A., Odak, M., Weaver, C.D., Meiler, J., Aschner, M., Bowman, A.B., 2014. Cellular manganese content is developmentally regulated in human dopaminergic neurons. Sci. Rep. 4.

Kwakye, G., Paoliello, M., Mukhopadhyay, S., Bowman, A., Aschner, M., 2015. Manganese-induced parkinsonism and Parkinson's disease: shared and distinguishable features. Int. J. Environ. Res. Public Health 12, 7519–7540.
Lao, Y., Dion, L.-A., Gilbert, G., Bouchard, M., Rocha, G., Wang, Y., Leporé, N., Saint-Amour, D., 2017. Mapping the basal ganglia alterations in children chronically exposed to manganese. Sci. Rep. 7, 41804.
Leo, D., Sukhanov, I., Zoratto, F., Illiano, P., Caffino, L., Sanna, F., Messa, G., Emanuele, M., Esposito, A., Dorofeikova, M., Budygin, E.A., Mus, L., Efimova, E.V., Niello, M., Espinoza, S., Sotnikova, T.D., Hoener, M.C., Laviola, G., Fumagalli, F., Adriani, W., Gainetdinov, R.R., 2018. Pronounced hyperactivity, cognitive dysfunctions, and BDNF dysregulation in dopamine transporter knock-out rats. J. Neurosci. 38, 1959–1972.
Li, X., Allen, R., Earley, C., Liu, H., Cruz, T., Edden, R., Barker, P., van Zijl, P., 2016. Brain iron deficiency in idiopathic restless legs syndrome measured by quantitative magnetic susceptibility at 7 tesla. Sleep Med. 22, 75–82.
Lu, L., Zhang, L., Li, J., Guo, W., Liang, W., Zheng, W., 2005. Alteration of serum concentrations of manganese, iron, ferritin, and transferrin receptor following exposure to welding fumes among career welders. Neurotoxicology 26, 257–265.
Lucchini, R., Placidi, D., Cagna, G., Fedrighi, C., Oppini, M., Peli, M., Zoni, S., 2017. Manganese and developmental neurotoxicity. Adv. Neurobiol. 18, 13–34.
Maccani, J., Koestler, D., Houseman, A., Armstrong, D., Marsit, C., Kelsey, K., 2015. DNA methylation changes in the placenta are associated with fetal manganese exposure. Reprod. Toxicol. 57, 43–49.
Madejczyk, M., Ballatori, N., 2012. The iron transporter ferroportin can also function as a manganese exporter. Biochim. Biophys. Acta 1818, 651–657.
Madison, J., Wegrzynowicz, M., Aschner, M., Bowman, A., 2012. Disease-toxicant interactions in manganese exposed Huntington disease mice: early changes in striatal neuron morphology and dopamine metabolism. PLoS One 7, e31024.
Marshall, E.F., Perry, R.H., Perry, E.K., Piggott, M.A., Thompson, P., Jaros, E., Burn, D.J., 1997. Striatal dopaminergic loss without parkinsonism in a case of corticobasal degeneration. Acta Neurol. Scand. 95, 287–292.
McLean, E., Cogswell, M., Egli, I., Wojdyla, D., Benoist, B., 2008. Worldwide prevalence of anaemia, WHO vitamin and mineral nutrition information system, 1993–2005. Public Health Nutr. 12, 444–454.
Menezes-Filho, J., Carvalho-Vivas, C., Viana, G., Ferreira, J., Nunes, L., Mergler, D., Abreu, N., 2014. Elevated manganese exposure and school-aged children's behavior: a gender-stratified analysis. Neurotoxicology 45, 293–300.
Mitterling, T., Heidbreder, A., Stefani, A., Fritz, J., Ulmer, H., Poewe, W., Högl, B., 2015. Natural course of restless legs syndrome/Willis–Ekbom disease: long-term observation of a large clinical cohort. Sleep Med. 16, 1252–1258.
Molina, R.M., Donaghey, T., Schwob, J.E., Brain, J.D., 2007. Olfactory uptake of manganese requires DMT1 and is enhanced by anemia. FASEB J.
Montoya, A., Restrepo, S., Arias, J., Ochoa, W., 2018. Prevalence and clinical characteristics of the restless legs syndrome (RLS) in patients diagnosed with attention deficit hyperactivity disorder (ADHD) in antioquia. Int. J. Psychol. Res. 11, 58–69.
Moos, T., Morgan, E.H., 2000. Transferrin and transferrin receptor function in brain barrier systems. Cell. Mol. Neurobiol. 20 (1), 77–95.
Moos, T., Rosengren Nielsen, T., Skjørringe, T., Morgan, E.H., 2007. Iron trafficking inside the brain. J. Neurochem. 103, 1730–1740.
Newland, M.C., Ceckler, T., Kordower, J., Weiss, B., 1989. Visualizing manganese in the primate basal ganglia with magnetic resonance imaging. Exp. Neurol. 106, 251–258.
Norlander, N.B., 1954. Restless legs. Br. J. Phys. Med. 17, 160–162.

Ohayon, M., O'Hara, R., Vitiello, M., 2012. Epidemiology of restless legs syndrome: a synthesis of the literature. Sleep Med. Rev. 16, 283–295.

Ohtake, T., Negishi, K., Okamoto, K., Oka, M., Maesato, K., Moriya, H., Kobayashi, S., 2005. Manganese-induced parkinsonism in a patient undergoing maintenance hemodialysis. Am. J. Kidney Dis. 46, 749–753.

Olanow, C.W., 2004. Manganese-induced parkinsonism and Parkinson's disease. Ann. N. Y. Acad. Sci. 1012 (1), 209–223.

O'Neal, S., Lee, J.-W., Zheng, W., Cannon, J., 2014. Subacute manganese exposure in rats is a neurochemical model of early manganese toxicity. Neurotoxicology 44, 303–313.

O'Neal, S., Zheng, W., 2015. Manganese toxicity upon overexposure: a decade in review. Curr. Environ. Heal Reports 2, 315–328.

Oulhote, Y., Mergler, D., Barbeau, B., Bellinger, D., Bouffard, T., Brodeur, M.-È., Saint-Amour, D., Legrand, M., Sauvé, S., Bouchard, M., 2014. Neurobehavioral function in school-age children exposed to manganese in drinking water. Environ. Health Perspect. 122, 1343–1350.

Parenti, M., Flauto, C., Parati, E., Vescovi, A., Groppetti, A., 1986. Manganese neurotoxicity: effects of L-DOPA and pargyline treatments. Brain Res. 367, 8–13.

Parenti, M., Rusconi, L., Cappabianca, V., Parati, E., Groppetti, A., 1988. Role of dopamine in manganese neurotoxicity. Brain Res. 473, 236–240.

Patton, S., Ponnuru, P., Snyder, A., Podskalny, G., Connor, J., 2011. Hypoxia-inducible factor pathway activation in restless legs syndrome patients. Eur. J. Neurol. 18, 1329–1335.

Peeraully, T., Tan, E.-K., 2012. Linking restless legs syndrome with Parkinson's disease: clinical, imaging and genetic evidence. Transl. Neurodegener. 1 (1), 6.

Peres, T.V., Eyng, H., Lopes, S.C., Colle, D., Gonçalves, F.M., Venske, D.K.K., Lopes, M.W., Ben, J., Bornhorst, J., Schwerdtle, T., Aschner, M., Farina, M., Prediger, R.D., Leal, R.B., 2015. Developmental exposure to manganese induces lasting motor and cognitive impairment in rats. Neurotoxicology 50, 28–37.

Peres, T., Parmalee, N., Martinez-Finley, E., Aschner, M., 2016. Untangling the manganese-α-synuclein web. Front. Neurosci. 10, 364.

Pfalzer, A., Bowman, A., 2017. Relationships between essential manganese biology and manganese toxicity in neurological disease. Curr. Environ. Health Rep. 4, 223–228.

Picchietti, D., Eeden, S., Inoue, Y., Berger, K., 2017. Achievements, challenges, and future perspectives of epidemiologic research in restless legs syndrome (RLS). Sleep Med. 31, 3–9.

Pratesi, A., Vella, A., Pasini, E., Salvi, F., Mascalchi, M., 2008. Parkinsonism in polycythaemia vera probably due to manganism. Mov. Disord. 23, 2420–2421.

Pringsheim, T., Jette, N., Frolkis, A., Steeves, T., 2014. The prevalence of Parkinson's disease: a systematic review and meta-analysis. Mov. Disord. 29, 1583–1590.

Qi, Y., Zheng, Y., Li, Z., Xiong, L., 2017. Progress in genetic studies of Tourette's syndrome. Brain Sci. 7 (10), 134–166.

Racette, B.A., Tabbal, S.D., Jennings, D., Good, L., Perlmutter, J.S., Evanoff, B., 2005. Prevalence of parkinsonism and relationship to exposure in a large sample of Alabama welders. Neurology 64, 230–235.

Racette, B., Aschner, M., Guilarte, T., Dydak, U., Criswell, S., Zheng, W., 2012. Pathophysiology of manganese-associated neurotoxicity. Neurotoxicology 33, 881–886.

Racette, B., Nielsen, S., Criswell, S., Sheppard, L., Warden, M.N., Checkoway, H., 2017. Dose-dependent progression of parkinsonism in manganese-exposed welders. Neurology 88 (4), 344–351.

Reif, A., 2010. Is NOS1 a genetic link between RLS and ADHD? J. Psychiatr. Res. 44, 60–61.

Riva, D., Taddei, M., Bulgheroni, S., 2018. The neuropsychology of basal ganglia. Eur. J. Paediatr. Neurol. 22.

Rivera-Mancía, S., Ríos, C., Montes, S., 2011. Manganese accumulation in the CNS and associated pathologies. Biometals 24, 811–825.
Rivière, J.-B.B., Xiong, L., Levchenko, A., St-Onge, J., Gaspar, C., Dion, Y., Lespérance, P., Tellier, G., Richer, F., Chouinard, S., Rouleau, G.A., 2009. Association of intronic variants of the BTBD9 gene with Tourette syndrome. Arch. Neurol. 66, 1267–1272.
Robertson, M.M., 2000. Tourette syndrome, associated conditions and the complexities of treatment. Brain 123 (Pt. 3), 425–462.
Robertson, M., 2005. Tourette syndrome. Psychiatry 4, 92–97.
Roth, J., Garrick, M., 2003. Iron interactions and other biological reactions mediating the physiological and toxic actions of manganese. Biochem. Pharmacol. 66, 1–13.
Roth, J., Li, Z., Sridhar, S., Khoshbouei, H., 2013. The effect of manganese on dopamine toxicity and dopamine transporter (DAT) in control and DAT transfected HEK cells. Neurotoxicology 35, 121–128.
Ruppert, E., Bataillard, M., Namer, I., Tatu, L., Hacquard, A., Hugueny, L., Hubbard, J., Kilic-Huck, U., Wolff, V., Bourgin, P., 2017. Hyperdopaminergism in lenticulostriate stroke-related restless legs syndrome: an imaging study. Sleep Med. 30, 136–138.
Saputra, D., Chang, J., Lee, B.-J., Yoon, J.-H., Kim, J., Lee, K., 2016. Short-term manganese inhalation decreases brain dopamine transporter levels without disrupting motor skills in rats. J. Toxicol. Sci. 41, 391–402.
Schapira, A., 2004. Restless legs syndrome. Drugs 64, 149–158.
Scharf, J., Miller, L., Mathews, C., Ben-Shlomo, Y., 2012. Prevalence of Tourette syndrome and chronic tics in the population-based Avon longitudinal study of parents and children cohort. J. Am. Acad. Child Adolesc. Psychiatry 51, 192–201.e5.
Schimmelmann, B.G., Friedel, S., Nguyen, T.T., Sauer, S., Vogel, C.I., Konrad, K., Wilhelm, C., Sinzig, J., Renner, T.J., Romanos, M., Palmason, H., Dempfle, A., Walitza, S., Freitag, C., Meyer, J., Linder, M., Schäfer, H., Warnke, A., Lesch, K.P., Herpertz-Dahlman, B., Hinney, A., Hebebrand, J., 2009. Exploring the genetic link between RLS and ADHD. J. Psychiatr. Res. 43, 941–945.
Schormair, B., Plag, J., Kaffe, M., Groß, N., Czamara, D., Samtleben, W., Lichtner, P., Ströhle, A., Stefanidis, I., Vainas, A., Dardiotis, E., Sakkas, G., Gieger, C., Müller-Myhsok, B., Meitinger, T., Heemann, U., Hadjigeorgiou, G., Oexle, K., Winkelmann, J., 2011. MEIS1 and BTBD9: genetic association with restless leg syndrome in end stage renal disease. J. Med. Genet. 48, 462–466.
Schulte, E., Winkelmann, J., 2015. Movement Disorder Genetics. Springer.
Sharma, A., Couture, J., 2014. A review of the pathophysiology, etiology, and treatment of attention-deficit hyperactivity disorder (ADHD). Ann. Pharmacother. 48, 209–225.
Sikk, K., Haldre, S., Aquilonius, S.-M., Taba, P., 2011. Manganese-induced parkinsonism due to ephedrone abuse. Adv. Neurol. 2011, 865319.
Sistrunk, S.C., Ross, M.K., Filipov, N.M., 2007. Direct effects of manganese compounds on dopamine and its metabolite Dopac: an in vitro study. Environ. Toxicol. Pharmacol. 23, 286–296.
Skjørringe, T., Burkhart, A., Johnsen, K.B., Moos, T., 2015. Divalent metal transporter 1 (DMT1) in the brain: implications for a role in iron transport at the blood-brain barrier, and neuronal and glial pathology. Front. Mol. Neurosci. 8, 19.
Smith, E., Newland, P., Bestwick, K., Ahmed, N., 2012. Increased whole blood manganese concentrations observed in children with iron deficiency anaemia. J. Trace Elem. Med. Biol. 27, 65–69.
Sørensen, E., Grau, K., Berg, T., Simonsen, A., Magnussen, K., Erikstrup, C., Hansen, M., Ullum, H., 2012. A genetic risk factor for low serum ferritin levels in Danish blood donors. Transfusion 52, 2585–2589.
Søvik, E., Perry, C., LaMora, A., Barron, A., Ben-Shahar, Y., 2015. Negative impact of manganese on honeybee foraging. Biol. Lett. 11, 20140989.

Spangler, A., Spangler, J., 2009. Groundwater manganese and infant mortality rate by County in North Carolina: an ecological analysis. Ecohealth 6, 596–600.

Spieler, D., Kaffe, M., Knauf, F., Bessa, J., Tena, J., Giesert, F., Schormair, B., Tilch, E., Lee, H., Horsch, M., Czamara, D., Karbalai, N., von Toerne, C., Waldenberger, M., Gieger, C., Lichtner, P., Claussnitzer, M., Naumann, R., Müller-Myhsok, B., Torres, M., Garrett, L., Rozman, J., Klingenspor, M., Gailus-Durner, V., Fuchs, H., de Angelis, M., Beckers, J., Hölter, S., Meitinger, T., Hauck, S., Laumen, H., Wurst, W., Casares, F., Gómez-Skarmeta, J., Winkelmann, J., 2014. Restless legs syndrome-associated intronic common variant in *Meis1* alters enhancer function in the developing telencephalon. Genome Res. 24, 592–603.

Stefansson, H., Rye, D.B., Hicks, A., Petursson, H., Ingason, A., Thorgeirsson, T.E., Palsson, S., Sigmundsson, T., Sigurdsson, A.P., Eiriksdottir, I., Soebech, E., Bliwise, D., Beck, J.M., Rosen, A., Waddy, S., Trotti, L.M., Iranzo, A., Thambisetty, M., Hardarson, G.A., Kristjansson, K., Gudmundsson, L.J., Thorsteinsdottir, U., Kong, A., Gulcher, J.R., Gudbjartsson, D., Stefansson, K., 2007. A genetic risk factor for periodic limb movements in sleep. N. Engl. J. Med. 357, 639–647.

Stepens, A., Logina, I., Liguts, V., Aldins, P., Eksteina, I., Platkājis, A., Mārtinsone, I., Tērauds, E., Rozentāle, B., Donaghy, M., 2008. A Parkinsonian syndrome in methcathinone users and the role of manganese. N. Engl. J. Med. 358, 1009–1017.

Stredrick, D., Stokes, A., Worst, T., Freeman, W., Johnson, E., Lash, L., Aschner, M., Vrana, K., 2004. Manganese-induced cytotoxicity in dopamine-producing cells. Neurotoxicology 25, 543–553.

Stubbe, D., 2000. Attention-deficit/hyperactivity disorder overview. Historical perspective, current controversies, and future directions. Child Adolesc. Psychiatr. Clin. N. Am. 9, 469–479.

Sulzer, D., 2007. Multiple hit hypotheses for dopamine neuron loss in Parkinson's disease. Trends Neurosci. 30, 244–250.

Tai, Y., Chew, K., Tan, B., Lim, K.-L., Soong, T., 2016. Iron mitigates DMT1-mediated manganese cytotoxicity via the ASK1-JNK signaling axis: implications of iron supplementation for manganese toxicity. Sci. Rep. 6, 21113.

Thireau, J., Farah, C., Molinari, N., Bouilloux, F., Torreilles, L., Winkelmann, J., Scholz, S., Richard, S., Dauvilliers, Y., Marmigère, F., 2017. MEIS1 variant as a determinant of autonomic imbalance in restless legs syndrome. Sci. Rep. 7, 46620.

Tidball, A., Bryan, M., Uhouse, M., Kumar, K., Aboud, A., Feist, J., Ess, K., Neely, D., Aschner, M., Bowman, A., 2015. A novel manganese-dependent ATM-p53 signaling pathway is selectively impaired in patient-based neuroprogenitor and murine striatal models of Huntington's disease. Hum. Mol. Genet. 24, 1929–1944.

Tong, M., Dong, M., Monte, S., 2009. Brain insulin-like growth factor and neurotrophin resistance in Parkinson's disease and dementia with Lewy bodies: potential role of manganese neurotoxicity. J. Alzheimers Dis. 16, 585–599.

Trenkwalder, C., Högl, B., Winkelmann, J., 2009. Recent advances in the diagnosis, genetics and treatment of restless legs syndrome. J. Neurol. 256, 539–553.

Tripp, G., Wickens, J.R., 2009. Neurobiology of ADHD. Neuropharmacology 57 (7–8), 579–589.

Trotti, L.M., 2017. Restless legs syndrome and sleep-related movement disorders. Continuum (Minneap. Minn.) 23, 1005–1016.

Wang, L., Shiraki, A., Itahashi, M., Akane, H., Abe, H., Mitsumori, K., Shibutani, M., 2013. Aberration in epigenetic gene regulation in hippocampal neurogenesis by developmental exposure to manganese chloride in mice. Toxicol. Sci. 136, 154–165.

Wang, J.-W., Zhang, Y.-Q., Zhang, X.-H., Wang, Y.-P., Li, J.-P., Li, Y.-J., 2017. Deep brain stimulation of pedunculopontine nucleus for postural instability and gait disorder after Parkinson disease: a meta-analysis of individual patient data. World Neurosurg. 102, 72–78.

Wasserman, G., Liu, X., Parvez, F., Factor-Litvak, P., Ahsan, H., Levy, D., Kline, J., van Geen, A., Mey, J., Slavkovich, V., Siddique, A., Islam, T., Graziano, J., 2011. Arsenic and manganese exposure and children's intellectual function. Neurotoxicology 32, 450–457.
Weinstock, L.B., Walters, A.S., Paueksakon, P., 2012. Restless legs syndrome—theoretical roles of inflammatory and immune mechanisms. Sleep Med. Rev. 16, 341–354.
Wijemanne, S., Jankovic, J., 2015. Restless legs syndrome: clinical presentation diagnosis and treatment. Sleep Med. 16, 678–690.
Williams, B., Kwakye, G., Wegrzynowicz, M., Li, D., Aschner, M., Erikson, K., Bowman, A., 2010. Altered manganese homeostasis and manganese toxicity in a Huntington's disease striatal cell model are not explained by defects in the iron transport system. Toxicol. Sci. 117, 169–179.
Winkelmann, J., Schormair, B., Lichtner, P., Ripke, S., Xiong, L., Jalilzadeh, S., Fulda, S., Pütz, B., Eckstein, G., Hauk, S., Trenkwalder, C., Zimprich, A., Stiasny-Kolster, K., Oertel, W., Bachmann, C., Paulus, W., Peglau, I., Eisensehr, I., Montplaisir, J., Turecki, G., Rouleau, G., Gieger, C., Illig, T., Wichmann, H.-E., Holsboer, F., Müller-Myhsok, B., Meitinger, T., 2007. Genome-wide association study of restless legs syndrome identifies common variants in three genomic regions. Nat. Genet. 39, 1000–1006.
Winkelmann, J., Czamara, D., Schormair, B., Knauf, F., Schulte, E., Trenkwalder, C., Dauvilliers, Y., Polo, O., Högl, B., Berger, K., Fuhs, A., Gross, N., Stiasny-Kolster, K., Oertel, W., Bachmann, C., Paulus, W., Xiong, L., Montplaisir, J., Rouleau, G., Fietze, I., Vávrová, J., Kemlink, D., Sonka, K., Nevsimalova, S., Lin, S.-C., Wszolek, Z., Vilariño-Güell, C., Farrer, M., Gschliesser, V., Frauscher, B., Falkenstetter, T., Poewe, W., Allen, R., Earley, C., Ondo, W., Le, W.-D., Spieler, D., Kaffe, M., Zimprich, A., Kettunen, J., Perola, M., Silander, K., Cournu-Rebeix, I., Francavilla, M., Fontenille, C., Fontaine, B., Vodicka, P., Prokisch, H., Lichtner, P., Peppard, P., Faraco, J., Mignot, E., Gieger, C., Illig, T., Wichmann, H.-E., Müller-Myhsok, B., Meitinger, T., 2011. Genome-wide association study identifies novel restless legs syndrome susceptibility loci on 2p14 and 16q12.1. PLoS Genet. 7, e1002171.
Winkelmann, J., Schormair, B., Xiong, L., Dion, P., Rye, D., Rouleau, G., 2017. Genetics of restless legs syndrome. Sleep Med. 31, 18–22.
Xiong, L., Dion, P., Montplaisir, J., Levchenko, A., Thibodeau, P., Karemera, L., Rivière, J., St-Onge, J., Gaspar, C., Dubé, M., Desautels, A., Turecki, G., Rouleau, G., 2007. Molecular genetic studies of DMT1 on 12q in French-Canadian restless legs syndrome patients and families. Am. J. Med. Genet. B Neuropsychiatr. Genet. 144B, 911–917.
Xiong, L., Catoire, H., Dion, P., Gaspar, C., Lafrenière, R., Girard, S., Levchenko, A., Rivière, J.-B., Fiori, L., St-Onge, J., Bachand, I., Thibodeau, P., Allen, R., Earley, C., Turecki, G., Montplaisir, J., Rouleau, G., 2009. MEIS1 intronic risk haplotype associated with restless legs syndrome affects its mRNA and protein expression levels. Hum. Mol. Genet. 18, 1065–1074.
Yang, Q., Li, L., Chen, Q., Foldvary-Schaefer, N., Ondo, W., Wang, Q., 2011. Association studies of variants in MEIS1, BTBD9, and MAP2K5/SKOR1 with restless legs syndrome in a US population. Sleep Med. 12, 800–804.
Yeh, P., Walters, A.S., Tsuang, J.W., 2012. Restless legs syndrome: a comprehensive overview on its epidemiology, risk factors, and treatment. Sleep Breath. 16, 987–1007.
Zecca, L., Youdim, M., Riederer, P., Connor, J., Crichton, R., 2004. Iron, brain ageing and neurodegenerative disorders. Nat. Rev. Neurosci. 5, 863–873.
Zhang, D., Kanthasamy, A., Anantharam, V., Kanthasamy, A., 2011. Effects of manganese on tyrosine hydroxylase (TH) activity and TH-phosphorylation in a dopaminergic neural cell line. Toxicol. Appl. Pharm. 254, 65–71.

CHAPTER FOUR

Developmental Neurotoxicity of Alcohol: Effects and Mechanisms of Ethanol on the Developing Brain

Xiaolu Zhang[*,†], Joel G. Hashimoto[†], Marina Guizzetti[†,1]
[*]Department of Pharmaceutical Science, Northern Jiangsu People's Hospital, Yangzhou, China
[†]Department of Behavioral Neuroscience, Oregon Health & Science University and VA Portland Health Care System, Portland, OR, USA
[1]Corresponding author: e-mail address: guizzett@ohsu.edu

Contents

Advances in Neurotoxicology, Volume 2
ISSN 2468-7480
https://doi.org/10.1016/bs.ant.2018.03.006

1. INTRODUCTION

Prenatal alcohol exposure profoundly alters brain development leading to mental retardation and cognitive and behavioral dysfunction (Mattson et al., 2011; Streissguth et al., 1980). While the developmental effects of alcohol abuse during gestation have been recognized for centuries, with references to the effects of alcohol during pregnancy present in the Bible and in the work by the philosopher Aristotle (Calhoun and Warren, 2007), the recognition of these effects by the scientific community, occurred much later.

In 1968, a French study for the first time ascribed a list of abnormalities to prenatal alcohol exposure after observing them in the offspring of women who drunk alcohol heavily during pregnancy (Lemoine et al., 1968). A few years later, similar abnormalities were described in another cohort of children born from mothers who abused alcohol during pregnancy and the term "fetal alcohol syndrome (FAS)" was introduced for the first time to describe the common pattern of observed deficits (Jones and Smith, 1973; Jones et al., 1973). It soon became apparent that FAS did not describe the full spectrum of abnormalities caused by prenatal alcohol exposure, and other terms, including partial (p)FAS, alcohol-related neurodevelopmental disorders (ARND), and alcohol-related birth defects (ARBD) were introduced, which were subsequently grouped under the umbrella term fetal alcohol spectrum disorders (FASD) (Streissguth and O'Malley, 2000). The studies from Jones and colleagues triggered a new, intense wave of research focused on systematically characterizing the effects of alcohol on the developing brain, on identifying the mechanisms of brain development affected by alcohol, and on devising possible treatments and prevention strategies to reduce the impact of prenatal alcohol on the life of affected individuals.

Since there has not been definitive identification of a safe alcohol dose or duration of exposure in pregnancy, the recommendations by the US Centers for Disease Control and Prevention (CDC) and the US Surgeon General are that: "There is no known safe amount of alcohol to drink while pregnant. There is also no known safe time during pregnancy or safe type of alcohol." However, while even mild to moderate levels of *in utero* alcohol exposure has been associated with brain and behavioral alterations, the degree of CNS dysfunction worsens with increasing dose and exposure time (Flak et al., 2014; Marquardt and Brigman, 2016). In addition, the rate of alcohol use during pregnancy in the United States remains surprisingly high.

In this chapter, clinical studies, prospective human studies, and preclinical studies employing experimental animals and cells are discussed to provide an overall picture of the body of knowledge on FASD with regard to its manifestations, diagnosis, and mechanisms involved.

2. FETAL ALCOHOL SPECTRUM DISORDERS

The term FASD was adopted as a non-diagnostic umbrella term encompassing a wide range of effects caused by prenatal alcohol exposure and leading to lifelong physical, behavioral, and cognitive dysfunctions. The severity of structural and functional/behavioral alterations caused by ethanol in the developing brain depends on many factors including the timing and duration of alcohol consumption, the genetic background of the mother and the fetus, and the maternal nutritional status and overall heath. Because of these variables, the manifestations and severity of FASD vary widely (May et al., 2014; Sokol et al., 2003).

FASD includes the following diagnostic conditions: FAS, pFAS, ARND, and ARBD (Hoyme et al., 2005). Currently, there are two major published sets of diagnostic criteria which are used to evaluate children with known prenatal alcohol exposure: the 1996 Institute of Medicine (IOM) criteria and the CDC criteria (Bertrand et al., 2004; Hoyme et al., 2005; Institute of Medicine (IOM) et al., 1996).

2.1 Fetal Alcohol Syndrome

FAS was the first of the conditions caused by developmental alcohol exposure to be documented (Jones and Smith, 1973; Jones et al., 1973). The key diagnostic features of FAS are growth deficiency, a characteristic facial phenotype, central nervous system damage/dysfunction, and ascertained *in utero* alcohol exposure. Both the CDC and the IOM have published guidelines for FAS diagnosis (Bertrand et al., 2004; Hoyme et al., 2005; Institute of Medicine (IOM) et al., 1996). In this section, we will refer to the CDC guidelines.

Growth deficiency. Alcohol-induced growth deficiencies can be pre- and/or postnatal (failure to thrive) (Abel, 1998). Prenatal growth deficiencies can last through early childhood and are often reflected by low weight and stature, i.e., below the 10th percentile (after race, sex, gestational or postnatal age are considered).

Distinct facial appearance. Three facial features have diagnostic value in FAS: short palpebral fissures, smooth philtrum and thin upper lip (Astley, 2004). In addition, other facial characteristics such as ptosis of the eyelids, epicanthal folds, strabismus, low/flat nasal bridge, midface hypoplasia, abnormal palmar creases and railroad track ears can also be present, although they are not required for a FAS diagnosis. These characteristic facial features may disappear in older children and/or adult; thus, early diagnosis is highly recommended.

CNS dysfunction. CNS deficits are a major component of FAS. For diagnostic purposes, they can be structural/neurological abnormalities, such as microcephaly, defined as small head circumference (<10%) and agenesis of the corpus callosum, or functional deficits, which can be either intellectual or behavioral and include intellectual disability, impaired concentration, attention problems, and motor, cognitive or speech deficits. In the absence of a major structural abnormality, deficits in at least three CNS functional domains need to be present for a FAS diagnosis (Roszel, 2015; Shilko et al., 2010).

2.2 Partial (p)FAS

Children with pFAS can display the typical facial features as well as abnormalities in one of the other areas including growth and central nervous system. A diagnosis of pFAS can be rendered when a confirmed history of maternal alcohol exposure, evidence of two facial characteristics, and of growth retardation or of CNS deficits are presents (Hoyme et al., 2005).

2.3 Alcohol-Related Neurodevelopmental Disorders

Subsequent to the identification of FAS, it was observed that, in some cases, prenatal alcohol exposure caused cognitive and behavioral disabilities in the absence of growth retardation and of the characteristic facial features. The diagnostic term of alcohol-related neurodevelopmental disorders (ARND) was introduced to describe cognitive and behavioral disabilities in the presence of confirmed prenatal alcohol exposure (Bertrand et al., 2004; Hoyme et al., 2005). The ARND diagnosis does not require ethanol-induced growth retardation or facial features, but requires the presence of significant CNS structural or functional abnormalities. In the absence of major structural abnormalities, deficits in three or more CNS functional domains (intellectual and/or behavioral) need to be present for an ARND diagnosis including intellectual disability, impaired concentration, attention problems, poor academic performance, motor disorder, hand–eye coordination, impairment of executive functioning (planning, judgment, complex problem-solving), and social communication dysfunction (Lang, 2006). However, differentiating ARND from other complex neurodevelopmental disorders can be challenging.

2.4 Alcohol-Related Birth Defects

In utero alcohol exposure can also cause congenital abnormalities, most frequently in the heart, kidney, ear, eye and skeletal system in the absence or presence of apparent CNS dysfunction (Bertrand et al., 2004; Institute of Medicine (IOM) et al., 1996). ARBD is the diagnostic term used to describe congenital malformations due to fetal alcohol exposure (Manning and Hoyme, 2007).

3. NEUROBEHAVIORAL DEFICITS IN FASD

Outcomes of prenatal alcohol exposure vary widely with regards to the physical and neurobehavioral phenotypes. In particular, the neurobehavioral effects of ethanol range from decreased IQ, developmental delays, and deficits in learning, memory, social communication, executive functioning, attention span, and motor control (Reynolds et al., 2011; Riley and McGee, 2005). Understanding the cause of this considerable variability in neurobehavioral outcomes is a challenge. Major factors involved include the amount and patterns of alcohol consumption as well as the duration of exposure (Mattson et al., 2011). Indeed, studies indicate that high levels of prenatal alcohol exposure are associated with an increased risk for intellectual disability (Mattson et al., 1997). While intellectual deficit can occur in children without characteristic facial features, they are often more severe in patients with a diagnosis of FAS (Mattson and Riley, 1998).

Fetal alcohol can affect cognitive functions; children with a FAS diagnosis have in average lower IQs than children with pFAS or ARDN (Chasnoff et al., 2010). One study carried out on a small cohort of children heavily exposed to alcohol during gestation found that the average IQ score for individuals with FAS was 70, while for those with pFAS or ARND was 80 (Mattson et al., 1997). However, other studies showed that most individuals diagnosed with FAS are not intellectually disabled, with approximately 25% having IQ scores <70 (Mattson et al., 2011; Streissguth et al., 1997). For this reason, intellectual disability is not a necessary criterion for the diagnosis of FAS.

Cognitive abilities termed "executive functions" are highly affected in individuals with FASD (Hoyme et al., 2005; Riley and McGee, 2005).

The term "executive functions" include a variety of cognitive abilities, such as planning, response inhibition, working memory, and problem-solving (Pennington and Ozonoff, 1996; Rai et al., 2017). Executive function deficits can predict poorer social skills and greater behavioral problems (Schonfeld et al., 2006). Learning and memory are also affected in FASD, as individuals affected display deficits in the ability to remember and acquire information (Mattson et al., 1996a). *In utero* alcohol exposure is also associated with hyperactivity and attention deficits. Affected children show deficits in attention on neuropsychological tasks, including alertness, reaction time, and information processing. A significantly higher rate of attention deficit and hyperactivity disorders (ADHD) is found among individuals with FASD (Coles et al., 2002; Fryer et al., 2007).

Children with FAS also display motor impairments, tremors, weak grasp, poor hand/eye coordination, and gait and balance deficits (Jones et al., 1973; Roebuck et al., 1998b). Motor dysfunction in heavily exposed individuals persists through adulthood (Guerri et al., 2009; Mattson et al., 2011).

4. FASD AND BRAIN STRUCTURE

Children with FASD often display abnormalities in brain structure that likely contribute to behavioral and cognitive deficits. In general, the alterations of brain structure associated with prenatal alcohol exposure include brain size, shape, and symmetry, with a decrease in the volume of gray and white matter (McGee and Riley, 2006).

Early studies of brain malformations derived from autopsies of infants heavily exposed to alcohol prenatally revealed widespread damage throughout much of the brain, including microcephaly, errors in neuron migration, agenesis of the corpus callosum and anterior commissure, neuroglia heterotopias, and ventricle, basal ganglia, brain stem, and cerebellar anomalies (Jones and Smith, 1973; Riley and McGee, 2005). These malformations are the result of the most severe forms of FAS, as they were the results of autopsies of children who did not survive.

Structural magnetic resonance imaging (MRI) is a non-invasive imaging technique that allows for observation and measurement of brain structures in living subjects. Studies using MRI have identified specific alterations in brains of living individual with confirmed prenatal alcohol exposure with and without a diagnosis of FAS and have identified abnormalities in regional volume, cortical thickness, and shape (Lebel et al., 2011; Riley and McGee, 2005).

Postmortem studies and neuroimaging studies have both shown an overall brain size reduction associated with prenatal alcohol exposure. Shape abnormalities and asymmetry have also been found (Archibald et al., 2001; Wisniewski et al., 1983). Understanding the exact location of regional volume reduction and shape alterations may be more important than information on the overall brain size, as it enables inferences to be made about the brain systems and circuits affected by FASD (Nunez et al., 2011).

Researchers have demonstrated that cortical thickness and volume are reduced in FAS; however, not all areas of the cortex are equally affected by prenatal alcohol exposure. Higher gray matter density, thicker cortices, displacement of posterior regions, regional surface area reductions in anterior and posterior areas, and asymmetry were reported (Lebel et al., 2011; Norman et al., 2009; Riley and McGee, 2005; Sowell et al., 2001b). The volume of the white matter in the frontal lobe, parietal lobe, temporal lobe, and total lobe is decreased. Furthermore, children with heavy prenatal alcohol exposure have increases in gray matter and decreases in white matter in the perisylvian cortex of the temporal and parietal lobes, and white matter hypoplasia appears to be more severe than gray matter hypoplasia (Archibald et al., 2001; Sowell et al., 2001b). The effects of *in utero* alcohol exposure on gray matter and white matter are associated with the impairments in verbal learning and memory, executive functions, language, visuospatial functions and attention (Lebel et al., 2011; Norman et al., 2009; Riley and McGee, 2005).

The hippocampus is located in the medial temporal lobe and is involved in the storage of memories. Heavy prenatal alcohol exposure may affect the hippocampal structure, but data are controversial. A MRI study found that the absolute volume of the hippocampus in the left temporal lobe is smaller than that of the corresponding area in the right temporal lobe in children with FAS (Riikonen et al., 1999). More recently, a study reported an altered hippocampus shape in FAS children (Joseph et al., 2014). Furthermore, animal studies have also suggested that the hippocampus is affected by prenatal alcohol exposure (Berman and Hannigan, 2000). Conversely, among subcortical structures, the hippocampus was relatively preserved in FAS participants in comparison to the disproportionate effects on caudate nucleus, suggesting that the hippocampus was less affected by developmental alcohol exposure than some other brain areas (Archibald et al., 2001).

The cerebellum is known to be particularly sensitive to prenatal alcohol exposure. Several studies have found significant gray and white matter reductions in the cerebellum in children exposed to alcohol during fetal development with and without facial dysmorphology (McGee and Riley, 2006; Roebuck et al., 1998a). Alcohol exposure may cause disproportionate reductions in size and greater displacement of the anterior portion of the vermis (Sowell et al., 1996). Animal models also support this pattern of abnormal development of the cerebellum (Goodlett et al., 1990). The structural abnormalities of the cerebellum of children with FASD may be associated with learning and memory impairment, alterations in balance, movement, and coordination, distraction, and abnormalities in classically conditioned eye-blink (Lebel et al., 2011; Norman et al., 2009; Riley and McGee, 2005).

Structural abnormalities in the location, size, and shape of the corpus callosum in children, adolescents, young adults and adults with FASD were also reported (Bookstein et al., 2002; Lebel et al., 2011; Norman et al., 2009; Riley and McGee, 2005; Sowell et al., 2001b). Individuals with FASD showed a reduction in the corpus callosum, which is more pronounced in the posterior region (Nunez et al., 2011; Sowell et al., 2001a). Altered motor control, attention, verbal, learning and executive functions in FASD individuals are associated with corpus callosum abnormalities (Norman et al., 2009).

Basal ganglia, particularly the caudate nucleus, display disproportionate reductions in volume in FAS individuals (Mattson et al., 1994, 1996b). Because of the connections of the basal ganglia with the motor and frontal cortex, deficits in executive and motor functions and coordination found in individuals with FASD can be ascribed, at least in part, to an effect of ethanol on basal ganglia (Lebel et al., 2011; Mattson et al., 1994, 1996b; Norman et al., 2009; Riley and McGee, 2005).

5. ANIMAL MODELS FOR FASD

To study the effects of ethanol on basic mechanisms of brain development and to test the effects of treatments aimed at ameliorating the effects of developmental ethanol exposure several animal models of FASD have been developed. Many of the neurobehavioral outcomes observed in individuals with FASD including microcephaly, facial features, growth retardation, spatial learning disability, motor performance, attention, and social deficits have been recapitulated in animal models (Bayer et al., 1993; Cronise et al., 2001).

A number of animal models of developmental ethanol exposure have been established, including nonhuman primates, pigs, sheep, chicken embryos, rats, mice, drosophila, zebrafish, insects, and *C. elegans*. Rodent models (rats and mice) are by far the most commonly used in the vast majority of behavioral and mechanistic studies (Cole et al., 2012; Cudd, 2005;

Kelly et al., 2009). The most common methods used to administer ethanol in rodent models of FASD are: the addition of ethanol to liquid diet or the drinking water, artificial rearing following gastrostomy surgery, intragastric intubation, subcutaneous or intraperitoneal injection, and vapor inhalation (Heaton et al., 2006; Ikonomidou et al., 2000; Tran et al., 2007; West et al., 1986). Rodent models reproduced several of the brain abnormalities reported in FASD including microcephaly and structural alterations to the cerebral neocortex, the hippocampus, the cerebellum, the basal ganglia and the corpus callosum (Bonthius and West, 1990; Cao et al., 2014; Maier et al., 1997; Margret et al., 2005; Parnell et al., 2009).

A challenge in designing experiments that test the developmental effects of ethanol is the different timing of human and rodent brain development. The events of human brain development occurring during the first two trimesters of gestation also occur prenatally in rodents, while the equivalent of the third trimester of human gestation occurs for the most part during the first two postnatal weeks in rodents (Bayer et al., 1993).

The development of the central nervous system, both in human and rodents, begins with the formation of the neural plate, and continues through gestation and postnatal periods. Ethanol appears to interfere with all CNS developmental stages (Guerri, 1998).

Neural tube formation and generation, migration, and differentiation of forebrain neurons occurs primarily during gestation days 5(GD5) to GD11 in rats, which correspond to the first trimester of human gestation (Rice and Barone, 2000; White et al., 2015). Ethanol exposure during this period has profound effects on brain structure and on the number of newly generated neurons (Hashimoto-Torii et al., 2011; Miller and Potempa, 1990). Ethanol exposure during the first trimester-equivalent of human gestation also results in neural tube defects, facial defects and forebrain abnormalities, hypoplasia or aplasia of the corpus callosum, and alterations in the hippocampus and the anterior cingulated cortex consistent with holoprosencephaly (Sulik, 2005).

A second critical period of development occurs in rats between GD11–12 and GD18–21, equivalent to the second trimester of human pregnancy. This period is characterized by the differentiation of the central nervous system, which includes a high rate of neurogenesis and neuronal migration in the cerebral cortex and hippocampus (Guerri, 1998; Miranda, 2012; Rice and Barone, 2000; Richards et al., 2004). Ethanol exposure during this period decreases the thickness and cell proliferation rate within the ventricular zone, while increasing the thickness and proliferation within the sub-ventricular zone (Miller, 1989). During the second trimester-equivalent of human gestation, ethanol induces premature differentiation of radial glia into astrocytes (Gressens et al., 1992), while inhibiting the generation, survival, and migration of neurons in the neocortex, hippocampus, and the principal sensory nucleus, as well as inducing corpus callosum malformation (Miller, 1995).

The maturation that occurs during the third trimester of gestation in humans takes place primarily postnatally in rats as the rodent-equivalent of the third trimester of human gestation has been estimated to occur between GD18 and postnatal day (PD) 10 (Rice and Barone, 2000). During this developmental stage, the brain undergoes a rapid period of growth and functional development, which includes astrocyte and oligodendrocyte proliferation, the beginning of myelination and synaptogenesis and an enormous increase in dendrite arborization (Rice and Barone, 2000). Alcohol exposure during this period induces microcephaly, abnormal development of the cerebellum and hippocampus, neuronal apoptosis in the neocortex and hippocampus, reduced dendritic arborization, and behavioral dysfunctions (Bonthius and West, 1990; Gil-Mohapel et al., 2010; Ikonomidou et al., 2000; Melcer et al., 1994; West et al., 1986).

6. MECHANISMS INVOLVED IN FASD

Multiple cellular mechanisms involved in the developmental effects of alcohol have been described implicating both neurons and glial (macroglia and microglia) cells. Fig. 1 shows the main mechanisms and the cellular targets of developmental alcohol exposure; these mechanisms are described in details below.

6.1 Genetics and Early Development Gene Expression

The genetic background of the mother and/or the fetus may affect the outcome of prenatal alcohol exposure; in particular genetic variants in genes encoding for enzymes involved in alcohol metabolism appear to affect the fetus susceptibility to the developmental effects of alcohol (Alfonso-Loeches and Guerri, 2011). One of the genetic factors in humans is the alcohol dehydrogenase 1B (ADH1B) gene, which encodes for the enzyme that catalyzes the rate-limiting step in alcohol metabolism. The ADH1B*2 and ADH1B*3 alleles encode for proteins that are faster in metabolizing ethanol than the product of the ADH1B*1 allele. Individuals that are homozygous for ADH1B*1 have an increased risk for FASD, while the ADH1B*2 or

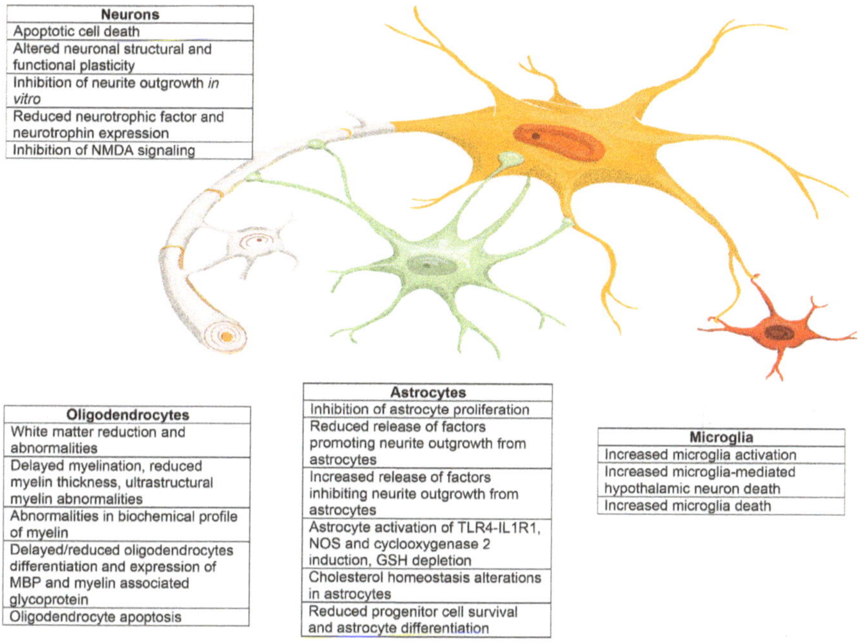

Fig. 1 Cellular targets of developmental alcohol exposure. Shown is the representation of the main effects exerted by ethanol on developing neurons, astrocytes, oligodendrocytes, and microglia that are likely to play a role in the neuropathology of FASD. *Reproduced with modifications from Guizzetti et al. (2014).*

ADH1B*3 alleles are associated with a protective effect toward FASD (Das et al., 2004; Green and Stoler, 2007; Jacobson et al., 2006).

Animal models of FASD have identified additional gene variants that are potentially involved in the susceptibility of the fetus to the effects of alcohol (McCarthy and Eberhart, 2014). It has been reported that binge alcohol exposure during the early phases of development reduces the expression of some homeobox genes, which direct the formation of many body structures during early development (Boncinelli, 1997; Halder et al., 2014; Rifas et al., 1997).

There is also some evidence that alcohol-induced abnormalities may be caused by deficiencies in retinoic acid, its receptors, or its signaling. Retinoic acid plays an important role in the regulation of homeobox gene expression during development (Chambon, 1996; Ross et al., 2000). It has been proposed that alcohol may compete with retinol (the retinoic acid precursor) for the binding to alcohol dehydrogenase (ADH), causing a reduction in the formation of retinoic acid (Duester et al., 1997).

6.2 Cell–Cell Interaction

Cell adhesion molecules (CAMs) are members of a subgroup of an immunoglobulin superfamily and, in the brain, have been established as key players in modulating cell–cell interaction throughout CNS development (Shapiro et al., 2007). Ethanol may alter brain development by inhibiting CAM-mediated cell–cell interaction. Neuronal CAM (NCAM) and L1CAM (L1) have been the most studied CAMs with regards to FASD as they are involved in several aspects of neural development including neuronal migration, axon and dendrite growth, and the formation of synaptic contacts on target neurons (Schmid and Maness, 2008).

L1 is a cell surface transmembrane glycoprotein that plays a critical role during normal neuronal development. Humans or rodents with genetic defects in L1 have extensive neuropathological defects and aberrant behaviors. Interestingly, it was found that FAS individuals show several similarities with the phenotype of individuals with L1 mutations, suggesting a common mechanism. This observation led to the investigation of the effects of neonatal alcohol exposure on L1CAM, which was shown to inhibit L1 mediated neurite outgrowth in cerebellar granule neurons (Bearer, 2001; Bearer et al., 1999). This effect if potentially mediated by alcohol-induced alterations in the processes of tyrosine phosphorylation/dephosphorylation of the L1 cytoplasmic domain (Littner et al., 2013).

NCAM, also called CD56, is a homophilic binding glycoprotein expressed in the brain on the surface of neurons and glia (Rutishauser et al., 1988). NCAM knockout mice show a reduction in brain size and persistent behavioral deficits (Cremer et al., 1994). During postnatal cerebral cortex development in rats, *in utero* alcohol exposure increased the levels of polysialylated NCAM (PSA-NCAM), associated with lower adhesive properties, while decreasing the levels of other NCAM isoforms (NCAM 140 and NCAM 180) (Alfonso-Loeches and Guerri, 2011; Minana et al., 2000).

6.3 Neuronal Apoptosis

Binge-like alcohol exposure during the neonatal period in rodents induces massive cell death in the forebrain, midbrain, cerebellum, brainstem, spinal cord and retina which was attributed to apoptosis (Creeley and Olney, 2013; Dikranian et al., 2005; Ikonomidou et al., 2000; Tenkova et al., 2003). Alcohol-induced apoptosis in neurons is mediated by multiple mechanisms. Ethanol increases the expression of the pro-apoptotic protein Bax, inhibits the phosphorylation (associated with activation) of antiapoptotic proteins extracellular signal-regulated protein kinases (ERK) and Akt, and disrupts the balance between Bax and BclXL protein. Bax induction by alcohol leads to mitochondrial damage, leaking of mitochondrial cytochrome *c*, and activation of caspases 3 and 8 (Olney et al., 2002a; Young et al., 2003, 2005, 2008). However, neuronal commitment to apoptotic cell death occurs before the activation of caspase 3 as caspase 3 knock-down does not prevent ethanol-induced apoptosis (Young et al., 2003, 2005). In addition, ethanol-induced inhibition of NMDA receptors and activation of GABA A receptors have been shown to be involved in apoptotic neurodegeneration in the developing brain (Olney et al., 2002b).

Several studies have also been carried out on the effects of ethanol on cerebellar neurons *in vivo* and *in vitro*. The two main neuronal populations in the cerebellum are Purkinje cells, which are mainly generated prenatally and undergo massive dendritic outgrowth and synaptogenesis during the neonatal period, and cerebellar granule cells, which are generated postnatally. Ethanol exposure during the neonatal period leads to the loss of both cerebellar Purkinje and granule cells by apoptosis (Dikranian et al., 2005; Luo, 2012). Both cell types are more sensitive to ethanol-induced cell death during the early stages of differentiation and maturation (Luo, 2012; Marcussen et al., 1994). For this reason, after neonatal ethanol exposure, cerebellar granule cells are more sensitive to alcohol than Purkinje cells, as at this developmental time-point granule cells are newly generated and are less mature than Purkinje cells (Maier and West, 2001; Wozniak et al., 2004). Mechanisms implicated in Purkinje cell loss include ethanol-induced alterations in neurotrophins, including brain-derived neurotrophic factor (BDNF) and neurotrophin-3 (NT3) (Light et al., 2002). Inhibition of anti-apoptotic NMDA signaling, neurotrophic factors (such as insulin-like growth factor, IGF, I and II) and neurotrophins (such as BDNF) have been implicated in cerebellar granule cell apoptosis (Luo, 2012).

Finally, early gestation exposure to alcohol, before neurogenesis occurs, leads to a massive increase in apoptotic degeneration of precursor cell populations of the developing brain and craniofacial region, leading to a reduction in the number of neurons generated later in gestation and to the development of facial features characteristic of FAS (Dunty et al., 2001; Sulik, 2005).

6.4 Glial Cells

It is now well established that glial cells play a major role during brain development and that ethanol affects several glial functions that contribute to the brain and behavioral deficits observed in FASD. Clinical and preclinical studies have shown that *in utero* alcohol exposure induces structural and functional abnormalities in glia, which affect glia-mediated trophic support to neurons, formation of mature synapses, and neuroplasticity (Phillips et al., 1991; Rubert et al., 2006; Saito et al., 2016).

Radial glia are a transient population of glial cells that are present in the developing cortex (in the retina and the cerebellum radial glia cells persist into adulthood and are called Muller cells and Bergman glia, respectively). Radial glia function as precursor cells generating neurons, astrocytes, oligodendrocytes, and ependymal cells. During embryogenesis alcohol affects the migration of young neurons along radial glial fibers and reduces the radial glia progenitor pool (Miller and Potempa, 1990; Perez-Torrero et al., 1997; Rubert et al., 2006). In studies of children with FAS, hypoplasia of the corpus callosum and anterior commissure, two areas originally generated by glia, has been observed (Guerri et al., 2001; Riley et al., 1995).

Astrocytes play several crucial roles in the mature and developing brain as they are involved in the regulation of ion homeostasis, reuptake and breakdown of neurotransmitters, maintenance of the blood–brain-barrier, regulation of blood flow, proliferation of stem cells, modulation of neurotransmission, and modulation of neuronal plasticity by inducing axon and dendrite outgrowth and synaptogenesis (Accorsi-Mendonca et al., 2013; Araque et al., 2014; Santello et al., 2012). Several early studies investigated the effects of ethanol on astrocyte functions. Under certain conditions (Goodlett et al., 1993, 1997) but not others (Ryabinin et al., 1995) neonatal ethanol exposure induces reactive astrogliosis measured as increased Glial Fibrillary Acidic Protein (GFAP) levels. In an *in vitro* study astrocyte cultures prepared from the brains of animals neonatally treated with ethanol showed a decrease in GFAP levels due to hypermethylation of DNA in the GFAP promoter region (Valles et al., 1997).

Ethanol inhibits the proliferation of astrocytes in culture induced by serum, IGF-I, the cholinergic agonist carbachol, and PDGF (Guizzetti and Costa, 1996; Kotter and Klein, 1999; Resnicoff et al., 1994). We have shown that ethanol inhibition of carbachol-induced proliferation is mediated by the inhibition of phospholipase D (PLD) signaling. Alcohol is a substrate of PLD that competes with water (the normal PLD substrate together with phosphatidylcholine); in the presence of ethanol, PLD catalyzes the formation of phosphatidyl ethanol instead of the physiological second messenger phosphatidic acid (Guizzetti and Costa, 2000, 2002; Guizzetti et al., 2003, 2004, 2014; Tsuji et al., 2003).

Ethanol also profoundly alters the astrocyte extracellular matrix (ECM) through multiple mechanisms. Ethanol can alter gene and/or protein expression of ECM proteins or can affect extracellular proteolysis of ECM proteins in astrocyte cultures. These effects of ethanol on astrocyte ECM profoundly affects neuronal development, as described in detail in Section 6.5 (Guizzetti et al., 2010, 2014; Wilhelm and Guizzetti, 2015; Zhang et al., 2014a).

Oligodendrocytes, which form myelin sheaths in the brain and provide trophic support for the maintenance of axonal integrity (Funfschilling et al., 2012), are also affected by ethanol. Oligodendrogenesis and the beginning of myelination occur during the third trimester of pregnancy in humans and prenatal alcohol exposure in humans is associated with white matter abnormalities in imaging studies (Lebel et al., 2011; Norman et al., 2009). Exposure to ethanol during the third trimester-equivalent of human gestation reduced the number of oligodendrocytes and proliferating oligodendrocyte precursors. These effects, though transient, lead to long-lasting changes in myelin protein composition and white matter injury (Newville et al., 2017). Additional studies also provide evidence of an ethanol-induced delay in oligodendrocyte differentiation and maturation, causing ultrastructural damage of myelin sheaths (Creeley et al., 2013; Guerri et al., 2001; Zoeller et al., 1994). Prenatal alcohol exposure in monkeys during the equivalent of the third trimester of human gestation induces apoptosis of oligodendrocytes that are starting to produce myelin proteins across white matter regions (Creeley and Olney, 2013; Creeley et al., 2013). Interestingly, an *in vitro* study found a protective effect of low-doses of proinflammatory cytokines toward ethanol-induced oligodendrocyte cytotoxicity (Benjamins et al., 2011).

Microglia are distributed across the CNS and are the resident innate immunity cells of the brain. Microglia are therefore the main cell types in the brain involved in control of infections and clearance of pathogens. Even in the absence of infection, microglia can be activated by neuronal cell death and other noxious stimuli. Activation of microglia involves the fast expression and release of several proinflammatory mediators leading to a neuroinflammatory response that can be harmful to neurons. Alcohol can induce microglial activation leading to the expression and release of inflammatory cytokines (TNFα, IL1β) and nitric oxide (Fernandez-Lizarbe et al., 2009; Zhang et al., 2016). The activation of microglia following alcohol exposure in the developing mouse brain is primarily mediated by BAX-induced neuronal degeneration *via* apoptosis (Ahlers et al., 2015; Boyadjieva and Sarkar, 2013a,b). Ethanol has also been shown to cause microglia cell death (Kane et al., 2011).

6.5 Neuronal Plasticity

Learning and memory deficits described in animal models of FASD (Berman and Hannigan, 2000; Clements et al., 2005; Girard et al., 2000; Goodlett and Johnson, 1997; Subbanna et al., 2013; Thomas et al., 2007) are due in part to the effects of developmental alcohol exposure on different forms of neuronal plasticity. Developmental alcohol exposure alters long term potentiation and long term depression (Izumi et al., 2005; Richardson et al., 2002; Servais et al., 2007; Sutherland et al., 1997), barrel cortex plasticity (Rema and Ebner, 1999), ocular dominance plasticity (Foxworthy and Medina, 2015) and reduces both expression and activity in AMPA receptors, which regulate dentate gyrus synaptic plasticity and hippocampal-sensitive learning (Staples et al., 2014). In a recent review article, the relationship between developmental alcohol-induced abnormalities in learning and memory and alterations in hippocampal synaptic plasticity was highlighted (Fontaine et al., 2016).

Several studies also investigated the effects of developmental alcohol exposure on structural plasticity. Dendritic arborization and dendritic spine density are decreased in spiny neurons of the nucleus accumbens and in pyramidal neurons of the cortex following developmental alcohol exposure (Cui et al., 2010; Hamilton et al., 2010a,b; Whitcher and Klintsova, 2008).

A new mechanism by which ethanol alters neuronal plasticity and development is by affecting the factors released by astrocytes. Astrocytes release numerous factors involved in neuronal development; in a proteomic study we found that the majority of the proteins released by astrocytes are ECM proteins or components of extracellular protease systems involved in the modulation of the ECM (Moore et al., 2009). Furthermore, *in utero* alcohol exposure strongly affects the release of neuritogenic factors by astrocytes, which, in turn, affect neuronal development and plasticity. Astrocytes prepared from animals that were prenatally exposed to ethanol displayed a reduced ability to foster neurite growth in co-cultured cortical neurons (Pascual and Guerri, 2007). Furthermore ethanol inhibits axonal growth of cerebellar neurons induced by the active fragment of the astrocyte-released activity-dependent neuroprotective protein NAPVSIPQ (NAP) (Chen and Charness, 2008).

We reported that carbachol-treated astrocytes induce neurite outgrowth in hippocampal neurons co-cultured with astrocytes. In this co-culture system, astrocytes and neurons are in close proximity, facing each other, but are not touching (Guizzetti et al., 2008). In the same system, co-treatment of astrocytes with carbachol and ethanol prevents the neuritogenic effect of carbachol-treated astrocytes (Guizzetti et al., 2010). We also found that carbachol increases the levels of neuritogenic ECM proteins laminin and fibronectin by increasing the levels of an inhibitor of the extracellular protease plasmin. Ethanol inhibits the effects of carbachol, thereby inhibiting neurite outgrowth (Guizzetti et al., 2008, 2010). Ethanol-treated astrocytes also displayed a reduced ability to foster neurite outgrowth when neurons are cultured on top of ethanol-treated astrocytes (Zhang et al., 2014a). This effect of ethanol is mediated by the upregulation of ECM proteins that act as inhibitors of neurite outgrowth, namely chondroitin sulfate proteoglycans (Zhang et al., 2014a) and by the upregulation of tissue-type plasminogen activator, which, through the activation of plasminogen to plasmin, increases the degradation of the neuritogenic ECM (Zhang et al., 2014b).

6.6 Epigenetics

We use the word epigenetics to refer to modification of histones, DNA, or RNA, resulting in changes in gene expression. The most common epigenetic mechanisms investigated with regards to FASD research are DNA methylation, histone posttranslational modification including histone acetylation and methylation, and microRNAs, a subset of non-coding RNAs that selectively degrades mRNAs before they are translated to proteins. Of particular relevance to FASD is the fact that, during early development, cell-type specification and differentiation is regulated by epigenetic mechanisms.

Recently, many studies have been carried out to explore the effects of developmental alcohol exposure on histone posttranslational modifications, DNA methylation, and microRNAs, including several genome-wide studies. Several excellent reviews have been published on this topic and we refer to these reviews that adequately summarize this field (Basavarajappa and Subbanna, 2016; Chater-Diehl et al., 2017; Gavin et al., 2017; Miranda, 2012; Resendiz et al., 2014).

6.7 Oxidative Stress

Several studies suggest ethanol exposure prenatally or neonatally in rodents triggers oxidative stress in the brain (Brocardo et al., 2011, 2012). Oxidative stress has been reported in the whole brain and in several brain regions such as hippocampus, hypothalamus, and cerebellum after prenatal and neonatal alcohol exposure. These effects can be long-lasting, and are involved in altered synaptic plasticity, learning and memory deficits, and neuronal cell death (Chu et al., 2007; Dembele et al., 2006; Dong et al., 2010; Patten et al., 2013; Ramachandran et al., 2001; Reyes et al., 1993).

Gestational or neonatal exposure to ethanol in rodents induces reactive oxygen species (ROS) production, increases lipid peroxidation, and decreases the levels of glutathione (GSH) and superoxide dismutase (SOD) (Brocardo et al., 2011; Dembele et al., 2006; Guerri, 1998). Ethanol also induces oxidative stress in astrocyte cultures by stimulating the formation of ROS, depleting GSH, increasing cyclooxygenase 2, and inducing nitric oxide synthase expression *via* NF-kB activation (Blanco et al., 2004; Montoliu et al., 1995).

In conclusion, published evidence indicates that developmental ethanol exposure induces ROS production, oxidative damage, and imbalance of the endogenous antioxidant system leading to oxidative stress, which is involved in the neuropathogenesis of FASD.

7. FASD TREATMENTS

An important area of FASD research is aimed at identifying interventions, including therapeutics, dietary supplements, and environmental enrichments, that can improve the consequences of prenatal alcohol exposure. Choline supplementation has emerged as a promising dietary intervention to ameliorate the effects of prenatal alcohol exposure. Prenatal or postnatal choline supplementation improves memory performance and neuronal plasticity, while prenatal choline deficiency detrimentally affects brain functions (Li et al., 2004; Meck et al., 2008; Zeisel and Niculescu, 2006). Prenatal or neonatal choline supplementation also reduces hyperactivity and improves spatial learning deficits associated with the hippocampus and behavioral abnormalities associated with the cerebellum in FASD rodent models (Thomas and Tran, 2012; Thomas et al., 2000, 2004, 2007, 2009, 2010). Choline supplementation may improve physiological and/or pathological (as in the case of FASD) brain development by increasing the levels of membrane phospholipids phosphatidylcholine and sphingomyelin that also can generate second messengers such as phosphatidic acid, diacylglycerol, and ceramide. Choline is also the precursor of two signaling lipids, platelet-activating factor and sphingosylphosphorylcholine, of the neurotransmitter acetylcholine, and of betaine, a methyl donor in DNA methylation (Blusztajn, 1998). Choline is currently being tested clinically for its effectiveness in treating FASD (Nguyen et al., 2016; Wozniak et al., 2015).

Several pharmacological compounds have been investigated as possible FASD interventions. These interventions include peptides that antagonize ethanol-induced inhibition of L1CAM in cell adhesion (Chen et al., 2005; Wilkemeyer et al., 2003); antioxidants and SOD (Kotch et al., 1995), vitamin C (Peng et al., 2005), vitamin E (Heaton et al., 2000), and green tea extract (Long et al., 2010) that protect against ethanol-induced oxidative stress in FASD animal models; and serotonin agonists buspirone and ipsapirone that protect against ethanol-induced serotoninergic neuron damage (Kim and Druse, 1996; Tajuddin and Druse, 1999).

Postnatal environment by neonatal handling (Weinberg et al., 1995), environmental enrichment (Hannigan et al., 1993), or complex motor training (Klintsova et al., 1998, 2000, 2002) can improve behavioral abnormalities caused by developmental alcohol exposure in rodents. These postnatal enrichment interventions may be the easiest to apply as therapies for FASD in clinical settings.

ACKNOWLEDGMENTS

This work was supported by VA Merit Review Award # I01BX001819 and by NIH/NIAAA R01AA021468, and R01AA022948 to M.G. and by facilities and resources at the VA Portland Health Care System (M.G.). The opinions expressed in this paper are solely those of the authors. The contents do not represent the views of the US Department of Veterans Affairs or the US government. The authors are extremely grateful to Mr. Jeff Frkonja for the graphic design of Fig. 1.

REFERENCES

Abel, E.L., 1998. Fetal alcohol syndrome: the 'American Paradox'. Alcohol Alcohol. 33, 195–201.

Accorsi-Mendonca, D., Zoccal, D.B., Bonagamba, L.G., Machado, B.H., 2013. Glial cells modulate the synaptic transmission of NTS neurons sending projections to ventral medulla of Wistar rats. Physiol. Rep. 1, e00080.

Ahlers, K.E., Karacay, B., Fuller, L., Bonthius, D.J., Dailey, M.E., 2015. Transient activation of microglia following acute alcohol exposure in developing mouse neocortex is primarily driven by BAX-dependent neurodegeneration. Glia 63, 1694–1713.

Alfonso-Loeches, S., Guerri, C., 2011. Molecular and behavioral aspects of the actions of alcohol on the adult and developing brain. Crit. Rev. Clin. Lab. Sci. 48, 19–47.

Araque, A., Carmignoto, G., Haydon, P.G., Oliet, S.H., Robitaille, R., Volterra, A., 2014. Gliotransmitters travel in time and space. Neuron 81, 728–739.

Archibald, S.L., Fennema-Notestine, C., Gamst, A., Riley, E.P., Mattson, S.N., Jernigan, T.L., 2001. Brain dysmorphology in individuals with severe prenatal alcohol exposure. Dev. Med. Child Neurol. 43, 148–154.

Astley, S.J., 2004. Diagnostic Guide for Fetal Alcohol Spectrum Disroders: The 4-Digit Diagnostic Code. University of Washington Publication Service, Seattle.

Basavarajappa, B.S., Subbanna, S., 2016. Epigenetic mechanisms in developmental alcohol-induced neurobehavioral deficits. Brain Sci. 6, 12.

Bayer, S.A., Altman, J., Russo, R.J., Zhang, X., 1993. Timetables of neurogenesis in the human brain based on experimentally determined patterns in the rat. Neurotoxicology 14, 83–144.
Bearer, C.F., 2001. L1 cell adhesion molecule signal cascades: targets for ethanol developmental neurotoxicity. Neurotoxicology 22, 625–633.
Bearer, C.F., Swick, A.R., O'Riordan, M.A., Cheng, G., 1999. Ethanol inhibits L1-mediated neurite outgrowth in postnatal rat cerebellar granule cells. J. Biol. Chem. 274, 13264–13270.
Benjamins, J.A., Nedelkoska, L., Lisak, R.P., Hannigan, J.H., Sokol, R.J., 2011. Cytokines reduce toxic effects of ethanol on oligodendroglia. Neurochem. Res. 36, 1677–1686.
Berman, R.F., Hannigan, J.H., 2000. Effects of prenatal alcohol exposure on the hippocampus: spatial behavior, electrophysiology, and neuroanatomy. Hippocampus 10, 94–110.
Bertrand J., Floyd R.L., Weber M. K., O'Connor M., Riley E. P., Johnson K. A., Cohen D. E., National Task Force on FAS/FAE. 2004. Fetal Alcohol Syndrome: Guidelines for Referral and Diagnosis. *In:* Centers for Disease Control and Prevention. (ed.). Atlanta, GA.
Blanco, A.M., Pascual, M., Valles, S.L., Guerri, C., 2004. Ethanol-induced iNOS and COX-2 expression in cultured astrocytes via NF-kappa B. Neuroreport 15, 681–685.
Blusztajn, J.K., 1998. Choline, a vital amine. Science 281, 794–795.
Boncinelli, E., 1997. Homeobox genes and disease. Curr. Opin. Genet. Dev. 7, 331–337.
Bonthius, D.J., West, J.R., 1990. Alcohol-induced neuronal loss in developing rats: increased brain damage with binge exposure. Alcohol. Clin. Exp. Res. 14, 107–118.
Bookstein, F.L., Streissguth, A.P., Sampson, P.D., Connor, P.D., Barr, H.M., 2002. Corpus callosum shape and neuropsychological deficits in adult males with heavy fetal alcohol exposure. Neuroimage 15, 233–251.
Boyadjieva, N.I., Sarkar, D.K., 2013a. Cyclic adenosine monophosphate and brain-derived neurotrophic factor decreased oxidative stress and apoptosis in developing hypothalamic neuronal cells: role of microglia. Alcohol. Clin. Exp. Res. 37, 1370–1379.
Boyadjieva, N.I., Sarkar, D.K., 2013b. Microglia play a role in ethanol-induced oxidative stress and apoptosis in developing hypothalamic neurons. Alcohol. Clin. Exp. Res. 37, 252–262.
Brocardo, P.S., Gil-Mohapel, J., Christie, B.R., 2011. The role of oxidative stress in fetal alcohol spectrum disorders. Brain Res. Rev. 67, 209–225.
Brocardo, P.S., Boehme, F., Patten, A., Cox, A., Gil-Mohapel, J., Christie, B.R., 2012. Anxiety- and depression-like behaviors are accompanied by an increase in oxidative stress in a rat model of fetal alcohol spectrum disorders: protective effects of voluntary physical exercise. Neuropharmacology 62, 1607–1618.
Calhoun, F., Warren, K., 2007. Fetal alcohol syndrome: historical perspectives. Neurosci. Biobehav. Rev. 31, 168–171.
Cao, W., Li, W., Han, H., O'Leary-Moore, S.K., Sulik, K.K., Allan Johnson, G., Liu, C., 2014. Prenatal alcohol exposure reduces magnetic susceptibility contrast and anisotropy in the white matter of mouse brains. Neuroimage 102 (Pt 2), 748–755.
Chambon, P., 1996. A decade of molecular biology of retinoic acid receptors. FASEB J. 10, 940–954.
Chasnoff, I.J., Wells, A.M., Telford, E., Schmidt, C., Messer, G., 2010. Neurodevelopmental functioning in children with FAS, pFAS, and ARND. J. Dev. Behav. Pediatr. 31, 192–201.
Chater-Diehl, E.J., Laufer, B.I., Singh, S.M., 2017. Changes to histone modifications following prenatal alcohol exposure: an emerging picture. Alcohol 60, 41–52.
Chen, S., Charness, M.E., 2008. Ethanol inhibits neuronal differentiation by disrupting activity-dependent neuroprotective protein signaling. Proc. Natl. Acad. Sci. U. S. A. 105, 19962–19967.

Chen, S.Y., Charness, M.E., Wilkemeyer, M.F., Sulik, K.K., 2005. Peptide-mediated protection from ethanol-induced neural tube defects. Dev. Neurosci. 27, 13–19.
Chu, J., Tong, M., de la Monte, S.M., 2007. Chronic ethanol exposure causes mitochondrial dysfunction and oxidative stress in immature central nervous system neurons. Acta Neuropathol. 113, 659–673.
Clements, K.M., Girard, T.A., Ellard, C.G., Wainwright, P.E., 2005. Short-term memory impairment and reduced hippocampal c-Fos expression in an animal model of fetal alcohol syndrome. Alcohol. Clin. Exp. Res. 29, 1049–1059.
Cole, G.J., Zhang, C., Ojiaku, P., Bell, V., Devkota, S., Mukhopadhyay, S., 2012. Effects of ethanol exposure on nervous system development in zebrafish. Int. Rev. Cell Mol. Biol. 299, 255–315.
Coles, C.D., Platzman, K.A., Lynch, M.E., Freides, D., 2002. Auditory and visual sustained attention in adolescents prenatally exposed to alcohol. Alcohol. Clin. Exp. Res. 26, 263–271.
Creeley, C.E., Olney, J.W., 2013. Drug-induced apoptosis: mechanism by which alcohol and many other drugs can disrupt brain development. Brain Sci. 3, 1153–1181.
Creeley, C.E., Dikranian, K.T., Johnson, S.A., Farber, N.B., Olney, J.W., 2013. Alcohol-induced apoptosis of oligodendrocytes in the fetal macaque brain. Acta Neuropathol. Commun. 1, 23.
Cremer, H., Lange, R., Christoph, A., Plomann, M., Vopper, G., Roes, J., Brown, R., Baldwin, S., Kraemer, P., Scheff, S., et al., 1994. Inactivation of the N-CAM gene in mice results in size reduction of the olfactory bulb and deficits in spatial learning. Nature 367, 455–459.
Cronise, K., Marino, M.D., Tran, T.D., Kelly, S.J., 2001. Critical periods for the effects of alcohol exposure on learning in rats. Behav. Neurosci. 115, 138–145.
Cudd, T.A., 2005. Animal model systems for the study of alcohol teratology. Exp. Biol. Med. (Maywood) 230, 389–393.
Cui, Z.J., Zhao, K.B., Zhao, H.J., Yu, D.M., Niu, Y.L., Zhang, J.S., Deng, J.B., 2010. Prenatal alcohol exposure induces long-term changes in dendritic spines and synapses in the mouse visual cortex. Alcohol Alcohol. 45, 312–319.
Das, U.G., Cronk, C.E., Martier, S.S., Simpson, P.M., McCarver, D.G., 2004. Alcohol dehydrogenase 2*3 affects alterations in offspring facial morphology associated with maternal ethanol intake in pregnancy. Alcohol. Clin. Exp. Res. 28, 1598–1606.
Dembele, K., Yao, X.H., Chen, L., Nyomba, B.L., 2006. Intrauterine ethanol exposure results in hypothalamic oxidative stress and neuroendocrine alterations in adult rat offspring. Am. J. Physiol. Regul. Integr. Comp. Physiol. 291, R796–802.
Dikranian, K., Qin, Y.Q., Labruyere, J., Nemmers, B., Olney, J.W., 2005. Ethanol-induced neuroapoptosis in the developing rodent cerebellum and related brain stem structures. Brain Res. Dev. Brain Res. 155, 1–13.
Dong, J., Sulik, K.K., Chen, S.Y., 2010. The role of NOX enzymes in ethanol-induced oxidative stress and apoptosis in mouse embryos. Toxicol. Lett. 193, 94–100.
Duester, G., Deltour, L., Ang, H.L., 1997. Evidence that class IV alcohol dehydrogenase may function in embryonic retinoic acid synthesis. Adv. Exp. Med. Biol. 414, 357–364.
Dunty Jr., W.C., Chen, S.Y., Zucker, R.M., Dehart, D.B., Sulik, K.K., 2001. Selective vulnerability of embryonic cell populations to ethanol-induced apoptosis: implications for alcohol-related birth defects and neurodevelopmental disorder. Alcohol. Clin. Exp. Res. 25, 1523–1535.
Fernandez-Lizarbe, S., Pascual, M., Guerri, C., 2009. Critical role of TLR4 response in the activation of microglia induced by ethanol. J. Immunol. 183, 4733–4744.
Flak, A.L., Su, S., Bertrand, J., Denny, C.H., Kesmodel, U.S., Cogswell, M.E., 2014. The association of mild, moderate, and binge prenatal alcohol exposure and child neuropsychological outcomes: a meta-analysis. Alcohol. Clin. Exp. Res. 38, 214–226.

Fontaine, C.J., Patten, A.R., Sickmann, H.M., Helfer, J.L., Christie, B.R., 2016. Effects of pre-natal alcohol exposure on hippocampal synaptic plasticity: sex, age and methodological considerations. Neurosci. Biobehav. Rev. 64, 12–34.

Foxworthy, W.A., Medina, A.E., 2015. Overexpression of serum response factor in neurons restores ocular dominance plasticity in a model of fetal alcohol spectrum disorders. Alcohol. Clin. Exp. Res. 39, 1951–1956.

Fryer, S.L., McGee, C.L., Matt, G.E., Riley, E.P., Mattson, S.N., 2007. Evaluation of psychopathological conditions in children with heavy prenatal alcohol exposure. Pediatrics 119, e733–e741.

Funfschilling, U., Supplie, L.M., Mahad, D., Boretius, S., Saab, A.S., Edgar, J., Brinkmann, B.G., Kassmann, C.M., Tzvetanova, I.D., Mobius, W., Diaz, F., Meijer, D., Suter, U., Hamprecht, B., Sereda, M.W., Moraes, C.T., Frahm, J., Goebbels, S., Nave, K.A., 2012. Glycolytic oligodendrocytes maintain myelin and long-term axonal integrity. Nature 485, 517–521.

Gavin, D.P., Grayson, D.R., Varghese, S.P., Guizzetti, M., 2017. Chromatin switches during neural cell differentiation and their dysregulation by prenatal alcohol exposure. Genes (Basel) 8 (e137), 1–31.

Gil-Mohapel, J., Boehme, F., Kainer, L., Christie, B.R., 2010. Hippocampal cell loss and neurogenesis after fetal alcohol exposure: insights from different rodent models. Brain Res. Rev. 64, 283–303.

Girard, T.A., Xing, H.C., Ward, G.R., Wainwright, P.E., 2000. Early postnatal ethanol exposure has long-term effects on the performance of male rats in a delayed matching-to-place task in the Morris water maze. Alcohol. Clin. Exp. Res. 24, 300–306.

Goodlett, C.R., Johnson, T.B., 1997. Neonatal binge ethanol exposure using intubation: timing and dose effects on place learning. Neurotoxicol. Teratol. 19, 435–446.

Goodlett, C.R., Marcussen, B.L., West, J.R., 1990. A single day of alcohol exposure during the brain growth spurt induces brain weight restriction and cerebellar Purkinje cell loss. Alcohol 7, 107–114.

Goodlett, C.R., Leo, J.T., O'Callaghan, J.P., Mahoney, J.C., West, J.R., 1993. Transient cortical astrogliosis induced by alcohol exposure during the neonatal brain growth spurt in rats. Brain Res. Dev. Brain Res. 72, 85–97.

Goodlett, C.R., Peterson, S.D., Lundahl, K.R., Pearlman, A.D., 1997. Binge-like alcohol exposure of neonatal rats via intragastric intubation induces both Purkinje cell loss and cortical astrogliosis. Alcohol. Clin. Exp. Res. 21, 1010–1017.

Green, R.F., Stoler, J.M., 2007. Alcohol dehydrogenase 1B genotype and fetal alcohol syndrome: a HuGE minireview. Am. J. Obstet. Gynecol. 197, 12–25.

Gressens, P., Lammens, M., Picard, J.J., Evrard, P., 1992. Ethanol-induced disturbances of gliogenesis and neuronogenesis in the developing murine brain: an in vitro and in vivo immunohistochemical and ultrastructural study. Alcohol Alcohol. 27, 219–226.

Guerri, C., 1998. Neuroanatomical and neurophysiological mechanisms involved in central nervous system dysfunctions induced by prenatal alcohol exposure. Alcohol. Clin. Exp. Res. 22, 304–312.

Guerri, C., Pascual, M., Renau-Piqueras, J., 2001. Glia and fetal alcohol syndrome. Neurotoxicology 22, 593–599.

Guerri, C., Bazinet, A., Riley, E.P., 2009. Foetal alcohol spectrum disorders and alterations in brain and behaviour. Alcohol Alcohol. 44, 108–114.

Guizzetti, M., Costa, L.G., 1996. Inhibition of muscarinic receptor-stimulated glial cell proliferation by ethanol. J. Neurochem. 67, 2236–2245.

Guizzetti, M., Costa, L.G., 2000. Possible role of protein kinase C ζ in muscarinic receptor-induced proliferation of astrocytoma cells. Biochem. Pharmacol. 60, 1457–1466.

Guizzetti, M., Costa, L.G., 2002. Effect of ethanol on protein kinase C ζ and p70S6 kinase activation by carbachol: a possible mechanism for ethanol-induced inhibition of glial cell proliferation. J. Neurochem. 82, 38–46.

Guizzetti, M., Bordi, F., Dieguez-Acuna, F.J., Vitalone, A., Madia, F., Woods, J.S., Costa, L.G., 2003. Nuclear factor kappaB activation by muscarinic receptors in astroglial cells: effect of ethanol. Neuroscience 120, 941–950.

Guizzetti, M., Thompson, B.D., Kim, Y., VanDeMark, K., Costa, L.G., 2004. Role of phospholipase D signaling in ethanol-induced inhibition of carbachol-stimulated DNA synthesis of 1321N1 astrocytoma cells. J. Neurochem. 90, 646–653.

Guizzetti, M., Moore, N.H., Giordano, G., Costa, L.G., 2008. Modulation of neuritogenesis by astrocyte muscarinic receptors. J. Biol. Chem. 283, 31884–31897.

Guizzetti, M., Moore, N.H., Giordano, G., VanDeMark, K.L., Costa, L.G., 2010. Ethanol inhibits neuritogenesis induced by astrocyte muscarinic receptors. Glia 58, 1395–1406.

Guizzetti, M., Zhang, X., Goeke, C., Gavin, D.P., 2014. Glia and neurodevelopment: focus on fetal alcohol spectrum disorders. Front. Pediatr. 2, 123.

Halder, D., Park, J.H., Choi, M.R., Chai, J.C., Lee, Y.S., Mandal, C., Jung, K.H., Chai, Y.G., 2014. Chronic ethanol exposure increases goosecoid (GSC) expression in human embryonic carcinoma cell differentiation. J. Appl. Toxicol. 34, 66–75.

Hamilton, D.A., Akers, K.G., Rice, J.P., Johnson, T.E., Candelaria-Cook, F.T., Maes, L.I., Rosenberg, M., Valenzuela, C.F., Savage, D.D., 2010a. Prenatal exposure to moderate levels of ethanol alters social behavior in adult rats: relationship to structural plasticity and immediate early gene expression in frontal cortex. Behav. Brain Res. 207, 290–304.

Hamilton, G.F., Whitcher, L.T., Klintsova, A.Y., 2010b. Postnatal binge-like alcohol exposure decreases dendritic complexity while increasing the density of mature spines in mPFC layer II/III pyramidal neurons. Synapse 64, 127–135.

Hannigan, J.H., Berman, R.F., Zajac, C.S., 1993. Environmental enrichment and the behavioral effects of prenatal exposure to alcohol in rats. Neurotoxicol. Teratol. 15, 261–266.

Hashimoto-Torii, K., Kawasawa, Y.I., Kuhn, A., Rakic, P., 2011. Combined transcriptome analysis of fetal human and mouse cerebral cortex exposed to alcohol. Proc. Natl. Acad. Sci. U. S. A. 108, 4212–4217.

Heaton, M.B., Mitchell, J.J., Paiva, M., 2000. Amelioration of ethanol-induced neurotoxicity in the neonatal rat central nervous system by antioxidant therapy. Alcohol. Clin. Exp. Res. 24, 512–518.

Heaton, M.B., Paiva, M., Madorsky, I., Siler-Marsiglio, K., Shaw, G., 2006. Effect of bax deletion on ethanol sensitivity in the neonatal rat cerebellum. J. Neurobiol. 66, 95–101.

Hoyme, H.E., May, P.A., Kalberg, W., Kodituwakku, P., Gossage, J.P., Trujillo, P.M., Buckley, D., Miller, J.H., Aragon, A.S., Khaole, N., Viljoen, D.L., Jones, K.L., Robinson, L.K., 2005. A practical clinical approach to diagnosis of fetal alcohol spectrum disorders: clarification of the 1996 Institute of Medicine criteria. Pediatrics 115, 37–49.

Ikonomidou, C., Bittigau, P., Ishimaru, M.J., Wozniak, D.F., Koch, C., Genz, K., Price, M.T., Stefovska, V., Horster, F., Tenkova, T., Dikranian, K., Olney, J.W., 2000. Ethanol-induced apoptotic neurodegeneration and fetal alcohol syndrome. Science 287, 1056–1060.

Institute of Medicine (IOM), Stratton, K., Howe, C., Battaglia, F., Committee to Study Fetal Alcohol Syndrome, Division of Biobehavioral Sciences and Mental Disorders, 1996. Fetal Alcohol Syndrome: Diagnosis, Epidemiology, Prevention, and Treatment. National Academy Press, Washington, DC.

Izumi, Y., Kitabayashi, R., Funatsu, M., Izumi, M., Yuede, C., Hartman, R.E., Wozniak, D.F., Zorumski, C.F., 2005. A single day of ethanol exposure during development has persistent effects on bi-directional plasticity, N-methyl-D-aspartate receptor function and ethanol sensitivity. Neuroscience 136, 269–279.

Jacobson, S.W., Carr, L.G., Croxford, J., Sokol, R.J., Li, T.K., Jacobson, J.L., 2006. Protective effects of the alcohol dehydrogenase-ADH1B allele in children exposed to alcohol during pregnancy. J. Pediatr. 148, 30–37.

Jones, K.L., Smith, D.W., 1973. Recognition of the fetal alcohol syndrome in early infancy. Lancet 302, 999–1001.

Jones, K.L., Smith, D.W., Ulleland, C.N., Streissguth, P., 1973. Pattern of malformation in offspring of chronic alcoholic mothers. Lancet 1, 1267–1271.

Joseph, J., Warton, C., Jacobson, S.W., Jacobson, J.L., Molteno, C.D., Eicher, A., Marais, P., Phillips, O.R., Narr, K.L., Meintjes, E.M., 2014. Three-dimensional surface deformation-based shape analysis of hippocampus and caudate nucleus in children with fetal alcohol spectrum disorders. Hum. Brain Mapp. 35, 659–672.

Kane, C.J., Phelan, K.D., Han, L., Smith, R.R., Xie, J., Douglas, J.C., Drew, P.D., 2011. Protection of neurons and microglia against ethanol in a mouse model of fetal alcohol spectrum disorders by peroxisome proliferator-activated receptor-gamma agonists. Brain Behav. Immun. 25 (Suppl. 1), S137–45.

Kelly, S.J., Goodlett, C.R., Hannigan, J.H., 2009. Animal models of fetal alcohol spectrum disorders: impact of the social environment. Dev. Disabil. Res. Rev. 15, 200–208.

Kim, J.A., Druse, M.J., 1996. Protective effects of maternal buspirone treatment on serotonin reuptake sites in ethanol-exposed offspring. Brain Res. Dev. Brain Res. 92, 190–198.

Klintsova, A.Y., Cowell, R.M., Swain, R.A., Napper, R.M., Goodlett, C.R., Greenough, W.T., 1998. Therapeutic effects of complex motor training on motor performance deficits induced by neonatal binge-like alcohol exposure in rats. I. Behavioral results. Brain Res. 800, 48–61.

Klintsova, A.Y., Goodlett, C.R., Greenough, W.T., 2000. Therapeutic motor training ameliorates cerebellar effects of postnatal binge alcohol. Neurotoxicol. Teratol. 22, 125–132.

Klintsova, A.Y., Scamra, C., Hoffman, M., Napper, R.M., Goodlett, C.R., Greenough, W.T., 2002. Therapeutic effects of complex motor training on motor performance deficits induced by neonatal binge-like alcohol exposure in rats: II. A quantitative stereological study of synaptic plasticity in female rat cerebellum. Brain Res. 937, 83–93.

Kotch, L.E., Chen, S.Y., Sulik, K.K., 1995. Ethanol-induced teratogenesis: free radical damage as a possible mechanism. Teratology 52, 128–136.

Kotter, K., Klein, J., 1999. Ethanol inhibits astroglial cell proliferation by disruption of phospholipase D-mediated signaling. J. Neurochem. 73, 2517–2523.

Lang, J., 2006. Ten brain domains: a proposal for functional central nervous system parameters for fetal alcohol spectrum disorder diagnosis and follow-up. J. FAS Inst. 4, 1–11.

Lebel, C., Roussotte, F., Sowell, E.R., 2011. Imaging the impact of prenatal alcohol exposure on the structure of the developing human brain. Neuropsychol. Rev. 21, 102–118.

Lemoine, P., Harousseau, H., Borteyru, J.P., Menuet, J.C., 1968. Children of alcoholic parents: abnormalities observed in 127 cases. Ouest Med. 8, 476–482.

Li, Q., Guo-Ross, S., Lewis, D.V., Turner, D., White, A.M., Wilson, W.A., Swartzwelder, H.S., 2004. Dietary prenatal choline supplementation alters postnatal hippocampal structure and function. J. Neurophysiol. 91, 1545–1555.

Light, K.E., Brown, D.P., Newton, B.W., Belcher, S.M., Kane, C.J., 2002. Ethanol-induced alterations of neurotrophin receptor expression on Purkinje cells in the neonatal rat cerebellum. Brain Res. 924, 71–81.

Littner, Y., Tang, N., He, M., Bearer, C.F., 2013. L1 cell adhesion molecule signaling is inhibited by ethanol in vivo. Alcohol. Clin. Exp. Res. 37, 383–389.

Long, L., Li, Y., Wang, Y.D., He, Q.Y., Li, M., Cai, X.D., Peng, K., Li, X.P., Xie, D., Wen, Y.L., Yin de, L., Peng, Y., 2010. The preventive effect of oral EGCG in a fetal alcohol spectrum disorder mouse model. Alcohol. Clin. Exp. Res. 34, 1929–1936.

Luo, J., 2012. Mechanisms of ethanol-induced death of cerebellar granule cells. Cerebellum 11, 145–154.

Maier, S.E., West, J.R., 2001. Regional differences in cell loss associated with binge-like alcohol exposure during the first two trimesters equivalent in the rat. Alcohol 23, 49–57.

Maier, S.E., Chen, W.J., Miller, J.A., West, J.R., 1997. Fetal alcohol exposure and temporal vulnerability regional differences in alcohol-induced microencephaly as a function of the

timing of binge-like alcohol exposure during rat brain development. Alcohol. Clin. Exp. Res. 21, 1418–1428.
Manning, M.A., Hoyme, H.E., 2007. Fetal alcohol spectrum disorders: a practical clinical approach to diagnosis. Neurosci. Biobehav. Rev. 31, 230–238.
Marcussen, B.L., Goodlett, C.R., Mahoney, J.C., West, J.R., 1994. Developing rat Purkinje cells are more vulnerable to alcohol-induced depletion during differentiation than during neurogenesis. Alcohol 11, 147–156.
Margret, C.P., Li, C.X., Elberger, A.J., Matta, S.G., Chappell, T.D., Waters, R.S., 2005. Prenatal alcohol exposure alters the size, but not the pattern, of the whisker representation in neonatal rat barrel cortex. Exp. Brain Res. 165, 167–178.
Marquardt, K., Brigman, J.L., 2016. The impact of prenatal alcohol exposure on social, cognitive and affective behavioral domains: insights from rodent models. Alcohol 51, 1–15.
Mattson, S.N., Riley, E.P., 1998. A review of the neurobehavioral deficits in children with fetal alcohol syndrome or prenatal exposure to alcohol. Alcohol. Clin. Exp. Res. 22, 279–294.
Mattson, S.N., Riley, E.P., Jernigan, T.L., Garcia, A., Kaneko, W.M., Ehlers, C.L., Jones, K.L., 1994. A decrease in the size of the basal ganglia following prenatal alcohol exposure: a preliminary report. Neurotoxicol. Teratol. 16, 283–289.
Mattson, S.N., Riley, E.P., Delis, D.C., Stern, C., Jones, K.L., 1996a. Verbal learning and memory in children with fetal alcohol syndrome. Alcohol. Clin. Exp. Res. 20, 810–816.
Mattson, S.N., Riley, E.P., Sowell, E.R., Jernigan, T.L., Sobel, D.F., Jones, K.L., 1996b. A decrease in the size of the basal ganglia in children with fetal alcohol syndrome. Alcohol. Clin. Exp. Res. 20, 1088–1093.
Mattson, S.N., Riley, E.P., Gramling, L., Delis, D.C., Jones, K.L., 1997. Heavy prenatal alcohol exposure with or without physical features of fetal alcohol syndrome leads to IQ deficits. J. Pediatr. 131, 718–721.
Mattson, S.N., Crocker, N., Nguyen, T.T., 2011. Fetal alcohol spectrum disorders: neuropsychological and behavioral features. Neuropsychol. Rev. 21, 81–101.
May, P.A., Baete, A., Russo, J., Elliott, A.J., Blankenship, J., Kalberg, W.O., Buckley, D., Brooks, M., Hasken, J., Abdul-Rahman, O., Adam, M.P., Robinson, L.K., Manning, M., Hoyme, H.E., 2014. Prevalence and characteristics of fetal alcohol spectrum disorders. Pediatrics 134, 855–866.
McCarthy, N., Eberhart, J.K., 2014. Gene-ethanol interactions underlying fetal alcohol spectrum disorders. Cell. Mol. Life Sci. 71, 2699–2706.
McGee, C.L., Riley, E.P., 2006. Brain imaging and fetal alcohol spectrum disorders. Ann. Ist. Super. Sanita 42, 46–52.
Meck, W.H., Williams, C.L., Cermak, J.M., Blusztajn, J., 2008. Developmental periods of choline sensitivity provide an ontogenic mechanism for regulating memory capacity and age-related dementia. Front. Integr. Neurosci. 1, 1–11.
Melcer, T., Gonzalez, D., Barron, S., Riley, E.P., 1994. Hyperactivity in preweanling rats following postnatal alcohol exposure. Alcohol 11, 41–45.
Miller, M.W., 1989. Effects of prenatal exposure to ethanol on neocortical development: II. Cell proliferation in the ventricular and subventricular zones of the rat. J. Comp. Neurol. 287, 326–338.
Miller, M.W., 1995. Effect of pre- or postnatal exposure to ethanol on the total number of neurons in the principal sensory nucleus of the trigeminal nerve: cell proliferation and neuronal death. Alcohol. Clin. Exp. Res. 19, 1359–1363.
Miller, M.W., Potempa, G., 1990. Numbers of neurons and glia in mature rat somatosensory cortex: effects of prenatal exposure to ethanol. J. Comp. Neurol. 293, 92–102.
Minana, R., Climent, E., Barettino, D., Segui, J.M., Renau-Piqueras, J., Guerri, C., 2000. Alcohol exposure alters the expression pattern of neural cell adhesion molecules during brain development. J. Neurochem. 75, 954–964.

Miranda, R.C., 2012. MicroRNAs and fetal brain development: implications for ethanol teratology during the second trimester period of neurogenesis. Front. Genet. 3, 77.

Montoliu, C., Sancho-Tello, M., Azorin, I., Burgal, M., Valles, S., Renau-Piqueras, J., Guerri, C., 1995. Ethanol increases cytochrome P4502E1 and induces oxidative stress in astrocytes. J. Neurochem. 65, 2561–2570.

Moore, N.H., Costa, L.G., Shaffer, S.A., Goodlett, D.R., Guizzetti, M., 2009. Shotgun proteomics implicates extracellular matrix proteins and protease systems in neuronal development induced by astrocyte cholinergic stimulation. J. Neurochem. 108, 891–908.

Newville, J., Valenzuela, C.F., Li, L., Jantzie, L.L., Cunningham, L.A., 2017. Acute oligodendrocyte loss with persistent white matter injury in a third trimester equivalent mouse model of fetal alcohol spectrum disorder. Glia 65, 1317–1332.

Nguyen, T.T., Risbud, R.D., Mattson, S.N., Chambers, C.D., Thomas, J.D., 2016. Randomized, double-blind, placebo-controlled clinical trial of choline supplementation in school-aged children with fetal alcohol spectrum disorders. Am. J. Clin. Nutr. 104, 1683–1692.

Norman, A.L., Crocker, N., Mattson, S.N., Riley, E.P., 2009. Neuroimaging and fetal alcohol spectrum disorders. Dev. Disabil. Res. Rev. 15, 209–217.

Nunez, C.C., Roussotte, F., Sowell, E.R., 2011. Focus on: structural and functional brain abnormalities in fetal alcohol spectrum disorders. Alcohol Res. Health 34, 121–131.

Olney, J.W., Tenkova, T., Dikranian, K., Muglia, L.J., Jermakowicz, W.J., D'Sa, C., Roth, K.A., 2002a. Ethanol-induced caspase-3 activation in the in vivo developing mouse brain. Neurobiol. Dis. 9, 205–219.

Olney, J.W., Tenkova, T., Dikranian, K., Qin, Y.Q., Labruyere, J., Ikonomidou, C., 2002b. Ethanol-induced apoptotic neurodegeneration in the developing C57BL/6 mouse brain. Brain Res. Dev. Brain Res. 133, 115–126.

Parnell, S.E., O'Leary-Moore, S.K., Godin, E.A., Dehart, D.B., Johnson, B.W., Allan Johnson, G., Styner, M.A., Sulik, K.K., 2009. Magnetic resonance microscopy defines ethanol-induced brain abnormalities in prenatal mice: effects of acute insult on gestational day 8. Alcohol. Clin. Exp. Res. 33, 1001–1011.

Pascual, M., Guerri, C., 2007. The peptide NAP promotes neuronal growth and differentiation through extracellular signal-regulated protein kinase and Akt pathways, and protects neurons co-cultured with astrocytes damaged by ethanol. J. Neurochem. 103, 557–568.

Patten, A.R., Brocardo, P.S., Sakiyama, C., Wortman, R.C., Noonan, A., Gil-Mohapel, J., Christie, B.R., 2013. Impairments in hippocampal synaptic plasticity following prenatal ethanol exposure are dependent on glutathione levels. Hippocampus 23, 1463–1475.

Peng, Y., Kwok, K.H., Yang, P.H., Ng, S.S., Liu, J., Wong, O.G., He, M.L., Kung, H.F., Lin, M.C., 2005. Ascorbic acid inhibits ROS production, NF-kappa B activation and prevents ethanol-induced growth retardation and microencephaly. Neuropharmacology 48, 426–434.

Pennington, B.F., Ozonoff, S., 1996. Executive functions and developmental psychopathology. J. Child Psychol. Psychiatry 37, 51–87.

Perez-Torrero, E., Duran, P., Granados, L., Gutierez-Ospina, G., Cintra, L., Diaz-Cintra, S., 1997. Effects of acute prenatal ethanol exposure on Bergmann glia cells early postnatal development. Brain Res. 746, 305–308.

Phillips, D.E., Krueger, S.K., Rydquist, J.E., 1991. Short- and long-term effects of combined pre- and postnatal ethanol exposure (three trimester equivalency) on the development of myelin and axons in rat optic nerve. Int. J. Dev. Neurosci. 9, 631–647.

Rai, J.K., Abecassis, M., Casey, J.E., Flaro, L., Erdodi, L.A., Roth, R.M., 2017. Parent rating of executive function in fetal alcohol spectrum disorder: a review of the literature and new data on Aboriginal Canadian children. Child Neuropsychol. 23, 713–732.

Ramachandran, V., Perez, A., Chen, J., Senthil, D., Schenker, S., Henderson, G.I., 2001. In utero ethanol exposure causes mitochondrial dysfunction, which can result in apoptotic cell death in fetal brain: a potential role for 4-hydroxynonenal. Alcohol. Clin. Exp. Res. 25, 862–871.

Rema, V., Ebner, F.F., 1999. Effect of enriched environment rearing on impairments in cortical excitability and plasticity after prenatal alcohol exposure. J. Neurosci. 19, 10993–11006.

Resendiz, M., Mason, S., Lo, C.L., Zhou, F.C., 2014. Epigenetic regulation of the neural transcriptome and alcohol interference during development. Front. Genet. 5, 285.

Resnicoff, M., Rubini, M., Baserga, R., Rubin, R., 1994. Ethanol inhibits insulin-like growth factor-1-mediated signalling and proliferation of C6 rat glioblastoma cells. Lab. Investig. 71, 657–662.

Reyes, E., Ott, S., Robinson, B., 1993. Effects of in utero administration of alcohol on glutathione levels in brain and liver. Alcohol. Clin. Exp. Res. 17, 877–881.

Reynolds, J.N., Weinberg, J., Clarren, S., Beaulieu, C., Rasmussen, C., Kobor, M., Dube, M.P., Goldowitz, D., 2011. Fetal alcohol spectrum disorders: gene-environment interactions, predictive biomarkers, and the relationship between structural alterations in the brain and functional outcomes. Semin. Pediatr. Neurol. 18, 49–55.

Rice, D., Barone Jr., S., 2000. Critical periods of vulnerability for the developing nervous system: evidence from humans and animal models. Environ. Health Perspect. 108 (Suppl. 3), 511–533.

Richards, L.J., Plachez, C., Ren, T., 2004. Mechanisms regulating the development of the corpus callosum and its agenesis in mouse and human. Clin. Genet. 66, 276–289.

Richardson, D.P., Byrnes, M.L., Brien, J.F., Reynolds, J.N., Dringenberg, H.C., 2002. Impaired acquisition in the water maze and hippocampal long-term potentiation after chronic prenatal ethanol exposure in the guinea-pig. Eur. J. Neurosci. 16 (8), 1593.

Rifas, L., Towler, D.A., Avioli, L.V., 1997. Gestational exposure to ethanol suppresses msx2 expression in developing mouse embryos. Proc. Natl. Acad. Sci. U. S. A. 94, 7549–7554.

Riikonen, R., Salonen, I., Partanen, K., Verho, S., 1999. Brain perfusion SPECT and MRI in foetal alcohol syndrome. Dev. Med. Child Neurol. 41, 652–659.

Riley, E.P., McGee, C.L., 2005. Fetal alcohol spectrum disorders: an overview with emphasis on changes in brain and behavior. Exp. Biol. Med. (Maywood) 230, 357–365.

Riley, E.P., Mattson, S.N., Sowell, E.R., Jernigan, T.L., Sobel, D.F., Jones, K.L., 1995. Abnormalities of the corpus callosum in children prenatally exposed to alcohol. Alcohol. Clin. Exp. Res. 19, 1198–1202.

Roebuck, T.M., Mattson, S.N., Riley, E.P., 1998a. A review of the neuroanatomical findings in children with fetal alcohol syndrome or prenatal exposure to alcohol. Alcohol. Clin. Exp. Res. 22, 339–344.

Roebuck, T.M., Simmons, R.W., Richardson, C., Mattson, S.N., Riley, E.P., 1998b. Neuromuscular responses to disturbance of balance in children with prenatal exposure to alcohol. Alcohol. Clin. Exp. Res. 22, 1992–1997.

Ross, S.A., McCaffery, P.J., Drager, U.C., De Luca, L.M., 2000. Retinoids in embryonal development. Physiol. Rev. 80, 1021–1054.

Roszel, E.L., 2015. Central nervous system deficits in fetal alcohol spectrum disorder. Nurse Pract. 40, 24–33.

Rubert, G., Minana, R., Pascual, M., Guerri, C., 2006. Ethanol exposure during embryogenesis decreases the radial glial progenitorpool and affects the generation of neurons and astrocytes. J. Neurosci. Res. 84, 483–496.

Rutishauser, U., Acheson, A., Hall, A.K., Mann, D.M., Sunshine, J., 1988. The neural cell adhesion molecule (NCAM) as a regulator of cell-cell interactions. Science 240, 53–57.

Ryabinin, A.E., Cole, M., Bloom, F.E., Wilson, M.C., 1995. Exposure of neonatal rats to alcohol by vapor inhalation demonstrates specificity of microcephaly and Purkinje cell loss but not astrogliosis. Alcohol. Clin. Exp. Res. 19, 784–791.
Saito, M., Chakraborty, G., Hui, M., Masiello, K., 2016. Ethanol-induced neurodegeneration and glial activation in the developing brain. Brain Sci. 6, 31.
Santello, M., Cali, C., Bezzi, P., 2012. Gliotransmission and the tripartite synapse. Adv. Exp. Med. Biol. 970, 307–331.
Schmid, R.S., Maness, P.F., 2008. L1 and NCAM adhesion molecules as signaling coreceptors in neuronal migration and process outgrowth. Curr. Opin. Neurobiol. 18, 245–250.
Schonfeld, A.M., Paley, B., Frankel, F., O'Connor, M.J., 2006. Executive functioning predicts social skills following prenatal alcohol exposure. Child. Neuropsychol. 12, 439–452.
Servais, L., Hourez, R., Bearzatto, B., Gall, D., Schiffmann, S.N., Cheron, G., 2007. Purkinje cell dysfunction and alteration of long-term synaptic plasticity in fetal alcohol syndrome. Proc. Natl. Acad. Sci. U. S. A. 104, 9858–9863.
Shapiro, L., Love, J., Colman, D.R., 2007. Adhesion molecules in the nervous system: structural insights into function and diversity. Annu. Rev. Neurosci. 30, 451–474.
Shilko, V.I., Malakhova, Zh.L., Bubnov, A.A., 2010. Brain disorders in fetal alcohol syndrome. Bull. Exp. Biol. Med. 150, 89–91.
Sokol, R.J., Delaney-Black, V., Nordstrom, B., 2003. Fetal alcohol spectrum disorder. JAMA 290, 2996–2999.
Sowell, E.R., Jernigan, T.L., Mattson, S.N., Riley, E.P., Sobel, D.F., Jones, K.L., 1996. Abnormal development of the cerebellar vermis in children prenatally exposed to alcohol: size reduction in lobules I-V. Alcohol. Clin. Exp. Res. 20, 31–34.
Sowell, E.R., Mattson, S.N., Thompson, P.M., Jernigan, T.L., Riley, E.P., Toga, A.W., 2001a. Mapping callosal morphology and cognitive correlates: effects of heavy prenatal alcohol exposure. Neurology 57, 235–244.
Sowell, E.R., Thompson, P.M., Mattson, S.N., Tessner, K.D., Jernigan, T.L., Riley, E.P., Toga, A.W., 2001b. Voxel-based morphometric analyses of the brain in children and adolescents prenatally exposed to alcohol. Neuroreport 12, 515–523.
Staples, M.C., Porch, M.W., Savage, D.D., 2014. Impact of combined prenatal ethanol and prenatal stress exposures on markers of activity-dependent synaptic plasticity in rat dentate gyrus. Alcohol 48, 523–532.
Streissguth, A.P., O'Malley, K., 2000. Neuropsychiatric implications and long-term consequences of fetal alcohol spectrum disorders. Semin. Clin. Neuropsychiatry 5, 177–190.
Streissguth, A.P., Landesman-Dwyer, S., Martin, J.C., Smith, D.W., 1980. Teratogenic effects of alcohol in humans and laboratory animals. Science 209, 353–361.
Streissguth, A.P., Barr, H.M., Kogan, J., Bookstein, F.L., 1997. Primary and secondary disabilities. In: Streissguth, A.P., Kanter, J. (Eds.), Fetal Alcohol Syndrome in the Challenge of Fetal Alcohol Syndrome: Overcoming Secondary Disabilities. University of Washington Press, Seattle, WA.
Subbanna, S., Shivakumar, M., Psychoyos, D., Xie, S., Basavarajappa, B.S., 2013. Anandamide-CB1 receptor signaling contributes to postnatal ethanol-induced neonatal neurodegeneration, adult synaptic, and memory deficits. J. Neurosci. 33, 6350–6366.
Sulik, K.K., 2005. Genesis of alcohol-induced craniofacial dysmorphism. Exp. Biol. Med. (Maywood) 230, 366–375.
Sutherland, R.J., McDonald, R.J., Savage, D.D., 1997. Prenatal exposure to moderate levels of ethanol can have long-lasting effects on hippocampal synaptic plasticity in adult offspring. Hippocampus 7, 232–238.
Tajuddin, N.F., Druse, M.J., 1999. In utero ethanol exposure decreased the density of serotonin neurons. Maternal ipsapirone treatment exerted a protective effect. Brain Res. Dev. Brain Res. 117, 91–97.

Tenkova, T., Young, C., Dikranian, K., Labruyere, J., Olney, J.W., 2003. Ethanol-induced apoptosis in the developing visual system during synaptogenesis. Invest. Ophthalmol. Vis. Sci. 44, 2809–2817.

Thomas, J.D., Tran, T.D., 2012. Choline supplementation mitigates trace, but not delay, eyeblink conditioning deficits in rats exposed to alcohol during development. Hippocampus 22, 619–630.

Thomas, J.D., La Fiette, M.H., Quinn, V.R., Riley, E.P., 2000. Neonatal choline supplementation ameliorates the effects of prenatal alcohol exposure on a discrimination learning task in rats. Neurotoxicol. Teratol. 22, 703–711.

Thomas, J.D., Garrison, M., O'Neill, T.M., 2004. Perinatal choline supplementation attenuates behavioral alterations associated with neonatal alcohol exposure in rats. Neurotoxicol. Teratol. 26, 35–45.

Thomas, J.D., Biane, J.S., O'Bryan, K.A., O'Neill, T.M., Dominguez, H.D., 2007. Choline supplementation following third-trimester-equivalent alcohol exposure attenuates behavioral alterations in rats. Behav. Neurosci. 121, 120–130.

Thomas, J.D., Abou, E.J., Dominguez, H.D., 2009. Prenatal choline supplementation mitigates the adverse effects of prenatal alcohol exposure on development in rats. Neurotoxicol. Teratol. 31, 303–311.

Thomas, J.D., Idrus, N.M., Monk, B.R., Dominguez, H.D., 2010. Prenatal choline supplementation mitigates behavioral alterations associated with prenatal alcohol exposure in rats. Birth Defects Res. A Clin. Mol. Teratol. 88, 827–837.

Tran, T.D., Stanton, M.E., Goodlett, C.R., 2007. Binge-like ethanol exposure during the early postnatal period impairs eyeblink conditioning at short and long CS-US intervals in rats. Dev. Psychobiol. 49, 589–605.

Tsuji, R., Guizzetti, M., Costa, L.G., 2003. In vivo ethanol decreases phosphorylated MAPK and p70S6 kinase in the developing rat brain. Neuroreport 14, 1395–1399.

Valles, S., Pitarch, J., Renau-Piqueras, J., Guerri, C., 1997. Ethanol exposure affects glial fibrillary acidic protein gene expression and transcription during rat brain development. J. Neurochem. 69, 2484–2493.

Weinberg, J., Kim, C.K., Yu, W., 1995. Early handling can attenuate adverse effects of fetal ethanol exposure. Alcohol 12, 317–327.

West, J.R., Hamre, K.M., Cassell, M.D., 1986. Effects of ethanol exposure during the third trimester equivalent on neuron number in rat hippocampus and dentate gyrus. Alcohol. Clin. Exp. Res. 10, 190–197.

Whitcher, L.T., Klintsova, A.Y., 2008. Postnatal binge-like alcohol exposure reduces spine density without affecting dendritic morphology in rat mPFC. Synapse 62, 566–573.

White, S.A., Weber, J.N., Howard, C.D., Favero, C.B., 2015. Effects of binge ethanol exposure during first-trimester equivalent on corticothalamic neurons in Swiss Webster outbred mice. Neuroreport 26, 1083–1088.

Wilhelm, C.J., Guizzetti, M., 2015. Fetal alcohol spectrum disorders: an overview from the glia perspective. Front. Integr. Neurosci. 9, 65.

Wilkemeyer, M.F., Chen, S.Y., Menkari, C.E., Brenneman, D.E., Sulik, K.K., Charness, M.E., 2003. Differential effects of ethanol antagonism and neuroprotection in peptide fragment NAPVSIPQ prevention of ethanol-induced developmental toxicity. Proc. Natl. Acad. Sci. U. S. A. 100, 8543–8548.

Wisniewski, K., Dambska, M., Sher, J.H., Qazi, Q., 1983. A clinical neuropathological study of the fetal alcohol syndrome. Neuropediatrics 14, 197–201.

Wozniak, D.F., Hartman, R.E., Boyle, M.P., Vogt, S.K., Brooks, A.R., Tenkova, T., Young, C., Olney, J.W., Muglia, L.J., 2004. Apoptotic neurodegeneration induced by ethanol in neonatal mice is associated with profound learning/memory deficits in juveniles followed by progressive functional recovery in adults. Neurobiol. Dis. 17, 403–414.

Wozniak, J.R., Fuglestad, A.J., Eckerle, J.K., Fink, B.A., Hoecker, H.L., Boys, C.J., Radke, J.P., Kroupina, M.G., Miller, N.C., Brearley, A.M., Zeisel, S.H., Georgieff, M.K., 2015. Choline supplementation in children with fetal alcohol spectrum disorders: a randomized, double-blind, placebo-controlled trial. Am. J. Clin. Nutr. 102, 1113–1125.

Young, C., Klocke, B.J., Tenkova, T., Choi, J., Labruyere, J., Qin, Y.Q., Holtzman, D.M., Roth, K.A., Olney, J.W., 2003. Ethanol-induced neuronal apoptosis in vivo requires BAX in the developing mouse brain. Cell Death Differ. 10, 1148–1155.

Young, C., Roth, K.A., Klocke, B.J., West, T., Holtzman, D.M., Labruyere, J., Qin, Y.Q., Dikranian, K., Olney, J.W., 2005. Role of caspase-3 in ethanol-induced developmental neurodegeneration. Neurobiol. Dis. 20, 608–614.

Young, C., Straiko, M.M., Johnson, S.A., Creeley, C., Olney, J.W., 2008. Ethanol causes and lithium prevents neuroapoptosis and suppression of pERK in the infant mouse brain. Neurobiol. Dis. 31, 355–360.

Zeisel, S.H., Niculescu, M.D., 2006. Perinatal choline influences brain structure and function. Nutr. Rev. 64, 197–203.

Zhang, X., Bhattacharyya, S., Kusumo, H., Goodlett, C.R., Tobacman, J.K., Guizzetti, M., 2014a. Arylsulfatase B modulates neurite outgrowth via astrocyte chondroitin-4-sulfate: dysregulation by ethanol. Glia 62, 259–271.

Zhang, X., Kusumo, H., Sakharkar, A.J., Pandey, S.C., Guizzetti, M., 2014b. Regulation of DNA methylation by ethanol induces tissue plasminogen activator expression in astrocytes. J. Neurochem. 128, 344–349.

Zhang, J., Wu, J., Zeng, W., Yao, K., Zu, H., Zhao, Y., 2016. Function of thymosin beta-4 in ethanol-induced microglial activation. Cell. Physiol. Biochem. 38, 2230–2238.

Zoeller, R.T., Butnariu, O.V., Fletcher, D.L., Riley, E.P., 1994. Limited postnatal ethanol exposure permanently alters the expression of mRNAS encoding myelin basic protein and myelin-associated glycoprotein in cerebellum. Alcohol. Clin. Exp. Res. 18, 909–916.

CHAPTER FIVE

Possible Mechanisms of Developmental Neurotoxicity of Organophosphate Insecticides

Russell L. Carr[1], Navatha Alugubelly, Afzaal N. Mohammed
Center for Environmental Health Sciences, Department of Basic Sciences, College of Veterinary Medicine, Mississippi State University, Mississippi State, MS, USA
[1]Corresponding author: e-mail address: rlcarr@cvm.msstate.edu

Contents

Although organophosphates (OPs) were first synthesized in the 1800s through the efforts of Jean Louis Lassaigne, Philippe de Clermont, and others, the neurotoxicity of the OPs was not recognized until 1932 when Willy Lange and his student Gerda v. Krueger reported on the neurotoxic effects of tetraethyl pyrophosphate or TEPP (Petroianu, 2010). The insecticidal properties of the OPs were discovered by Gerhard Schrader, a German insecticide chemist working at I.G. Farbenindustrie. During this period Schrader developed many OPs with insecticidal properties, including parathion and methyl parathion, which garnered him the historical title of the father of modern OP insecticides. He also developed tabun which had such high toxicity that it attracted the attention of the Nazi regime. They directed Schrader into efforts to develop these compounds as chemical weapons leading to the production of soman and sarin. Thus, he is unfortunately also known as the

Advances in Neurotoxicology, Volume 2
ISSN 2468-7480
https://doi.org/10.1016/bs.ant.2018.03.004

father of the nerve agents. After the end of World War II, the OP insecticides quickly joined the arsenal to control insects and were proven to be very efficacious. However, many OPs had high acute mammalian toxicity and further development of these compounds as insecticides focused on creating safer compounds. This led to the development of relatively safer insecticides such as malathion in the 1950s and chlorpyrifos in the 1960s both of which combined low mammalian toxicity with high insecticidal activity. As insect resistance to the organochlorine insecticides increased, the OP insecticides rose to the forefront of insecticide use but the introduction of the synthetic pyrethroids eventually decreased the use of OPs. However, the OP compounds still currently make-up a significant portion of the insecticides currently used world-wide.

With respect to the developmental neurotoxicity of the OP insecticides, a significant amount of research has focused on chlorpyrifos and its active metabolite, chlorpyrifos-oxon. These comprise about 83% of the total studies related to OP developmental toxicity. The second most studied are the OP insecticides ethyl parathion and methyl parathion along with their respective oxons. These are followed by diazinon and its oxon and dichlorvos or DDVP. In many of the *in vivo* studies, the results obtained come with some caveats. The first caveat is the choice of route. For many of these studies, the route of exposure was a subcutaneous (sc) injection which does not represent an actual route of exposure in humans. However, while it may affect the toxicokinetics/toxicodynamics of a compound, it can be justified for use in investigative mechanistic studies. The second caveat is the use of dimethyl sulfoxide (DMSO) as a vehicle. DMSO enhances the permeability of cell membranes (Gurtovenko and Anwar, 2007). DMSO has also been reported to alter the toxicokinetics of the OP chlorpyrifos (Marty et al., 2007; Smith et al., 2009). Multiple studies have reported that DMSO can potentiate the effects of the test chemical, including some OPs, and can even exert toxic effects by itself (Ballough et al., 2008; Carr and Nail, 2008; Chen et al., 2011a; Hallare et al., 2004; Hanslick et al., 2009; Rosecrans et al., 1984; Sackerman et al., 2010; Sierra and Uphouse, 1986; Turner et al., 2012; Uphouse et al., 1982). The concerns about the intrinsic toxicity of DMSO, its influence on the absorption of chemicals, and its potential to enhance the toxicity of other chemicals make DMSO an unsuitable vehicle for any *in vivo* study not just a developmental study. However, the data obtained from studies that utilized DMSO as a vehicle can provide some beneficial information by identifying potential hazards of a chemical. Therefore, these studies were included. For identification, the route and solvent used for each *in vivo* study, when available, has been clearly stated.

Some early evidence about the age-related differences in susceptibility to OP insecticides actually originated from human poisonings. In 1959 in Singapore, 36 children and three adults consumed barley contaminated with parathion apparently during shipping. As a result, nine children died while none of the adults did (Kanagaratnam et al., 1960). It was calculated that the lethal dosage consumed by the children was 0.1 mg/kg. In a volunteer study, this dosage was given daily to human adults for 42 days with no signs of poisoning or symptoms of toxicity (Durham, 1967). An additional example was the poisoning of 79 people in Jamaica in 1976 through the consumption of parathion-contaminated flour (Diggory et al., 1977). Of the 17 people who died as a result, the highest incidence of lethality was in children <4 years of age. One of the first detailed laboratory animal studies that demonstrated age-related differences in the susceptibility to acute OP insecticide exposure compared the acute oral toxicity of multiple OP insecticides between 23 day old weanling and adult male rats. With most insecticides, the weanling rats were more susceptible than the adults (Brodeur and DuBois, 1963). Although there are some exceptions (Chakraborti et al., 1993; Lu et al., 1965; Moser, 1999; Padilla et al., 2000; Pope and Liu, 1997), many additional studies have reported that younger animals are more susceptible to the acute toxicity of many different OPs (Benke and Murphy, 1975; Gagne and Brodeur, 1972; Harbison, 1975; Long et al., 1986; Mendoza, 1976; Pope et al., 1991).

Brodeur and DuBois (1963) reported that of the insecticides tested, the phosphorothioates and phosphorodithioates exhibited the greatest differences in toxicity with less differences present in compounds that did not require metabolic activation. Further research demonstrated that the basis for this difference in toxicity was not due to age-related differences in the sensitivity of AChE to inhibition (Atterberry et al., 1997; Benke and Murphy, 1975; Mortensen et al., 1996; Moser and Padilla, 1998). In fact, the age-related differences in toxicity have been attributed to a lower level of detoxication in developing animals as compared to adults (Atterberry et al., 1997; Benke and Murphy, 1975; Gagne and Brodeur, 1972; Mortensen et al., 1996; Moser and Padilla, 1998). These reactions that inactivate the OPs included not only the cytochrome P450-mediated detoxication of the parent insecticide or its oxon (oxidative cleavage or dearylation) but also detoxication of the oxon by PON1 and destruction of the oxon by binding to non-target enzymes such as carboxylesterase and butyrylcholinesterase (see complete metabolism scheme in Fig. 1). These latter two enzymes have been suggested to play a very important role in the age-related differences in the toxicity. Using an *in vitro* system that measures the degradation of an insecticide by

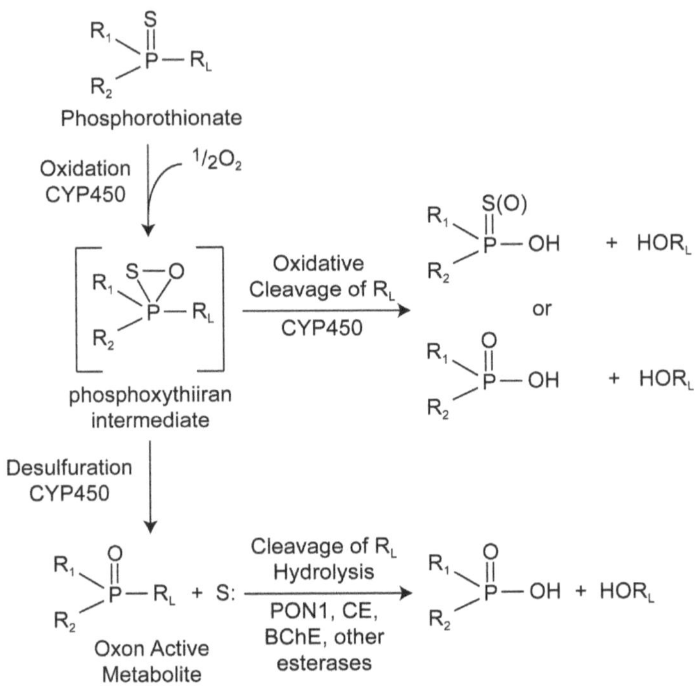

Fig. 1 Biotransformation of a phosphorothionate OP insecticide. The first step is cytochrome P450 (CYP450) mediated oxidation of the phosphorothionate to an unstable phosphoxythiiran intermediate. This is followed by either detoxication by oxidative cleavage of the leaving group (R_L) or activation by oxidative desulfuration to the oxon metabolite. The oxon can undergo detoxication by catalytic hydrolysis by paraoxonase (PON1) or stoichiometric detoxication by non-target esterases via inhibition of carboxylesterases (CE), butyrylcholinesterase (BChE), or other esterases.

the action of PON1 and the carboxylesterases via incubation with liver homogenates from adult and neonatal rats, chemicals that produced greater inhibition of AChE in neonates *in vivo* had considerably less detoxication in the presence of neonatal liver homogenates than in the presence of adult liver homogenates (Moser and Padilla, 2011; Padilla et al., 2000, 2004). However, when human liver was used in this system, this correlation was not straightforward (Moser and Padilla, 2016). Human liver exhibited differences in the detoxication patterns with different chemicals as compared to that with rat liver. There was also high individual variability in the samples from infants and children as compared to adults. This means that while inbred animal models may be useful in determining detoxification pathways, a better understanding of how the individual variability in these protective detoxification mechanisms in humans contributes to the sensitivity to specific compounds is needed before we can fully assess the role of PON1 and carboxylesterases in developmental human toxicity.

During the earliest stages of researching the developmental toxicity of organophosphate insecticides, a significant amount of the studies focused on teratogenicity using either the avian egg model or mammalian models (for review see Tyl, 1992). Many of these studies reported no evidence of teratology while others reported teratogenic effects at levels that induced maternal toxicity. However, the majority of these studies focused on embryotoxicity and the presence of physical malformations rather than measures of neurological function. Most of the early studies investigating the effects of OP exposure on neurological function in adults reported that high levels of AChE inhibition must be obtained prior to disrupting the ability of an animal to perform a behavior. In addition, gestational exposure of mice to parathion (oil) did not result in any significant changes in open field activity or shuttle box avoidance learning in the offspring (Al-Hachim and Fink, 1968a,b,c). Thus, the lack of observed behavioral effects following chronic low level exposures caused some to propose that such a gestational exposure would not produce any negative effects on neurological function. However, Spyker (1975) proposed that individuals are most vulnerable during development and subtle disturbances that occur during the developmental period could lead to changes in function that do not appear until later in life. To test this hypothesis, Spyker initiated one of the first studies that investigated the neurological effects induced by developmental exposure to an OP insecticide. Following daily gestational exposure of mice to low levels of diazinon by feeding, some subtle changes in neuromuscular function were observed in the offspring even after they reach adult ages (Spyker and Avery, 1977). While the changes in behavior were subtle, the authors suggested that the mechanism behind the toxicity could be the result of a direct action on the cholinergic system. However, they were also the first to suggest that the mechanism behind the toxicity could be due to an indirect interference on neuronal development.

Other early studies reported functional effects following gestational exposure to OPs. Daily oral gestational (oil) exposure to 1.0 mg/kg methyl parathion from gestational days 7–15 (GD7–15) produced effects in a simple maze transfer tests but did not in other behavioral tests (Crowder et al., 1980). The same dosage administered by feeding peanut butter from GD6–20 did not induce any signs of maternal toxicity but decreased amphetamine-stimulated locomotor activity (postnatal day 60 or PND60), decreased time of cage emergence (PND90), and impaired operant behavior (PND90–180) in the offspring (Gupta et al., 1985). However, other behaviors were not affected. The exposure did exert cholinergic deficits in that it resulted in decreased AChE activity and decreased muscarinic receptor levels around the time of birth. The study included a higher dosage of methyl parathion (po-oil) but there was no dose-response relationship in the effects on behavior prompting the authors to suggest U-shaped dose–response pattern. However, the method of administration between the two dosages differed making direct comparison of the two dosages difficult. The difference in time between the measurements of the neurochemical (AChE and receptors) and the behavioral parameters also made a relationship between the two difficult.

One of the initial studies investigating functional effects following postnatal exposure to an OP involved exposure to two dosages of parathion (sc-oil) from PND5–21 (Stamper et al., 1988). Both dosages resulted in decreased growth rate and reduced AChE activity and muscarinic receptor levels on PND21 and PND28. There was no effect on activity levels, neuromuscular coordination, or endurance (PND24) but spatial memory in a T-maze (PND24) and working memory in a radial arm maze (PND36) were altered. The behavioral tests were performed relatively close to the time of measurement of the neurochemical parameters suggesting that the negative functional effects were related to the impact of exposure on the cholinergic system.

In 1988, the US Congress requested that the National Academy of Sciences appoint a committee through its National Research Council to study science and policy concerning pesticides in the diets of infants and children. The resulting publication, Pesticides in the Diets of Infants and Children (National Research Council, 1993), pointed out the deficiencies in the understanding of how pesticides affect children and adults differently. The concern about this issue lead to the passage of the 1996 Food Quality Protection Act (FQPA), which established additional safety factors to account for age-related susceptibilities. Following the NRC publication, the research into the basis of the developmental toxicity of OP insecticides accelerated.

Historically, the acute toxicity of the OP insecticides is due to inhibition of acetylcholinesterase (EC 3.1.1.7; AChE) and that enzyme has long been considered the main toxicological target of organophosphates. AChE is a serine hydrolase that is imbedded in the postsynaptic membrane of neuromuscular junctions and synapses associated with cholinergic nerves. It is responsible for the termination of chemical signaling in the synapse through the hydrolysis of the neurotransmitter acetylcholine (ACh). The enzyme contains two sites of interest with respect to both hydrolysis of ACh and inhibition by OPs. The anionic site functions to attract the substrate/inhibitor to the enzyme and orient the molecule into the esteratic site where the molecule binds to a serine hydroxyl group. For ACh, this binding results in hydrolysis into choline and acetate. For OPs, this binding results in the phosphorylation of the esteratic site and the release of the leaving group which was associated with the anionic site. This phosphorylation inhibits the ability of the hydrolytic capacity of the enzyme resulting in the accumulation of ACh in the synapse. AChE is located close to the acetylcholine receptors, either muscarinic or nicotinic, on the postsynaptic membrane to ensure quick inactivation of ACh and prevent continual stimulation of those receptors. Thus, the inhibition of AChE and subsequent accumulation of ACh leads to excessive stimulation of these receptors. At high exposure levels yielding significant inhibition of AChE, the clinical effects observed are due to the excessive activation of the receptors in the autonomic and central nervous systems and at nicotinic receptors on skeletal muscles.

As stated earlier, there is an age-related difference in the toxicity of OP insecticides. While this is not directly due to any differences in AChE sensitivity, it requires much lower levels of exposure to induce the same effect in a juvenile than it does in an adult. However, the developing animal does have one advantage over adults. There is faster recovery of cholinergic biochemical parameters, especially AChE activity, in fetuses/neonates as compared to adults (Atterberry et al., 1997; Chakraborti et al., 1993; Lassiter et al., 1998). Why this occurs is not really clear but it may be that the fetuses/neonates are growing and the synthesis machinery is already in place. This ability for rapid synthesis of new proteins contributes to more effective and rapid recovery of AChE activity in neonates. While this faster recovery of activity probably does not play a role in the toxic effects of OPs, it does make estimating the level of maximum AChE inhibition more difficult in fetuses/neonates. To get an accurate measurement of the level of AChE in a fetus or neonate, AChE must be measured within an appropriate time following exposure.

Given that the traditional target of OP insecticides is the cholinergic system, a great deal of interest has been focused on the effects of this system. It is well established that the cholinergic system plays an important role in both prenatal and postnatal development (Abreu-Villaça et al., 2011). Therefore, disruption of the cholinergic system could result in negative effects on development. In fact, developmental exposure to OP insecticides has been demonstrated to induce both short term and long term alterations in the level of the components of the cholinergic system including choline acetyltransferase, vesicular acetylcholine transporter, high-affinity choline transporter, nicotinic receptors, and muscarinic receptors (Berge et al., 1986; Betancourt and Carr, 2003; Betancourt et al., 2006; Chakraborti et al., 1993; Chanda and Pope, 1996; Chanda et al., 1995; Dam et al., 1999; Dvergsten and Meeker, 1994; Eells and Brown, 2009; Fitzgerald and Costa, 1993; Guo-Ross et al., 2007; Gupta et al., 1985; Jett et al., 2001; Levin et al., 2010; Liu et al., 1999; Mehl et al., 2007; Qiao et al., 2002, 2004; Rhodes et al., 2004; Richardson and Chambers, 2003, 2005; Santos et al., 2004; Slotkin and Seidler, 2007a, 2015; Slotkin et al., 2001, 2004, 2006a, 2007a, 2008a, 2009a; Stamper et al., 1988; Stanton et al., 1994; Sun et al., 2003; Tang et al., 1999, 2003; Veronesi and Pope, 1990; Zhang et al., 2002; Zheng et al., 2000). These effects were observed following both gestational and postnatal exposures and frequently at dosages that inhibited brain AChE. At higher levels of AChE inhibition, the changes in components of the cholinergic system would be expected given the body's attempt to alter its physiology to compensate for the excessive stimulation. However, in several studies, effects on these components were observed when the level of AChE inhibition was not high enough to induce excessive stimulation. This suggested that the changes were not due to a physiological adaptation to AChE inhibition and was due to some other factor altering the levels of these components.

There have been multiple reports of altered behavior including learning and memory related tasks in children exposed developmentally to OP insecticides suggesting effects on neurological function in humans (for review see Burke et al., 2017; Burns et al., 2013; Eskenazi et al., 2008; Jurewicz and Hanke, 2008; Li et al., 2012; Marsillach et al., 2016; Mink et al., 2012; Polańska et al., 2013; Reiss et al., 2015). There have also been a plethora of animal behavioral studies investigating the effects of developmental OP exposure on neurological function. These involved monitoring either the short-term or persistent behavioral effects following gestational exposure, postnatal exposure, or a combination. These studies included a wide variety of behavioral tests and a wide variety of OP insecticides (for review see Abreu-Villaça and Levin, 2017; Burke et al., 2017; Li et al., 2012). These studies report effects at dosages that induced either a high level of brain AChE inhibition, a border-line level of brain AChE inhibition, or, for a few, no significant brain AChE inhibition. The neurochemical and behavioral effects observed without high levels of brain AChE inhibition led to the speculation that the developmental exposure to low levels of OP insecticides exerted their toxicity through mechanisms other than the inhibition of AChE. While many studies did not measure AChE activity at the appropriate time following exposure or did not measure AChE at all, the significant amount of literature demonstrating negative effects following very low dosages of OP insecticides during development has led to a general consensus that OPs, especially during repeated low level exposures, can exert toxic effects through a non-cholinergic mechanism.

In searching for the unknown non-cholinergic mechanism in mammals, there are several areas of research that have garnered quite a bit of attention in mammalian-based systems. Some of these studies have reported effects at very low levels of exposure. However, the basis for these effects remains unknown. In the following sections, some of the most popular hypotheses are presented as well as some hypotheses that have not been investigated to the extent that they should have been.

1. EFFECTS ON DNA

One of the first reports of non-cholinergic effects was decreased DNA and protein synthesis in the brains of rat pups following acute subcutaneous (sc-DMSO) injection of either 2 mg/kg chlorpyrifos on PND1 or 11 mg/kg chlorpyrifos on PND11 (Whitney et al., 1995). A follow-up study measured markers for cell packing density (DNA concentration), cell number (DNA content), and cell size (protein/DNA ratio) shortly after repeated postnatal exposure to chlorpyrifos (sc-DMSO) on PND1–4 and on PND11–14 and measured parameters 1 and 5 days post-exposure. Administration of higher dosages (5 and 25 mg/kg) lead to significant effects but also produced significant mortality. Reducing the dosages to either 1 mg/kg on PND1–4 or 5 mg/kg on PND11–14 eliminated the mortality and resulted in varied effects (including both increases and decreases) in the markers. The authors attributed the changes observed to chlorpyrifos-induced cell loss but no accompanying histological analysis was included to verify these changes (Campbell et al., 1997). No AChE measurement was performed for comparison in either of these studies. However, subsequent studies demonstrated that the earlier exposure paradigm (1 mg/kg from PND1–4, sc-DMSO) produced significant inhibition (~75%) of AChE at 2 h following exposure (Dam et al., 2000) and the late exposure paradigm (5 mg/kg from PND11–14, sc-DMSO) produced significant inhibition (~65%) of AChE on PND15 following exposure (Song et al., 1997). For both paradigms, significant inhibition was still present on PND10 and PND20, respectively (Song et al., 1997). At these levels of AChE inhibition and the persistence of significant inhibition for at least 5 days in a developing brain, cell loss is a possible outcome. Using these same exposure paradigms, further work reported decreased RNA concentration which was suggested to serve as a biomarker for adverse effects on brain cell development (Johnson et al., 1998). An additional study, using the early exposure paradigm, reported decreased DNA synthesis but no effects on RNA or protein synthesis. It was proposed that the chlorpyrifos-induced cell loss was targeting glia or a population of late-developing neurons since the majority of neurogenesis would have already occurred by that point (Dam et al., 1998).

Gestational studies were conducted to measure the same markers determined above by Campbell et al. (1997). Exposure to a range of dosages of chlorpyrifos (sc-DMSO) occurred during the periods corresponding to organogenesis and neural tube formation (GD9–12) and neurogenesis in the brainstem and forebrain (GD17–20). The GD9–12 exposure did not alter any parameter in forebrain on GD17 or GD21 but did decrease cell size (protein/DNA ratio) in the brainstem on GD21 at 2 mg/kg but not at 1 or 5 mg/kg. On GD21 following the GD17–20 exposure, no parameter in forebrain or brainstem was altered with the exception of cell size which was increased at 1 and 2 mg/kg but not at the higher dosages (Qiao et al., 2002). This increased cell size following exposure to 1 or 5 mg/kg on GD17–20 was still detectable in the cerebral cortex and hippocampus on PND30 and PND60. At these two time points, membrane protein concentration was also decreased in these two regions (Qiao et al., 2003). Unfortunately, the actual fetal brain AChE inhibition that occurs following this gestational exposure paradigm has not been adequately determined.

In supporting *in vitro* studies, DNA and RNA synthesis was decreased in undifferentiated PC12 cells following exposure to varying concentrations of chlorpyrifos with the threshold for effects at 1.5 μg/mL (Song et al., 1998). Using a C6 glioma cell line, DNA synthesis but not RNA synthesis was decreased following exposure to varying concentrations of chlorpyrifos (Garcia et al., 2001). In addition, this effect on DNA synthesis was not blocked by the presence of the muscarinic receptor antagonist atropine or the nicotinic receptor antagonist mecamylamine suggesting that cholinergic stimulation was not involved. A subsequent *in vitro* study reported that DNA synthesis in both PC12 and C6 glioma cells was inhibited by diazinon, chlorpyrifos, and its active metabolite chlorpyrifos-oxon, the active metabolite of chlorpyrifos, with the C6 cells being more sensitive (Qiao et al., 2001).

Interestingly, trichloropyridinol (TCP), the major catabolic product of chlorpyrifos, also inhibited DNA synthesis in both cell lines and physostigmine, a carbamate cholinesterase inhibitor, inhibited DNA synthesis in C6 cells only.

The *in vitro* data indicate that DNA synthesis can be inhibited by chlorpyrifos, chlorpyrifos-oxon, diazinon, physostigmine, and even TCP which has no cholinergic effects. Other chemicals have demonstrated similar effects in these *in vitro* systems such as tobacco smoke, plasticizers, silver, the organochlorine dieldrin, and divalent nickel (Chen et al., 2011b; Powers et al., 2010; Slotkin et al., 2007b, 2015a). Thus, the alteration of DNA dynamics may not be due to a specific interaction of an OP with the macromolecules responsible for DNA synthesis and content. It is possible that the altered DNA dynamics are the result of increased oxidative stress or some other factors that are simply by-products of an exposure to an exogenous chemical. Another possibility is that the altered DNA dynamics are due to epigenetic mechanisms such as non-coding RNAs, DNA methylation, or histone modifications. A small amount of work has addressed these possibilities (for review see Burke et al., 2017).

2. FREE RADICAL FORMATION

As with most environmental exposures, the production of free radicals is frequently reported following exposure to OP insecticides (Nurulain et al., 2013). The developing brain is highly susceptible to both reactive oxygen species (ROS) and reactive nitrogen species (RNS) because of its low concentration of antioxidants, high rate of oxygen consumption, populations of immature cells, and other factors related to the targets and molecules that assist in free radical formation (Ikonomidou and Kaindl, 2011; Wilhelm et al., 2016). Thus, the production of free radicals has been put forward as a mechanism of or at least a contributor to developmental OP toxicity. The first *in vivo* evidence came following acute exposure to chlorpyrifos on PND19 that resulted in decreased activity of catalase, the enzyme that converts hydrogen peroxide to oxygen and water (Jett and Navoa, 2000). In a subsequent *in vivo* study, thiobarbituric acid-reactive species (TBARS), a marker for lipid peroxidation, was determined following chlorpyrifos (sc-DMSO) exposure either at a range of dosages (1–40 mg/kg) from GD17–20 and assessed at GD21, 1 mg/kg from PND1–4 and assessed on PND5, or 5 mg/kg from PND11–14 and assessed on PND15. Only the latter postnatal exposure period produced increased TBARS levels in the forebrain and cerebellum of males but not females on PND15 suggesting a sexual-dimorphic effect and greater oxidative stress effects during late postnatal exposure (Slotkin et al., 2005). However, exposure to 1 mg/kg chlorpyrifos or 2 mg/kg diazinon (sc-DMSO) on PND1–4 induced changes in genes related to the generation of oxidative stress on PND5 suggesting that the earlier ages may be susceptible as well (Slotkin and Seidler, 2007a). In another gene expression study, acute exposure of PND7 rats to 0.1, 0.5, 1.0 or 2.0 mg/kg chlorpyrifos (oral-oil) altered the canonical pathways associated with oxidative stress at 24 h post-exposure but only at the highest dosage (Ray et al., 2010). Gestational exposure (GD6–17) to 2, 4, 10, 12 or 15 mg/kg chlorpyrifos (sc-oil) did not result in changes in the enriched categories of gene expression related to oxidative stress at any dosage on GD18 (Moreira et al., 2010). Lactational exposure to 20, 60, or 200 mg/kg malathion (sc-saline) from PND1–21 resulted in significant brain AChE inhibition in the offspring on PND22 but no changes in the activities of the antioxidant enzymes glutathione peroxidase or glutathione reductase and no changes in TBARS or in glutathione levels (da Silva et al., 2006). In contrast, lactational exposure to 200 mg/kg malathion (oral-corn oil) yielded significant AChE inhibition but also increased malondialdehyde (MDA) production and decreased superoxide dismutase and catalase activities in the brain on both PND21 and PND51 (Selmi et al., 2012). Thus, for *in vivo* developmental exposures, there is some evidence for increased free radical production following developmental OP exposure but there are many negative reports as well.

Most of the speculation for the induction of oxidative stress by OP insecticides comes from *in vitro* studies. *In vitro* exposure of neural progenitor cells grown as neurospheres to malathion at levels that inhibited AChE did not result in any changes in measures of oxidative stress (Salama et al., 2015). *In vitro* evidence of increased ROS was observed in PC12 cells and in C6 glioma cells within 10 min following exposure to 50 μg/mL chlorpyrifos. For the PC12 cells, ROS production following chlorpyrifos-oxon exposure was determined to see if the effects were mediated through AChE inhibition. The amount of ROS following chlorpyrifos-oxon was lower than that obtained with chlorpyrifos but the concentration of the oxon (10 μg/mL) was five times lower than of the parent compound (Crumpton et al., 2000; Garcia et al., 2001). ROS production measured by the level of TBARS can also be induced in both differentiating and non-differentiating PC12 cells following exposure to 14 μM chlorpyrifos suggesting that both phases of cell replication and differentiation are susceptible to the effects (Qiao et al., 2005). In contrast, a dose–response time-course study measuring oxidative stress in PC12 cell exposed to ranges of CPF from 1 to 5000 μg/mL reported that effects are only observed after 4–5 h incubation with the dosages exhibiting effects being 500 μg/mL or greater (Geter et al., 2008). Thus, both positive and negative results have been reported. The discrepancies could be due to differences in the experimental design as well as the markers measured as an indicator of oxidative stress.

A good indicator that oxidative stress can play a role in the toxicity of OP insecticides comes from more recent reports investigating the ability of antioxidants to protect *in vitro* effects. In PC12 cells, the presence of antioxidants (ascorbic acid) during incubation with 30 μM or 50 μM of chlorpyrifos or several non-OPs (fipronil, perfluorooctanesulfonamide, or dieldrin) decreased the amount of oxidative stress produced but did not prevent cell loss. The addition of Vitamin E provided greater protection against oxidative stress but also did not prevent cell loss induced by fipronil, perfluorooctanesulfonamide, and dieldrin. Interestingly, the level of oxidative stress was much higher with fipronil and perfluorooctanesulfonamide than chlorpyrifos suggesting that the induction of oxidative stress in this system is not selective to chlorpyrifos (Slotkin and Seidler, 2010). Vitamin E with chlorpyrifos was not tested but previous work performed by this same laboratory had reported that incubation of Vitamin E with 30 μM chlorpyrifos can partially prevent PC12 cell loss (Slotkin et al., 2007b). Vitamin E has also shown to attenuate the production of ROS, lipid peroxidation, and cytotoxicity that are induced in PC12 cells by 40 μM acephate, methamidophos, chloramidophos, malathion, or malaoxon (Lu et al., 2012). Exposure to 100 μM chlorpyrifos induced oxidative stress, MAPK activation, and cytotoxicity in PC12 cells and incubation with the antioxidant *N*-acetyl cysteine effectively blocked the MAPK activation and cytotoxicity indicating that at high concentrations, chlorpyrifos induced oxidative stress was responsible for the MAPK activation mediated cytotoxicity (Lee et al., 2012).

In other cell types, the role of oxidative stress in cytotoxicity has been established. In oligodendrocyte progenitor cells, cytotoxicity following exposure to 60 μM and 100 μM chlorpyrifos was prevented by incubation with the antioxidant Trolox, a water soluble analog of Vitamin E (Saulsbury et al., 2009). Cytotoxicity of primary hippocampal neurons was observed following 72 h incubation with 0, 1, 10, 20, 50, and 100 μM chlorpyrifos with 10 μM exerting effects only after prolonged incubation. Co-incubation with the antioxidant Trolox reduced the cytotoxicity but did not totally eliminate it (Salyha, 2013). However, oxidative stress was only observed in mature cerebellar granule neurons at 500 μM chlorpyrifos, a concentration that induced significant cytotoxicity (Amani et al., 2016). In immature cerebellar granule neurons, oxidative stress was observed at a sub-lethal concentration of 100 μM (20% cytotoxicity) but not at lower concentrations. The developing neurons were the most sensitive to cytotoxicity but oxidative stress levels were not determined in this population.

The presence of ROS in the developing brain is necessary for appropriate development (Wilson et al., 2017). The levels of ROS must be tightly controlled since concentrations above or below certain levels can be detrimental. Therefore, increased oxidative stress certainly could factor into the developmental neurotoxicity of the OPs. However, the vast majority of the research performed in this area has been using *in vitro* models with limited *in vivo* studies and those having about equal numbers of positive and negative results. It would be beneficial to have a little more focus on *in vivo* effects. In adult *in vivo* studies, there are multiple reports of OP insecticide exposure, although many at dosages that yield AChE inhibition, inducing alterations in markers of oxidative stress and these alterations have been decreased by the presence of chemicals with antioxidant properties (Ahmed et al., 2013; Ambali et al., 2010; Canales-Aguirre et al., 2012; Coban et al., 2015; Singh and Panwar, 2014; Yilmaz et al., 2012). It might be useful to determine if the negative effects, including alteration of oxidative stress markers, following low level developmental OP exposure can be ameliorated by the presence of antioxidants. However, since ROS are necessary for normal development, the level of antioxidant used must not alter this development.

3. NEURITE OUTGROWTH AND EFFECTS ON THE CYTOSKELETON

The process of neurogenesis in the developing brain involves neuronal proliferation and differentiation of neural progenitors. This proliferation can occur by a variety of types of cell division: symmetric proliferative that produces two progenitors; symmetric differentiative that produces two daughter cells of the same fate (e.g., two neurons); asymmetric mono-differentiative that produces on progenitor and one daughter; and asymmetric bi-differentiative that produces two daughters destined to different fates (e.g., neurons or glia) (Huttner and Kosodo, 2005). Early during development, the brain is primarily in symmetric proliferative cell division to ensure enough progenitors are present for correct tissue formation. As time progresses, there is a switch from symmetric cell division during development to asymmetric cell division (Knoblich, 2008). During this phase both progenitors and cells required to undergo differentiation are produced requiring specific mechanisms that coordinate between cell division, cell cycle exit, and differentiation (Hardwick et al., 2015). The actual number of times a progenitor cell will reenter the cell cycle is determined by the size of the brain. Animals with larger brains require higher numbers of neural progenitor cells to renter the cell cycle (Ohnuma and Harris, 2003). As cells exit the cell cycle, they differentiate into either neuronal or glial cells with their fates being determined by when they exit the cell cycle in order to generate the range of functional cells required to be present at the correct time and place. As development progresses, the neurogenesis switches from asymmetric cell division to symmetric differentiative cell division where there is a need to increase the differentiation of cells into neurons. Although this reduces the number of progenitor cells, a population of progenitor cells remains in the brain even through adulthood. In addition to differentiation, neuronal migration is another main feature of mammalian brain development. The majority of neurons are generated at sites that are different from the sites they preside in the adult brain. Neuronal cell migration is a biological necessity that enables the communication between early and late forming neurons before they start forming the synaptic connections (Rakic, 1990). It is thought that the cell division, proliferation, migration, and differentiation that occur during development involve regulation by cytoskeletal components and any interference with those components can disrupt the process (Lian and Sheen, 2015; Marin et al., 2010).

The cytoskeleton often refers to the complex and interlinking network of filaments and tubules that extended throughout the cytoplasm giving structure, shape, and movement facilitation to the cell. The cytoskeleton plays an important role in nervous system development not only by controlling important phenomena such as neuronal differentiation and migration, but also dendritic branching, spine formation, and axon outgrowth (Geraldo and Gordon-Weeks, 2009) and even maintenance of synaptic plasticity once the synapses are formed (Gordon-Weeks and Fournier, 2014). The formation of these neural circuits is necessary for nervous system function and any alteration of this formation would be detrimental. Thus, effects on the cytoskeleton could be a potential target of developmental neurotoxicants such as OP insecticides. The first reports of OPs affecting the cytoskeletal components reported altered structure in the neurites of neuroblastoma 2-A (N2a) cells exposed to high levels of tri-*o*-tolyl-phosphate (TOTP), diisopropylfluorophosphate (DFP) and dicrotophos (Harvey and Sharma, 1980) and of cultured chick dorsal root ganglia cells exposed to fenthion and paraoxon (Tuler and Bowen, 1989). In PC12 cells that were stimulated to differentiate with nerve growth factor (NGF), exposure to tricresyl phosphate inhibited the growth and maintenance of neurites (Flaskos et al., 1994). Other studies involving NGF-stimulated PC12 cells also reported inhibition of neurite-like outgrowth when exposed to OPs. Chlorpyrifos-oxon did so at concentrations affecting cell viability but mipafox did so at concentrations not affecting cell viability (Li and Casida, 1998). Low levels of chlorpyrifos and chlorpyrifos-oxon inhibited outgrowth in PC12 cells but so did the chlorpyrifos-metabolite TCP at comparable concentrations (Das and Barone, 1999). Effects have been observed in other neuronal-related cell lines exposed to OPs. Inhibition of neurite outgrowth in N2a cells was reported following acute exposure to diazinon, pirimiphos methyl, phosmet, and chlorpyrifos. Interestingly, the pyrethroid pyrethrum also inhibited neurite outgrowth and was more potent than the OPs diazinon and chlorpyrifos (Axelrad et al., 2002, 2003). Mipafox decreased neurite length in differentiated SY5Y human neuroblastoma cells while paraoxon did not (Hong et al., 2003). Chlorpyrifos and chlorpyrifos-oxon, but not TCP, inhibited axonal growth in primary neurons from the superior cervical ganglia at concentrations that did not inhibit AChE. At higher concentrations, all three actually enhanced dendritic growth (Howard et al., 2005). In general, there was no correlation between the inhibition of axonal/neurite growth and the inhibition of AChE in many of these studies suggesting that OPs exert these effects through other mechanisms. However, many other chemicals that are not OPs have been reported to inhibit neurite outgrowth (Abdulla et al., 1995; Abdulla and Campbell, 1993; Attoff et al., 2016; Axelrad et al., 2002, 2003; Chiou and Westhead, 1992; Parran et al., 2001; Popova et al., 2017; Smith et al., 1997, 2001). This type of scenario makes it difficult to definitely state that the effects on outgrowth was an effect specific just to the OP compounds. However, more recently, alteration of cytoskeletal proteins has been hypothesized as a potential target for the OP insecticides.

In vivo evidence of the alteration of cytoskeletal proteins by OP exposure was first reported in adult rats exposed daily for 14 days to 10.0, 18.0, or 25.0 mg/kg chlorpyrifos (sc-oil) (Terry et al., 2003). Deficits in retrograde and axonal transport were observed in the sciatic nerve obtained from these animals and persisted for up to 20 days. This was thought to be the result of direct interference with kinesin-dependent intracellular transport as demonstrated *in vitro* with chlorpyrifos, chlorpyrifos-oxon, or DFP (Gearhart et al., 2007). These studies were followed up by an *in vitro* study in cultured neurons that observed that chlorpyrifos and chlorpyrifos-oxon altered mitochondrial transport in axons and mitochondrial length and number

(Middlemore-Risher et al., 2011). This effect was observed at levels below those that inhibited AChE; could not be blocked with cholinergic receptor antagonists; did not alter mitochondrial energy production; and did not produce elevated ROS. Subsequent *in vitro* studies utilizing fluorescent-tagged membrane bound organelles in cultured neurons observed that DPF, chlorpyrifos, and chlorpyrifos-oxon decreased both anterograde and retrograde velocity of movement of the membrane bound organelle in the axon (Gao et al., 2016, 2017). Using magnetic resonance imaging of the optic nerve, decreased axonal transport was observed after repeated *in vivo* exposure to dosages of chlorpyrifos (sc-oil) that significantly inhibited AChE, both immediately following exposure and after a 30-day washout period (Hernandez et al., 2015). This interference with the transport of organelles and other required material in axons could be detrimental since this process is essential for both development of the nervous system and maintenance of fully developed neurons. However, it is not clear if these effects are the result of direct effects on the transport motors themselves or due to the alteration of tubulin structure which impairs the rate at which the transport occurs.

About the same time that alteration of cytoskeletal proteins following adult OP exposure was observed (Terry et al., 2003), the connection between the effects on neurite outgrowth and effects on the cytoskeleton components was first implicated. Chlorpyrifos and leptophos inhibited outgrowth of axon-like processes in N2a cells but also reduced the levels of the neurofilament heavy chain (NFH) subunit and the growth-associated protein GAP-43 and increased the levels of the heat-shock protein HSP-70 (Sachana et al., 2001, 2003). The carbamate carbaryl also inhibited neurite outgrowth but did not disrupt NFH or GAP-43 levels. Chlorpyrifos-oxon also produced similar effects on outgrowth, NFH, and GAP-43 in N2a cells (Flaskos et al., 2011). Diazinon also inhibited outgrowth and reduced the levels of NFH subunits in N2a cells but also reduced in the levels of the microtubule associated protein MAP1B (Flaskos et al., 2007). Surprisingly, the effects on HSP-70 levels (reduced) with diazinon were the opposite of that observed with chlorpyrifos and leptophos. The effects of diazinon-oxon were similar to diazinon in that it inhibited outgrowth but differed in that it did not alter NFH levels and actually increased HSP-70 levels (Sidiropoulou et al., 2009a). The authors stated that the effects did not correlate with AChE inhibition but all concentrations used significantly inhibited AChE. These effects on outgrowth and cytoskeletal components have been observed in glial cell lines as well. Cell process outgrowth was inhibited by chlorpyrifos-oxon and mipafox in C6 glioma cells (Li and Casida, 1998). Both chlorpyrifos and chlorpyrifos-oxon inhibited extension outgrowth in C6 glioma cells and decreased the levels of the microtubule-associated protein MAP1B and tubulin but not MAP2c (Sachana et al., 2008). While extension outgrowth was not affected by diazinon in C6 glioma cells (Flaskos et al., 2007), diazinon-oxon decreased extension outgrowth in C6 glioma cells and the levels of glial fibrillary acidic protein (GFAP) and MAP1B but with no effects on MAP2c (Sidiropoulou et al., 2009a). The neurofilaments are structural components found in all parts of the neuron but are particularly abundant in axons and are essential for the growth of axons (Yuan et al., 2012). MAP1B is critically important in terms of neurogenesis since it is one of the first microtubule associated protein expressed in neurons during development and is intricately involved in the axonal elongation (Gordon-Weeks and Fischer, 2000; Schoenfeld et al., 1989; Tortosa et al., 2011). Growth-associated protein 43 (GAP43) is a phosphoprotein that mediates neurite outgrowth in developing neurons (Benowitz and Routtenberg, 1997; He et al., 1997; Strittmatter et al., 1995). Altered GAP3 expression in heterogeneous mice results in abnormal cortex organization (McIlvain et al., 2003). While these parameters are vital to cellular outgrowth/axon extension and their levels are clearly impacted by the presence of OPs, the effects observed vary between different OPs and can even differ between the parent insecticide and its active metabolite. This makes identifying a single target for all OPs difficult. In addition, it is unclear if these reductions in cytoskeletal components are the primary effect of the OPs or are the result of the OPs affecting other parameters leading to the reductions as a secondary effect.

Along a related line of study, persistent chlorpyrifos oxon exposure decreased microtubule-associated protein-2 (MAP-2) levels in organotypic hippocampal slice cultures which was associated with decreased neuronal viability in those cultures (Prendergast et al., 2007). Chlorpyrifos-oxon also reduced tubulin polymerization in a purified tubulin preparation which when combined with the effects observed in the organotypic slice cultures suggested impaired microtubule synthesis and/or function. Further research reported that chlorpyrifos-oxon can covalently bind to 17 tyrosine sites on tubulin and prevents the polymerization of tubulin to form microtubules (Grigoryan and Lockridge, 2009; Grigoryan et al., 2008, 2009). The phosphorylation of tubulin has been observed following *in vivo* exposure of adult mice by either daily administration of low dosage of chlorpyrifos that did not inhibit AChE or acute administration of chlorpyrifos-oxon that inhibited AChE. Both exposures resulted in phosphorylation of a single tyrosine residue (281) which is lower than the number of phosphorylation sites observed *in vitro* suggesting differences between *in vivo* and *in vitro* effects. In addition, the acute exposure also resulted in phosphorylation of a single serine residue (338) which was not observed *in vitro* (Jiang et al., 2010). The increased phosphorylation of tubulin suggests that *in vivo* OP exposure can affect the assembly and stability of microtubules. Increased phosphorylation of NFH has also been observed with higher concentrations of diazinon-oxon in N2a cells lending support to the phosphorylation of cytoskeletal components as a mechanism (Sidiropoulou et al., 2009b).

The research in this area clearly demonstrates that OPs can alter the neuronal cytoskeleton structure either by phosphorylation of tubulin, altered transport of molecules necessary for structural assembly, or both. In fact, phosphorylation of other cellular cytoskeletal components may occur as well (Flaskos, 2014). These events could definitely negatively affect the proper development of the brain but such events could also negatively affect the proper development of other organs as well. If possible, it would be interesting to determine the potencies of different OPs with respect to their propensity to phosphorylate cytoskeletal components and determine how those potencies relate to the effects of the same OPs on transport of molecules and axonal/neurite outgrowth. In addition, the vast majority of these studies were conducted in *in vitro* systems which are appropriate but do not possess the full dynamics of an *in vivo* system. The *in vivo* studies that have been performed have focused on the effect of OPs in adult animals. It would be interesting to determine if these events occur during an *in vivo* low level developmental OP exposure.

4. NEED FOR *IN VITRO* STUDIES TO INCLUDE ASTROCYTES

Astrocytes have essential roles in function and development of the brain. They also play a role in developmental toxicity of OP insecticides. Developmental exposure of rat pups to levels of chlorpyrifos that inhibited AChE has resulted in changes in the levels/expression of the astrocyte marker GFAP in the brain of treated animals (Betancourt et al., 2006; Garcia et al., 2002; Zhang et al., 2015). Increased GFAP levels typically indicate activation of astrocytes (Eng et al., 2000) which produce proinflammatory cytokines and leads to inflammation (Benveniste, 1998). Increased GFAP can precede neuronal loss (Dell'Anna et al., 1995). However, astrocytes also provide a protective role for neurons (Bélanger and Magistretti, 2009; Liu et al., 2017). A large volume of research investigating the effects of OP insecticides has utilized *in vitro* systems which contain a single cell line or primary neurons. The elimination of astrocytes from the cultures can totally change the dynamics of the results. For example, in neuron enriched cultures of rat telencephalon, both chlorpyrifos and parathion exhibited cytotoxicity but when tested in an aggregate culture containing both neurons and astrocytes, the cytotoxicity of both parathion and chlorpyrifos was reduced (Zurich et al., 2004). Exposure of hippocampal neurons to diazinon and diazoxon increased indicators of increased oxidative stress and inhibited neurite outgrowth. Addition of antioxidants ameliorated the inhibition of neurite outgrowth suggesting that the outgrowth was mediated by oxidative stress. When the neurons exposed in the presence of astrocytes, there was no inhibition of neuronal neurite outgrowth and this neuroprotection was demonstrated to be due to the production of glutathione by the astrocytes (Pizzurro et al., 2014). In human neural progenitor cells, chlorpyrifos inhibited the neurite length and neurite number. However, when astrocytes were included in the neuronal culture, the effects of chlorpyrifos were attenuated. The addition of the cytochrome P450 inhibitor SKF525A to the incubation partially prevented the protection mediated by the presence of the astrocytes. This indicated that the astrocytes can biotransform chlorpyrifos into inactive and less neurotoxic metabolites thereby providing neuroprotection (Wu et al., 2017). Given the protective contributions of astrocytes, aggregate co-cultures containing astrocytes would seem to better mimic the cellular heterogeneity and interactions in the central nervous system and would provide a more realistic scenario with respect to toxicity. Cultures of single cell types can be appropriate for identifying very specific mechanisms but they are not appropriate to assess toxicity. This is especially true for studies that attempt to relate *in vitro* concentrations with *in vivo* exposure levels which is often done.

5. ALTERING THE MORPHOGENETIC ACTION OF AChE

In cultured neurons, excessive amounts of acetylcholine inhibit neurite outgrowth when applied directly to cultured neurons (Lauder and Schambra, 1999) suggesting that the increased level of acetylcholine that occurs as a result of AChE inhibition could have inhibitory effects on development. However, in cultured primary cortical neurons, the cytotoxicity of chlorpyrifos and diazinon was not altered by the presence of muscarinic or nicotinic receptor antagonists but was actually attenuated by the addition of acetylcholine suggesting that the inhibition of AChE did not play a role in the cytotoxicity (Rush et al., 2010). It has also been demonstrated that

AChE has morphogenetic effects on development and this effect is not due to the catalytic activity of the enzyme (Bigbee et al., 1999; Brimijoin and Koenigsberger, 1999; Grisaru et al., 1999). Using dorsal root ganglia neurons from AChE+/+ mice, chlorpyrifos and chlorpyrifos-oxon produced deficits in axonal outgrowth at concentrations that did not decrease cell viability or inhibit the catalytic activity of AChE. In dorsal root ganglia neurons from AChE−/− mice, the axonal growth was lower than the AChE+/+ neurons but neither chlorpyrifos nor chlorpyrifos-oxon inhibited axonal growth. Transfection of AChE into the AChE−/− neurons restored the normal axonal growth and these neurons were now sensitive to the inhibitory effects of the OPs on axonal growth. Thus, deficits in axonal outgrowth induced by chlorpyrifos and chlorpyrifos-oxon were only present in cultures that expressed intact AChE suggesting that AChE is necessary for normal axonal growth and the alteration of that growth by OP compounds is due an interaction with AChE that is not related to its catalytic activity (Yang et al., 2008). This is an interesting phenomenon but unfortunately, only limited work has been done in this area.

6. MONOAMINE NEUROTRANSMITTER SYSTEMS

An area that has garnered a great deal of interest is the effect of developmental OP exposure on the monoamine neurotransmitter systems. The earliest work in this area reported decreased serotonin levels in rabbits exposed to 9 mg/kg dichlorvos (po) either repeatedly from PND6–15 or acutely on PND6. Both scenarios significantly inhibited AChE but the acute exposure increased serotonin levels while the repeated dosage decreased serotonin levels (Maślińska et al., 1981). Acute exposure to either 2 mg/kg (PND15) or 3 mg/kg (PND21) methyl parathion (ip) followed by sacrifice 48 h later reported decreased serotonin (5HT), dopamine, and norepinephrine levels in several brain regions of rat pups. Both these dosages inhibited AChE (Mahaboob Basha et al., 2001). With respect to effects on the serotonergic system following postnatal exposure to OP insecticides, the vast majority of the work in this area used two exposure paradigms and concentrated the efforts on the effects of chlorpyrifos. The early exposure paradigm involved administration (sc-DMSO) of 1 mg/kg from PND1–4 and the late exposure paradigm involved administration (sc-DMSO) of 5 mg/kg from PND11–14. Both paradigms result in significant inhibition of AChE (>65%) that persists at least 6 days following the last exposure. Both of these exposure scenarios also induce altered levels of the cholinergic components (Slotkin et al., 2002).

With respect to the levels of the functional components of the serotonergic system, exposure to chlorpyrifos during both the early exposure paradigm (PND1–4) and the late exposure paradigm (PND11–14) resulted in alterations in $5HT_{1A}$ receptor levels, $5HT_2$ receptor levels, 5HT high-affinity transporter levels, 5HT content and turnover, and adenylyl cyclase activation stimulated by 5HT with effects observed at neonatal, adolescent, and adult ages (Aldridge et al., 2003, 2004, 2005b,c; Raines et al., 2001; Slotkin and Seidler, 2005; Slotkin et al., 2015b). The effects varied based on the exposure paradigm and the age tested. In addition, sex differences were observed in many of these studies. The effects of diazinon and parathion was determined using the early exposure paradigm (PND1–4) and changes were observed in the levels of the $5HT_{1A}$ and $5HT_2$ receptors and the 5HT high-affinity transporter at neonatal, adolescent, and adult ages (Levin et al., 2010; Slotkin et al., 2006b, 2008b, 2009b). When similar parameters measured at similar time points were compared, there were differences in the effects of the three insecticides suggesting that these three organophosphates have different effects on serotonergic system following developmental exposure. An additional study reported that postnatal (PND3–9) exposure to methamidophos (sc-DMSO) produced different effects on the levels of the $5HT_{1A}$ and $5HT_2$ receptors and the 5HT high-affinity transporter than was observed with the other three OP insecticides (Lima et al., 2013). Differences between compounds were also observed in behavioral studies where the response of treated rats in identical tasks was different. In particular, a serotonergic-related behavior response was observed in chlorpyrifos treated rats but not in diazinon or parathion treated rats identified by administration of 5HT2 receptor antagonist ketanserin (Aldridge et al., 2005a; Levin et al., 2010; Timofeeva et al., 2008a,b). Gestational exposure to chlorpyrifos (sc-DMSO) from either GD9–12 or GD17–20 also produced changes in the levels of the $5HT_{1A}$ and $5HT_2$ receptors, the 5HT high-affinity transporter, and adenylyl cyclase activation by 5HT at neonatal, adolescent, and adult ages (Aldridge et al., 2003, 2004, 2005c; Slotkin and Seidler, 2007b). As with the postnatal exposures, the effects varied based on the exposure paradigm and the age tested and sex differences were observed in many of the studies. Overall, developmental OP insecticide exposure leads to an alteration in serotonergic components thereby altering synaptic function. This results in behavioral abnormalities that are usually associated with 5HT deficiencies.

With respect to the other catecholamines, developmental OP exposure has been shown to exert effects on the dopaminergic and noradrenergic systems. The main effect observed has been on neurotransmitter levels, metabolite levels, and turnover (Aldridge et al., 2005c; Chen et al., 2011c; Dam et al., 1999; Mahaboob Basha et al., 2001; Slotkin and Seidler, 2007b; Slotkin et al., 2002, 2009c, 2017). The majority of these studies used the same exposure paradigms (or very similar) as described for the serotonergic studies and used the same route and vehicle (sc-DMSO). There are several studies that report no effects on β-adrenergic receptor levels following postnatal exposure (sc-DMSO) to chlorpyrifos, parathion, or diazinon exposure (Slotkin et al., 2017; Song et al., 1997) but this depends on the time measured. Altered β-adrenergic receptor levels have been reported in neonates and adults following gestational exposure to chlorpyrifos (Meyer et al., 2003) and in adults exposed to chlorpyrifos using the late exposure paradigm (Meyer et al., 2004). However, very little, if any, information is available on the effects of developmental OP exposure on dopaminergic receptors. One study, using the late exposure paradigm, has reported that exposure results in a significant reduction of dopaminergic neurons at 16 and 46 days and a significant increase in the expression of microglia and astrocytes (Zhang et al., 2015). This study suggests that the dopaminergic neurons could be directly targeted by chlorpyrifos exposure which is an interesting finding. However, this is a single study and more studies demonstrating this observation are needed to give more support to this finding. In addition, this study did not measure AChE inhibition. In an earlier study, the same dosage of chlorpyrifos administered using the same route and at the same age yielded 65% inhibition of brain AChE (Song et al., 1997) and significantly reduced body weight (Campbell et al., 1997). While the dosage may be devoid of any systemic toxicity such as cholinergic hyperstimulation, the reduced body weight combined with the high inhibition of brain AChE suggests that exposure to 5 mg/kg of chlorpyrifos (sc-DMSO) from PND11–14 is exerting a systemic toxic effect.

In summary, these data clearly suggest that developmental exposure to OPs can induce changes in the components of the monoaminergic systems suggesting the possibility of architectural miswiring of the circuits. The deficits resulting from the dysfunction of these circuits can lead to abnormal behavior. In fact, many behaviors that have been reported to be disrupted by developmental OP exposure, such as anxiety, emotionality, and social behavior, are regulated by the monoaminergic neurotransmitters especially serotonin. However, the experimental evidence was not direct and it was difficult to conclude that the changes in the components of the monoaminergic systems were the result of direct actions of the OP on the monoaminergic system. This is a possibility but it is more likely that the reduction in the components of these neurotransmitter systems is the downstream effects on the action of the OP on another target.

7. ENDOCANNABINOID SYSTEM

First observed in adult mice, high level acute exposure of adults to OP insecticides results in greater inhibition of fatty-acid amide hydrolase (FAAH) and monoacylglycerol lipase (MAGL) which are responsible for the breakdown of the endocannabinoids arachidonoylethanolamide (AEA or anandamide) and 2-arachidonoylglycerol (2-AG), respectively. The level of inhibition of these enzymes was greater than the level of inhibition of AChE (Quistad et al., 2001, 2002b, 2006). Additional studies demonstrated that this acute *in vivo* exposure of adult mice resulted in increased levels of 2-AG and AEA in the brain (Nomura and Casida, 2011; Nomura et al., 2008). The studies were focusing on acute toxicity in adults and the general conclusion was that inhibition of endocannabinoid deactivation enzymes and the resulting accumulation of the endocannabinoids would not play a role in the acute toxicity of the OP insecticides unless it is administered at very high dose levels. Because the endocannabinoid system plays a pivotal role in normal brain development (Anavi-Goffer and Mulder, 2009; Harkany et al., 2008), abnormal levels of the endocannabinoids could disrupt appropriate signaling processes involved in brain development.

With respect to developmental OP exposure, exposure to 1, 2.5, or 5 mg/kg chlorpyrifos (po-oil) from PND10–16 resulted in inhibition of FAAH, MAGL, and AChE in a dose-related manner with the extent of enzyme inhibition from highest to lowest level being FAAH > AChE > MAGL (Carr et al., 2011). A subsequent time course study reported that exposure to chlorpyrifos using the same exposure paradigm resulted in the accumulation of AEA and 2-AG in the brain (Carr et al., 2013). When the dosages used in the study was lowered to 0.5 mg/kg, the exposure resulted in no measurable inhibition of brain AChE or MAGL but significant brain FAAH inhibition and a significant increase in AEA levels but not 2-AG (Carr et al., 2014). Emotional-related behavior is modulated by the endocannabinoid system and is susceptible to developmental OP toxicity but it was not clear if these low dosages would result in effects on emotional-related behavior.

Using a light–dark emergence test (PND25), rats treated with 0.5, 0.75, or 1.0 mg/kg chlorpyrifos (PND10–16) emerged faster from a dark container into a highly illuminated arena than did controls suggesting decreased emotional reactivity (Carr et al., 2017). Neither 0.5 nor 0.75 mg/kg inhibit brain AChE. In a follow-up study, these same dosages of chlorpyrifos administered using the same exposure paradigm resulted in increased social interactions in adolescent rats (PND35). In addition, a specific inhibitor for FAAH was included as a positive control and the behavioral results were very similar to that obtained for chlorpyrifos (Carr et al., 2015).

While these data suggest that altering FAAH activity and thereby inducing indirect signaling in the endocannabinoid system may be a measurable "non-cholinergic" effect of low dosages of chlorpyrifos, additional work needs to be done to clarify this possibility. For example, FAAH degrades other neuroactive chemicals such as the fatty-acid amides oleoylethanolamide and palmitoylethanolamide. Alterations of the levels of other neuroactive chemicals induced by OP inhibition of FAAH during the maturation of the endocannabinoid system could lead to alterations in that maturation process but less is known about the roles of these compounds in brain development. More work needs to be done in this area.

Chlorpyrifos, as well as other insecticides, have also been reported to displace the binding of ligands from cannabinoid receptor 1 (CB1) in adult mice brain (Quistad et al., 2002a). For chlorpyrifos-oxon, the *in vitro* sensitivity of the CB1 receptor is greater than that of FAAH as determined by IC50s. However, following acute *in vivo* exposure to chlorpyrifos, the level of inhibition of FAAH activity is much greater than the inhibition of ligand binding to the CB1 receptor (Quistad et al., 2001). Activation of the CB1 receptor inhibits adenylyl cyclase (Turu and Hunyady, 2010) and developmental exposure to chlorpyrifos results in the inhibition of adenylyl cyclase (Song et al., 1997) so it could be possible that direct activation of the CB1 receptor by chlorpyrifos-oxon could be responsible for the decreased adenylyl cyclase activity. However, it is not known if this displacement of ligand binding stimulates or inhibits the activity of the CB1 receptor. Thus, at this time, it is unclear if the action of chlorpyrifos-oxon on the CB1 receptor plays a role in the developmental toxicity of chlorpyrifos.

8. DISRUPTION OF EMOTIONAL/SOCIAL BEHAVIOR

While most behavioral findings were not presented here, the reported effects on emotional/social behaviors tends to connect the biochemical results especially at exposure to low levels of an OP which would be more reflective of what would likely occur in an actual environmental exposure. The first report of alteration of social behavior involved gestational exposure of rats to 5, 10, or 15 mg/kg Sumithion (active ingredient fenitrothion, po-oil) from GD7–15 which resulted in increased social interaction in the two higher treatment groups (Lehotzky et al., 1989). Exposure of mice to either 1 or 3 mg/kg chlorpyrifos (sc-DMSO) from PND1–4 did not alter behavior while exposure from PND11–14 increased activity levels (PND25) and the higher dosage altered novelty seeking (PND35). Males (PND45) from both ages and treatment groups exhibited increased aggressive behavior toward an intruder (Ricceri et al., 2003). Similar results on aggressive behavior in adult male mice (PND60) exposed to chlorpyrifos postnatally were observed in a subsequent study. This study also observed reduced anxiety in adult females measured using the elevated plus-maze and altered maternal behavior (Ricceri et al., 2006). Gestational exposure (GD15–18) of mice to 3 or 6 mg/kg of chlorpyrifos (po-oil) decreased number and duration of ultrasonic vocalizations in both sexes (PND10) (Venerosi et al., 2009) and increased social interactions and ultrasonic vocalizations in females (PND120) (Venerosi et al., 2006). Exposure of mice to 3 mg/kg chlorpyrifos (sc-oil) from PND11–14 had no effect on preference for social novelty as adolescents but decreased anxiety levels and maternal aggressive behavior and increased social investigatory behavior in adults (Venerosi et al., 2008). Similar negative results on ultrasonic vocalizations and social novelty were observed in mice following gestational exposure to chlorpyrifos-oxon (Mullen et al., 2013). Gestational exposure of mice (GD15–18) to 6 mg/kg of chlorpyrifos (po-oil) did not alter forced swimming behavior in adults but did decrease the aggressive attacks of nesting females toward an intruder unfamiliar male. Administration of the 5HT transporter inhibitor fluvoxamine produced changes in the behavior of the control animals but had no effect on the behavior of the chlorpyrifos-treated animals suggesting that the treatment resulted in changes in the response of the serotonergic system thus altering social–emotional behavior (Venerosi et al., 2010). Gestational exposure of mice (GD14–17) to 6 mg/kg chlorpyrifos did not show any effects of treatment on the response to either social or non-social odors and no difference in social discrimination but females did exhibit a higher level of social investigatory behavior (De Felice et al., 2014). Further demonstration of increased social investigatory behavior during a social recognition test in mice of both sexes has been reported following exposure to 6 mg/kg chlorpyrifos from GD15–PND14 (Venerosi et al., 2015). The majority of these studies originated from a single laboratory and used only one strain of mice (CD1). It would be better to have different laboratories report similar results or observe the same behavioral responses in other strains of mice or in other species. However, several of these studies measured the same behavioral parameters and reported the same results thereby verifying earlier studies which greatly strengthens the idea that effects on social behavior is a consistent finding. In general, these studies observed a lack of effect on social novelty/discrimination accompanied by reduced anxiety and increased social investigatory behaviors. Others have reported decreased anxiety in an emergence test and increased social interactions and play in adolescent rats following postnatal exposure to low levels of chlorpyrifos (Carr et al., 2015, 2017).

As these studies progressed, these findings suggested that chlorpyrifos exposure altered the neuroendocrine signaling pathways involved in regulating social and emotional responses. In a combined gestational and postnatal study where pregnant mice received 3 or 6 mg/kg chlorpyrifos (po-oil) on GD15–18 and/or the offspring received 1 or 3 mg/kg chlorpyrifos (sc-oil) on PND11–14, increased oxytocin (OT) and decreased arginine vasopressin (AVP) expression was observed in the hypothalamus of males (PND150) following the gestational exposure to 6 mg/kg. No effects on prolactin (PRL) expression were observed (Tait et al., 2009). Using the same exposure paradigm, the level of serum thyroxin (T4) and triiodothyronine (T3) was measured in the serum of dams on GD18 and pups on PND150. T4 was decreased in the dams treated with 6 mg/kg. Reduction of T3 levels in both males and females and T4 levels in the male pups treated postnatally with 3 mg/kg was observed regardless of whether or not gestational exposure occurred. (De Angelis et al., 2009). Using the same exposure paradigm, it was reported that the exposures to chlorpyrifos resulted in changes in the cytochrome P450s responsible for steroid hormone metabolism which could affect the levels of the hormones (Buratti et al., 2011). Gestational exposure of mice (GD14–17) to 6 mg/kg chlorpyrifos increased expression of genes for the estrogen receptor β in the hypothalamus, decreased OT expression in amygdala of adult males, and increased vasopressin 1a receptor expression in the amygdala of both sexes (Venerosi et al., 2015).

The presence of sexually-dimorphic findings in a large number of developmental OP studies strongly suggests that these compounds alter the maturation of important sexually-dimorphic neuroendocrine pathways. However, it is actually neurochemical signals that control the release of the neuroendocrine molecules which makes dissecting the neurochemical effects reported from the neuroendocrine effects reported a challenge. For example, from the above described studies, alterations in social behaviors are a frequent outcome of developmental OP exposure. Social behavior is modulated by AVP and OT (Choleris et al., 2009; Gabor et al., 2012). Most of AVP and OT secreting neurons are located in the supraoptic nucleus (SON) and paraventricular nucleus (PVN) of the hypothalamus. The neurosecretory functions of the SON and PVN are modulated by several transmitter systems including the cholinergic and monoaminergic systems (Daftary et al., 1998; Ho et al., 2007; Li et al., 2001), both of which are altered by developmental OP exposure. Endocannabinoid signaling, which is altered by developmental OP exposure, also modulates social behaviors (Achterberg et al., 2016; Wei et al., 2017) and endocannabinoid signaling modulates the activity of the AVP and OT (Luce et al., 2014; McDonald et al., 2008). Thus, developmental exposure to OP insecticides induces endocrine-related effects but how they do this is not clear. As the research continues to correlate changes in neuroendocrine molecules with changes with neurochemical signaling molecules with changes in behavioral outcomes, the picture will hopefully become clearer.

Some epidemiological studies have implicated that developmental exposure to chlorpyrifos may be associated with autism spectrum disorder (Eskenazi et al., 2007; Landrigan, 2010; Rauh et al., 2006). A recent publication reported that male offspring of C57BL/6J (B6) mice exposed gestationally to 2.5 or 5 mg/kg chlorpyrifos (po-oil) from GD12–15 exhibited decreased social preference (Lan et al., 2017). These authors suggested that their results could represent an autism spectrum disorder-like behavioral profile in male mice following gestational exposure to chlorpyrifos. However, following gestational exposure to 6 mg/kg chlorpyrifos (GD14–17) in the BTBR T+tf/J mouse strain, a validated model of idiopathic autism based on their behavior, this atypical behavioral phenotype was not worsened in the offspring exposed to chlorpyrifos (De Felice et al., 2015). In addition, a recent review of 13 behavioral studies of rats and mice treated during the perinatal period with chlorpyrifos does not support the association between gestational chlorpyrifos exposure and autism (Williams and DeSesso, 2014). While some epidemiological studies correlated OP exposure with autism, the majority of animal studies do not support this.

9. CONCLUSIONS

As the research literature research was perused for this review, there are issues that kept appearing and we felt that these needed to be addressed especially in regard to AChE inhibition.

(1) First, the *in vivo* literature is full of phrases such as "without biologically significant cholinesterase inhibition" or "dosages below those which cause appreciable brain cholinesterase inhibition." This leads to the question "What is appreciable or biologically significant cholinesterase inhibition?" The level of inhibition in these studies is certainly below that which causes enough inhibition to induce cholinergic crisis but there usually is statistically significant cholinesterase inhibition. In a developing brain where everything is in a constant flux, could 20% cholinesterase inhibition be biologically relevant? This is really difficult to answer but still remains a possibility.

(2) Second, when developing a model for studying developmental OP insecticide toxicity or when implementing someone else's model, it is necessary to perform a time course AChE inhibition study. Many studies never determine the level of AChE inhibition and simply rely on previous paper's claims about the effect of the dosage on cholinesterase activity. In several instances, different studies used the same exposure paradigm and dosage and the original phrase "without biologically significant cholinesterase inhibition" that was in the initial publication morphed into "no cholinesterase inhibition" in later publications. Determining a time course for inhibition is essential because of the faster recovery of AChE activity in the fetus/neonate. Several gestational studies have reported that the dosage they used yielded "no cholinesterase inhibition." However, they actually only assayed for AChE inhibition at a single time point, frequently at birth. It has been demonstrated that following gestational exposure (GD14–18) to 6 mg/kg chlorpyrifos (po-oil), the peak inhibition of AChE in the fetus occurred 5–10 h post-exposure on GD18 with significant recovery of activity by 24 h and total recovery by 48 h (Lassiter et al., 1998). Thus, only measuring AChE at birth, several days after the last administration of the OP, is an inaccurate assessment of brain AChE inhibition. This inaccuracy actually hampers the search for non-cholinergic mechanisms since the goal is to identify effects that will occur without any brain AChE inhibition.

In conclusion, the mechanism behind the developmental neurotoxicity of OP insecticides especially at very low dosages remains unknown. There are some very promising leads that need to be explored further. The recent report that the activation of microglia and astrocytes and the accompanying increase in inflammatory mediators may play a role in the developmental toxicity of OP insecticides is an interesting hypothesis (Tian et al., 2015; Zhang et al., 2015). However, both studies utilized the 5 mg/kg chlorpyrifos (sc-DMSO) from PND11–14 exposure paradigm. While this exposure paradigm does not cause signs of systemic toxicity, it does induce 65% AChE inhibition with significant inhibition persisting up to 5 days. Thus, there is a need for future work to focus on using dosages that have been verified to not inhibit brain AChE inhibition. It is hopeful that very soon, research will be able to unlock this mystery. Once an understanding of the mechanism(s) is obtained, it will allow the development of safer insecticides to protect future generations and, hopefully, allow therapeutic approaches to be developed for children affected by these low level OP insecticide exposures.

REFERENCES

Abdulla, E.M., Campbell, I.C., 1993. Use of neurite outgrowth as an *in vitro* method of assessing neurotoxicity. Ann. N. Y. Acad. Sci. 679, 276–279.

Abdulla, E.M., Calaminici, M., Campbell, I.C., 1995. Comparison of neurite outgrowth with neurofilament protein subunit levels in neuroblastoma cells following mercuric oxide exposure. Clin. Exp. Pharmacol. Physiol. 22, 362–363.

Abreu-Villaça, Y., Levin, E.D., 2017. Developmental neurotoxicity of succeeding generations of insecticides. Environ. Int. 99, 55–77.

Abreu-Villaça, Y., Filgueiras, C.C., Manhães, A.C., 2011. Developmental aspects of the cholinergic system. Behav. Brain Res. 221, 367–378.

Achterberg, E.J.M., van Swieten, M.M.H., Driel, N.V., Trezza, V., Vanderschuren, L.J.M.J., 2016. Dissociating the role of endocannabinoids in the pleasurable and motivational properties of social play behaviour in rats. Pharmacol. Res. 110, 151–158.

Ahmed, M.A.E., Ahmed, H.I., El-Morsy, E.M., 2013. Melatonin protects against diazinon-induced neurobehavioral changes in rats. Neurochem. Res. 38, 2227–2236.

Aldridge, J.E., Seidler, F.J., Meyer, A., Thillai, I., Slotkin, T.A., 2003. Serotonergic systems targeted by developmental exposure to chlorpyrifos: effects during different critical periods. Environ. Health Perspect. 111, 1736–1743.

Aldridge, J.E., Seidler, F.J., Slotkin, T.A., 2004. Developmental exposure to chlorpyrifos elicits sex-selective alterations of serotonergic synaptic function in adulthood: critical periods and regional selectivity for effects on the serotonin transporter, receptor subtypes, and cell signaling. Environ. Health Perspect. 112, 148–155.

Aldridge, J.E., Levin, E.D., Seidler, F.J., Slotkin, T.A., 2005a. Developmental exposure of rats to chlorpyrifos leads to behavioral alterations in adulthood, involving serotonergic mechanisms and resembling animal models of depression. Environ. Health Perspect. 113, 527–531.

Aldridge, J.E., Meyer, A., Seidler, F.J., Slotkin, T.A., 2005b. Developmental exposure to terbutaline and chlorpyrifos: pharmacotherapy of preterm labor and an environmental neurotoxicant converge on serotonergic systems in neonatal rat brain regions. Toxicol. Appl. Pharmacol. 203, 132–144.

Aldridge, J.E., Meyer, A., Seidler, F.J., Slotkin, T.A., 2005c. Alterations in central nervous system serotonergic and dopaminergic synaptic activity in adulthood after prenatal or neonatal chlorpyrifos exposure. Environ. Health Perspect. 113, 1027–1031.

Al-Hachim, G.M., Fink, G.B., 1968a. Effect of DDT or parathion on condition avoidance response of offspring from DDT or parathion treated mothers. Psychopharmacologia 12, 424–427.

Al-Hachim, G.M., Fink, G.B., 1968b. Effect of DDT or parathion on the minimal electroshock seizure threshold of offspring from DDT or parathion-treated mothers. Psychopharmacologia 13, 408–412.

Al-Hachim, G.M., Fink, G.B., 1968c. Effect of DDT or parathion on open-field behavior of offspring from DDT- or parathion-treated mothers. Psychol. Rep. 22, 1193–1196.

Amani, N., Soodi, M., Daraei, B., Dashti, A., 2016. Chlorpyrifos toxicity in mouse cultured cerebellar granule neurons at different stages of development: additive effect on glutamate-induced excitotoxicity. Cell J. 18, 464–472.
Ambali, S.F., Idris, S.B., Onukak, C., Shittu, M., Ayo, J.O., 2010. Ameliorative effects of vitamin C on short-term sensorimotor and cognitive changes induced by acute chlorpyrifos exposure in Wistar rats. Toxicol. Ind. Health 26, 547–558.
Anavi-Goffer, S., Mulder, J., 2009. The polarised life of the endocannabinoid system in CNS development. Chembiochem 10, 1591–1598.
Atterberry, T.T., Burnett, W.T., Chambers, J.E., 1997. Age-related differences in parathion and chlorpyrifos toxicity in male rats: target and nontarget esterase sensitivity and cytochrome P450-mediated metabolism. Toxicol. Appl. Pharmacol. 147, 411–418.
Attoff, K., Kertika, D., Lundqvist, J., Oredsson, S., Forsby, A., 2016. Acrylamide affects proliferation and differentiation of the neural progenitor cell line C17.2 and the neuroblastoma cell line SH-SY5Y. Toxicol. In Vitro 35, 100–111.
Axelrad, J.C., Howard, C.V., McLean, W.G., 2002. Interactions between pesticides and components of pesticide formulations in an *in vitro* neurotoxicity test. Toxicology 173, 259–268.
Axelrad, J.C., Howard, C.V., McLean, W.G., 2003. The effects of acute pesticide exposure on neuroblastoma cells chronically exposed to diazinon. Toxicology 185, 67–78.
Ballough, G.P., Kan, R.K., Nicholson, J.D., Fath, D.M., Tompkins, C.P., Moffa, G.M., Filbert, M.G., 2008. Brain damage from soman-induced seizures is greatly exacerbated by dimethyl sulfoxide (DMSO): modest neuroprotection by 2-aminoethyl diphenylborinate (2-APB), a transient receptor potential channel inhibitor and inositol 1,4,5-triphosphate receptor antag. J. Med. Chem. Biol. Radiol. Def. 6, 1–21.
Bélanger, M., Magistretti, P.J., 2009. The role of astroglia in neuroprotection. Dialogues Clin. Neurosci. 11, 281–295.
Benke, G.M., Murphy, S.D., 1975. The influence of age on the toxicity and metabolism of methyl parathion and parathion in male and female rats. Toxicol. Appl. Pharmacol. 31, 254–269.
Benowitz, L.I., Routtenberg, A., 1997. GAP-43: an intrinsic determinant of neuronal development and plasticity. Trends Neurosci. 20, 84–91.
Benveniste, E.N., 1998. Cytokine actions in the central nervous system. Cytokine Growth Factor Rev. 9, 259–275.
Berge, G.N., Nafstad, I., Fonnum, F., 1986. Prenatal effects of trichlorfon on the guinea pig brain. Arch. Toxicol. 59, 30–35.
Betancourt, A.M., Carr, R.L., 2003. The effect of chlorpyrifos and chlorpyrifos-oxon on brain cholinesterase, muscarinic receptor binding, and neurotrophin levels in rats following early postnatal exposure. Toxicol. Sci. 77, 63–71.
Betancourt, A.M., Burgess, S.C., Carr, R.L., 2006. Effect of developmental exposure to chlorpyrifos on the expression of neurotrophin growth factors and cell-specific markers in neonatal rat brain. Toxicol. Sci. 92, 500–506.
Bigbee, J.W., Sharma, K.V., Gupta, J.J., Dupree, J.L., 1999. Morphogenic role for acetylcholinesterase in axonal outgrowth during neural development. Environ. Health Perspect. 107, 81–87.
Brimijoin, S., Koenigsberger, C., 1999. Cholinesterases in neural development: new findings and toxicologic implications. Environ. Health Perspect. 107 (Suppl. 1), 59–64.
Brodeur, J., DuBois, K.P., 1963. Comparison of acute toxicity of anticholinesterase insecticides to weanling and adult male rats. Proc. Soc. Exp. Biol. Med. 114, 509–511.
Buratti, F.M., De Angelis, G., Ricceri, L., Venerosi, A., Calamandrei, G., Testai, E., 2011. Foetal and neonatal exposure to chlorpyrifos: biochemical and metabolic alterations in the mouse liver at different developmental stages. Toxicology 280, 98–108.

Burke, R.D., Todd, S.W., Lumsden, E., Mullins, R.J., Mamczarz, J., Fawcett, W.P., Gullapalli, R.P., Randall, W.R., Pereira, E.F.R., Albuquerque, E.X., 2017. Developmental neurotoxicity of the organophosphorus insecticide chlorpyrifos: from clinical findings to preclinical models and potential mechanisms. J. Neurochem. 142, 162–177.

Burns, C.J., McIntosh, L.J., Mink, P.J., Jurek, A.M., Li, A.A., 2013. Pesticide exposure and neurodevelopmental outcomes: review of the epidemiologic and animal studies. J. Toxicol. Environ. Health B Crit. Rev. 16, 127–283.

Campbell, C.G., Seidler, F.J., Slotkin, T.A., 1997. Chlorpyrifos interferes with cell development in rat brain regions. Brain Res. Bull. 43, 179–189.

Canales-Aguirre, A.A., Gomez-Pinedo, U.A., Luquin, S., Ramírez-Herrera, M.A., Mendoza-Magaña, M.L., Feria-Velasco, A., 2012. Curcumin protects against the oxidative damage induced by the pesticide parathion in the hippocampus of the rat brain. Nutr. Neurosci. 15, 62–69.

Carr, R.L., Nail, C.A., 2008. Effect of different administration paradigms on cholinesterase inhibition following repeated chlorpyrifos exposure in late preweanling rats. Toxicol. Sci. 106, 186–192.

Carr, R.L., Borazjani, A., Ross, M.K., 2011. Effect of developmental chlorpyrifos exposure, on endocannabinoid metabolizing enzymes, in the brain of juvenile rats. Toxicol. Sci. 122, 112–120.

Carr, R.L., Adams, A.L., Kepler, D.R., Ward, A.B., Ross, M.K., 2013. Induction of endocannabinoid levels in juvenile rat brain following developmental chlorpyrifos exposure. Toxicol. Sci. 135, 193–201.

Carr, R.L., Graves, C.A., Mangum, L.C., Nail, C.A., Ross, M.K., 2014. Low level chlorpyrifos exposure increases anandamide accumulation in juvenile rat brain in the absence of brain cholinesterase inhibition. Neurotoxicology 43, 82–89.

Carr, R.L., de Leon, K.A., Loyant, L., Mohammed, A.N., Nail, C.A., 2015. Juvenile rat emotional behavior and social play are altered by preweanling inhibitors of FAAH. Toxicologist 144, 2719.

Carr, R.L., Armstrong, N.H., Buchanan, A.T., Eells, J.B., Mohammed, A.N., Ross, M.K., Nail, C.A., 2017. Decreased anxiety in juvenile rats following exposure to low levels of chlorpyrifos during development. Neurotoxicology 59, 183–190.

Chakraborti, T.K., Farrar, J.D., Pope, C.N., 1993. Comparative neurochemical and neurobehavioral effects of repeated chlorpyrifos exposures in young and adult rats. Pharmacol. Biochem. Behav. 46, 219–224.

Chanda, S.M., Pope, C.N., 1996. Neurochemical and neurobehavioral effects of repeated gestational exposure to chlorpyrifos in maternal and developing rats. Pharmacol. Biochem. Behav. 53, 771–776.

Chanda, S.M., Harp, P., Liu, J., Pope, C.N., 1995. Comparative developmental and maternal neurotoxicity following acute gestational exposure to chlorpyrifos in rats. J. Toxicol. Environ. Health 44, 189–202.

Chen, T.-H., Wang, Y.-H., Wu, Y.-H., 2011a. Developmental exposures to ethanol or dimethylsulfoxide at low concentrations alter locomotor activity in larval zebrafish: implications for behavioral toxicity bioassays. Aquat. Toxicol. 102, 162–166.

Chen, T., Yang, W., Li, Y., Chen, X., Xu, S., 2011b. Mono-(2-ethylhexyl) phthalate impairs neurodevelopment: inhibition of proliferation and promotion of differentiation in PC12 cells. Toxicol. Lett. 201, 34–41.

Chen, X.-P., Wang, X., Dong, J.-Y., 2011c. Different reaction patterns of dopamine content to prenatal exposure to chlorpyrifos in different periods. J. Appl. Toxicol. 31, 355–359.

Chiou, J.Y., Westhead, E.W., 1992. Okadaic acid, a protein phosphatase inhibitor, inhibits nerve growth factor-directed neurite outgrowth in PC12 cells. J. Neurochem. 59, 1963–1966.

Choleris, E., Clipperton-Allen, A.E., Phan, A., Kavaliers, M., 2009. Neuroendocrinology of social information processing in rats and mice. Front. Neuroendocrinol. 30, 442–459.
Coban, F.K., Ince, S., Kucukkurt, I., Demirel, H.H., Hazman, O., 2015. Boron attenuates malathion-induced oxidative stress and acetylcholinesterase inhibition in rats. Drug Chem. Toxicol. 38, 391–399.
Crowder, L.A., Lanzaro, G.C., Whitson, R.S., 1980. Behavioral effects of methyl parathion and toxaphene exposure in rats. J. Environ. Sci. Health B 15, 365–378.
Crumpton, T.L., Seidler, F.J., Slotkin, T.A., 2000. Is oxidative stress involved in the developmental neurotoxicity of chlorpyrifos? Dev. Brain Res. 121, 189–195.
da Silva, A.P., Meotti, F.C., Santos, A.R.S., Farina, M., 2006. Lactational exposure to malathion inhibits brain acetylcholinesterase in mice. Neurotoxicology 27, 1101–1105.
Daftary, S.S., Boudaba, C., Szabó, K., Tasker, J.G., 1998. Noradrenergic excitation of magnocellular neurons in the rat hypothalamic paraventricular nucleus via intranuclear glutamatergic circuits. J. Neurosci. 18, 10619–10628.
Dam, K., Seidler, F.J., Slotkin, T.A., 1998. Developmental neurotoxicity of chlorpyrifos: delayed targeting of DNA synthesis after repeated administration. Dev. Brain Res. 108, 39–45.
Dam, K., Garcia, S.J., Seidler, F.J., Slotkin, T.A., 1999. Neonatal chlorpyrifos exposure alters synaptic development and neuronal activity in cholinergic and catecholaminergic pathways. Dev. Brain Res. 116, 9–20.
Dam, K., Seidler, F.J., Slotkin, T.A., 2000. Chlorpyrifos exposure during a critical neonatal period elicits gender-selective deficits in the development of coordination skills and locomotor activity. Dev. Brain Res. 121, 179–187.
Das, K.P., Barone, S., 1999. Neuronal differentiation in PC12 cells is inhibited by chlorpyrifos and its metabolites: is acetylcholinesterase inhibition the site of action? Toxicol. Appl. Pharmacol. 160, 217–230.
De Angelis, S., Tassinari, R., Maranghi, F., Eusepi, A., Di Virgilio, A., Chiarotti, F., Ricceri, L., Pesciolini, A.V., Gilardi, E., Moracci, G., Calamandrei, G., Olivieri, A., Mantovani, A., 2009. Developmental exposure to chlorpyrifos induces alterations in thyroid and thyroid hormone levels without other toxicity signs in CD1 mice. Toxicol. Sci. 108, 311–319.
De Felice, A., Venerosi, A., Ricceri, L., Sabbioni, M., Scattoni, M.L., Chiarotti, F., Calamandrei, G., 2014. Sex-dimorphic effects of gestational exposure to the organophosphate insecticide chlorpyrifos on social investigation in mice. Neurotoxicol. Teratol. 46, 32–39.
De Felice, A., Scattoni, M.L., Ricceri, L., Calamandrei, G., 2015. Prenatal exposure to a common organophosphate insecticide delays motor development in a mouse model of idiopathic autism. PLoS One 10, e0121663.
Dell'Anna, M.E., Geloso, M.C., Draisci, G., Luthman, J., 1995. Transient changes in Fos and GFAP immunoreactivity precede neuronal loss in the rat hippocampus following neonatal anoxia. Exp. Neurol. 131, 144–156.
Diggory, H.J., Landrigan, P.J., Latimer, K.P., Ellington, A.C., Kimbrough, R.D., Liddle, J.A., Cline, R.E., Smrek, A.L., 1977. Fatal parathion poisoning caused by contamination of flour in international commerce. Am. J. Epidemiol. 106, 145–153.
Durham, W.F., 1967. The interaction of pesticides with other factors. In: Gunther, F.A. (Ed.), Residue Reviews: Residues of Pesticides and Other Foreign Chemicals in Foods and Feeds. In: vol. 18. Springer New York, New York, NY, pp. 21–103.
Dvergsten, C., Meeker, R.B., 1994. Muscarinic cholinergic receptor regulation and acetylcholinesterase inhibition in response to insecticide exposure during development. Int. J. Dev. Neurosci. 12, 63–75.
Eells, J.B., Brown, T., 2009. Repeated developmental exposure to chlorpyrifos and methyl parathion causes persistent alterations in nicotinic acetylcholine subunit mRNA

expression with chlorpyrifos altering dopamine metabolite levels. Neurotoxicol. Teratol. 31, 98–103.

Eng, L.F., Ghirnikar, R.S., Lee, Y.L., 2000. Glial fibrillary acidic protein: GFAP-thirty-one years (1969-2000). Neurochem. Res. 25, 1439–1451.

Eskenazi, B., Marks, A.R., Bradman, A., Harley, K., Barr, D.B., Johnson, C., Morga, N., Jewell, N.P., 2007. Organophosphate pesticide exposure and neurodevelopment in young Mexican-American children. Environ. Health Perspect. 115, 792–798.

Eskenazi, B., Rosas, L.G., Marks, A.R., Bradman, A., Harley, K., Holland, N., Johnson, C., Fenster, L., Barr, D.B., 2008. Pesticide toxicity and the developing brain. Basic Clin. Pharmacol. Toxicol. 102, 228–236.

Fitzgerald, B.B., Costa, L.G., 1993. Modulation of muscarinic receptors and acetylcholinesterase activity in lymphocytes and in brain areas following repeated organophosphate exposure in rats. Fundam. Appl. Toxicol. 20, 210–216.

Flaskos, J., 2014. The neuronal cytoskeleton as a potential target in the developmental neurotoxicity of organophosphorothionate insecticides. Basic Clin. Pharmacol. Toxicol. 115, 201–208.

Flaskos, J., McLean, W.G., Hargreaves, A.J., 1994. The toxicity of organophosphate compounds towards cultured PC12 cells. Toxicol. Lett. 70, 71–76.

Flaskos, J., Harris, W., Sachana, M., Muñoz, D., Tack, J., Hargreaves, A.J., 2007. The effects of diazinon and cypermethrin on the differentiation of neuronal and glial cell lines. Toxicol. Appl. Pharmacol. 219, 172–180.

Flaskos, J., Nikolaidis, E., Harris, W., Sachana, M., Hargreaves, A.J., 2011. Effects of sublethal neurite outgrowth inhibitory concentrations of chlorpyrifos oxon on cytoskeletal proteins and acetylcholinesterase in differentiating N2a cells. Toxicol. Appl. Pharmacol. 256, 330–336.

Gabor, C.S., Phan, A., Clipperton-Allen, A.E., Kavaliers, M., Choleris, E., 2012. Interplay of oxytocin, vasopressin, and sex hormones in the regulation of social recognition. Behav. Neurosci. 126, 97–109.

Gagne, J., Brodeur, J., 1972. Metabolic studies on the mechanisms of increased susceptibility of weaning rats to parathion. Can. J. Physiol. Pharmacol. 50, 902–915.

Gao, J., Naughton, S.X., Wulff, H., Singh, V., Beck, W.D., Magrane, J., Thomas, B., Kaidery, N.A., Hernandez, C.M., Terry, A.V., 2016. Diisopropylfluorophosphate impairs the transport of membrane-bound organelles in rat cortical axons. J. Pharmacol. Exp. Ther. 356, 645–655.

Gao, J., Naughton, S.X., Beck, W.D., Hernandez, C.M., Wu, G., Wei, Z., Yang, X., Bartlett, M.G., Terry, A.V., 2017. Chlorpyrifos and chlorpyrifos oxon impair the transport of membrane bound organelles in rat cortical axons. Neurotoxicology 62, 111–123.

Garcia, S.J., Seidler, F.J., Crumpton, T.L., Slotkin, T.A., 2001. Does the developmental neurotoxicity of chlorpyrifos involve glial targets? Macromolecule synthesis, adenylyl cyclase signaling, nuclear transcription factors, and formation of reactive oxygen in C6 glioma cells. Brain Res. 891, 54–68.

Garcia, S.J., Seidler, F.J., Qiao, D., Slotkin, T.A., 2002. Chlorpyrifos targets developing glia: effects on glial fibrillary acidic protein. Dev. Brain Res. 133, 151–161.

Gearhart, D., Sickles, D., Buccafusco, J., Prendergast, M.A., Terry, A.V., 2007. Chlorpyrifos, chlorpyrifos-oxon, and diisopropylfluorophosphate inhibit kinesin-dependent microtubule motility. Toxicol. Appl. Pharmacol. 218, 20–29.

Geraldo, S., Gordon-Weeks, P.R., 2009. Cytoskeletal dynamics in growth-cone steering. J. Cell Sci. 122, 3595–3604.

Geter, D.R., Kan, H.L., Lowe, E.R., Rick, D.L., Charles, G.D., Gollapudi, B.B., Mattsson, J.L., 2008. Investigations of oxidative stress, antioxidant response, and protein binding in chlorpyrifos exposed rat neuronal PC12 cells. Toxicol. Mech. Methods 18, 17–23.

Gordon-Weeks, P.R., Fischer, I., 2000. MAP1B expression and microtubule stability in growing and regenerating axons. Microsc. Res. Tech. 48, 63–74.
Gordon-Weeks, P.R., Fournier, A.E., 2014. Neuronal cytoskeleton in synaptic plasticity and regeneration. J. Neurochem. 129, 206–212.
Grigoryan, H., Lockridge, O., 2009. Nanoimages show disruption of tubulin polymerization by chlorpyrifos oxon: implications for neurotoxicity. Toxicol. Appl. Pharmacol. 240, 143–148.
Grigoryan, H., Schopfer, L.M., Thompson, C.M., Terry, A.V., Masson, P., Lockridge, O., 2008. Mass spectrometry identifies covalent binding of soman, sarin, chlorpyrifos oxon, diisopropyl fluorophosphate, and FP-biotin to tyrosines on tubulin: a potential mechanism of long term toxicity by organophosphorus agents. Chem. Biol. Interact. 175, 180–186.
Grigoryan, H., Schopfer, L.M., Peeples, E.S., Duysen, E.G., Grigoryan, M., Thompson, C.M., Lockridge, O., 2009. Mass spectrometry identifies multiple organophosphorylated sites on tubulin. Toxicol. Appl. Pharmacol. 240, 149–158.
Grisaru, D., Sternfeld, M., Eldor, A., Glick, D., Soreq, H., 1999. Structural roles of acetylcholinesterase variants in biology and pathology. Eur. J. Biochem. 264, 672–686.
Guo-Ross, S.X., Chambers, J.E., Meek, E.C., Carr, R.L., 2007. Altered muscarinic acetylcholine receptor subtype binding in neonatal rat brain following exposure to chlorpyrifos or methyl parathion. Toxicol. Sci. 100, 118–127.
Gupta, R.C., Rech, R.H., Lovell, K.L., Welsch, F., Thornburg, J.E., 1985. Brain cholinergic, behavioral, and morphological development in rats exposed in utero to methylparathion. Toxicol. Appl. Pharmacol. 77, 405–413.
Gurtovenko, A.A., Anwar, J., 2007. Modulating the structure and properties of cell membranes: the molecular mechanism of action of dimethyl sulfoxide. J. Phys. Chem. B 111, 10453–10460.
Hallare, A.V., Köhler, H.-R., Triebskorn, R., 2004. Developmental toxicity and stress protein responses in zebrafish embryos after exposure to diclofenac and its solvent, DMSO. Chemosphere 56, 659–666.
Hanslick, J.L., Lau, K., Noguchi, K.K., Olney, J.W., Zorumski, C.F., Mennerick, S., Farber, N.B., 2009. Dimethyl sulfoxide (DMSO) produces widespread apoptosis in the developing central nervous system. Neurobiol. Dis. 34, 1–10.
Harbison, R.D., 1975. Comparative toxicity of some selected pesticides in neonatal and adult rats. Toxicol. Appl. Pharmacol. 32, 443–446.
Hardwick, L.J.A., Ali, F.R., Azzarelli, R., Philpott, A., 2015. Cell cycle regulation of proliferation versus differentiation in the central nervous system. Cell Tissue Res. 359, 187–200.
Harkany, T., Keimpema, E., Barabás, K., Mulder, J., 2008. Endocannabinoid functions controlling neuronal specification during brain development. Mol. Cell. Endocrinol. 286, S84–S90.
Harvey, M.J., Sharma, R.P., 1980. Organophosphate cytotoxicity: the effects on protein metabolism in cultured neuroblastoma cells. J. Environ. Pathol. Toxicol. 3, 423–436.
He, Q., Dent, E.W., Meiri, K.F., 1997. Modulation of actin filament behavior by GAP-43 (neuromodulin) is dependent on the phosphorylation status of serine 41, the protein kinase C site. J. Neurosci. 17, 3515–3524.
Hernandez, C.M., Beck, W.D., Naughton, S.X., Poddar, I., Adam, B.-L., Yanasak, N., Middleton, C., Terry, A.V., 2015. Repeated exposure to chlorpyrifos leads to prolonged impairments of axonal transport in the living rodent brain. Neurotoxicology 47, 17–26.
Ho, S.S.N., Chow, B.K.C., Yung, W.-H., 2007. Serotonin increases the excitability of the hypothalamic paraventricular nucleus magnocellular neurons. Eur. J. Neurosci. 25, 2991–3000.

Hong, M.S., Hong, S.J., Barhoumi, R., Burghardt, R.C., Donnelly, K.C., Wild, J.R., Venkatraj, V., Tiffany-Castiglioni, E., 2003. Neurotoxicity induced in differentiated SK-N-SH-SY5Y human neuroblastoma cells by organophosphorus compounds. Toxicol. Appl. Pharmacol. 186, 110–118.

Howard, A.S., Bucelli, R., Jett, D.A., Bruun, D., Yang, D., Lein, P.J., 2005. Chlorpyrifos exerts opposing effects on axonal and dendritic growth in primary neuronal cultures. Toxicol. Appl. Pharmacol. 207, 112–124.

Huttner, W.B., Kosodo, Y., 2005. Symmetric versus asymmetric cell division during neurogenesis in the developing vertebrate central nervous system. Curr. Opin. Cell Biol. 17, 648–657.

Ikonomidou, C., Kaindl, A.M., 2011. Neuronal death and oxidative stress in the developing brain. Antioxid. Redox Signal. 14, 1535–1550.

Jett, D.A., Navoa, R.V., 2000. *In vitro* and *in vivo* effects of chlorpyrifos on glutathione peroxidase and catalase in developing rat brain. Neurotoxicology 21, 141–145.

Jett, D.A., Navoa, R.V., Beckles, R.A., McLemore, G.L., 2001. Cognitive function and cholinergic neurochemistry in weanling rats exposed to chlorpyrifos. Toxicol. Appl. Pharmacol. 174, 89–98.

Jiang, W., Duysen, E.G., Hansen, H., Shlyakhtenko, L., Schopfer, L.M., Lockridge, O., 2010. Mice treated with chlorpyrifos or chlorpyrifos oxon have organophosphorylated tubulin in the brain and disrupted microtubule structures, suggesting a role for tubulin in neurotoxicity associated with exposure to organophosphorus agents. Toxicol. Sci. 115, 183–193.

Johnson, D.E., Seidler, F.J., Slotkin, T.A., 1998. Early biochemical detection of delayed neurotoxicity resulting from developmental exposure to chloropyrifos. Brain Res. Bull. 45, 143–147.

Jurewicz, J., Hanke, W., 2008. Prenatal and childhood exposure to pesticides and neurobehavioral development: review of epidemiological studies. Int. J. Occup. Med. Environ. Health 21, 121–132.

Kanagaratnam, K., Wong, H.B., Tan, K.H., 1960. Parathion poisoning from contaminated barley. Lancet 1, 538–542.

Knoblich, J.A., 2008. Mechanisms of asymmetric stem cell division. Cell 132, 583–597.

Lan, A., Kalimian, M., Amram, B., Kofman, O., 2017. Prenatal chlorpyrifos leads to autism-like deficits in C57Bl6/J mice. Environ. Health 16, 43.

Landrigan, P.J., 2010. What causes autism? Exploring the environmental contribution. Curr. Opin. Pediatr. 22, 219–225.

Lassiter, T.L., Padilla, S., Mortensen, S.R., Chanda, S.M., Moser, V.C., Barone, S., 1998. Gestational exposure to chlorpyrifos: apparent protection of the fetus? Toxicol. Appl. Pharmacol. 152, 56–65.

Lauder, J.M., Schambra, U.B., 1999. Morphogenetic roles of acetylcholine. Environ. Health Perspect. 107, 65–69.

Lee, J.E., Park, J.H., Shin, I.C., Koh, H.C., 2012. Reactive oxygen species regulated mitochondria-mediated apoptosis in PC12 cells exposed to chlorpyrifos. Toxicol. Appl. Pharmacol. 263, 148–162.

Lehotzky, K., Szeberenyi, M.J., Kiss, A., 1989. Behavioral consequences of prenatal exposure to the organophosphate insecticide sumithion. Neurotoxicol. Teratol. 11, 321–324.

Levin, E.D., Timofeeva, O.A., Yang, L., Petro, A., Ryde, I.T., Wrench, N., Seidler, F.J., Slotkin, T.A., 2010. Early postnatal parathion exposure in rats causes sex-selective cognitive impairment and neurotransmitter defects which emerge in aging. Behav. Brain Res. 208, 319–327.

Li, W., Casida, J.E., 1998. Organophosphorus neuropathy target esterase inhibitors selectively block outgrowth of neurite-like and cell processes in cultured cells. Toxicol. Lett. 98, 139–146.

Li, D.-P., Pan, Y.-Z., Pan, H.-L., 2001. Acetylcholine attenuates synaptic GABA release to supraoptic neurons through presynaptic nicotinic receptors. Brain Res. 920, 151–158.
Li, A.A., Lowe, K.A., McIntosh, L.J., Mink, P.J., 2012. Evaluation of epidemiology and animal data for risk assessment: chlorpyrifos developmental neurobehavioral outcomes. J. Toxicol. Environ. Health Part B 15, 109–184.
Lian, G., Sheen, V.L., 2015. Cytoskeletal proteins in cortical development and disease: actin associated proteins in periventricular heterotopia. Front. Cell. Neurosci. 9, 99.
Lima, C.S., Dutra-Tavares, A.C., Nunes, F., Nunes-Freitas, A.L., Ribeiro-Carvalho, A., Filgueiras, C.C., Manhães, A.C., Meyer, A., Abreu-Villaça, Y., 2013. Methamidophos exposure during the early postnatal period of mice: immediate and late-emergent effects on the cholinergic and serotonergic systems and behavior. Toxicol. Sci. 134, 125–139.
Liu, J., Olivier, K., Pope, C.N., 1999. Comparative neurochemical effects of repeated methyl parathion or chlorpyrifos exposures in neonatal and adult rats. Toxicol. Appl. Pharmacol. 158, 186–196.
Liu, B., Teschemacher, A.G., Kasparov, S., 2017. Neuroprotective potential of astroglia. J. Neurosci. Res. 95, 2126–2139.
Long, G.G., Scheidt, A.B., Everson, R.J., Oehme, F.W., 1986. Age related susceptibility of newborn pigs to the cutaneous application of chlorpyrifos. Vet. Hum. Toxicol. 28, 297–299.
Lu, F.C., Jessup, D.C., Lavallee, A., 1965. Toxicity of pesticides in young versus adult rats. Food Cosmet. Toxicol. 3, 591–596.
Lu, X.T., Ma, Y., Wang, C., Zhang, X.F., Jin, D.Q., Huang, C.J., 2012. Cytotoxicity and DNA damage of five organophosphorus pesticides mediated by oxidative stress in PC12 cells and protection by vitamin E. J. Environ. Sci. Health Part B 47, 445–454.
Luce, V., Fernandez Solari, J., Rettori, V., De Laurentiis, A., 2014. The inhibitory effect of anandamide on oxytocin and vasopressin secretion from neurohypophysis is mediated by nitric oxide. Regul. Pept. 188, 31–39.
Mahaboob Basha, P., Begum, S., Nayeemunnisa, 2001. Methyl parathion induced alterations in monoaminergic system of developing rat pups. Indian. J. Exp. Biol. 39, 276–279.
Marin, O., Valiente, M., Ge, X., Tsai, L.-H., 2010. Guiding neuronal cell migrations. Cold Spring Harb. Perspect. Biol. 2, a001834.
Marsillach, J., Costa, L.G., Furlong, C.E., 2016. Paraoxonase-1 and early-life environmental exposures. Ann. Glob. Health 82, 100–110.
Marty, M.S., Domoradzki, J.Y., Hansen, S.C., Timchalk, C., Bartels, M.J., Mattsson, J.L., 2007. The effect of route, vehicle, and divided doses on the pharmacokinetics of chlorpyrifos and its metabolite trichloropyridinol in neonatal Sprague-Dawley rats. Toxicol. Sci. 100, 360–373.
Maślińska, D., Lewandowska, I., Prokopczyk, J., 1981. Effect of prolonged acetylcholinesterase inhibition on postnatal brain development in rabbit. I. Level of serotonin in different brain regions. Acta Neuropathol. Suppl. 7, 52–55.
McDonald, N., Kuzmiski, J., Naderi, N., Schwab, Y., Pittman, Q., 2008. Endogenous modulators of synaptic transmission: cannabinoid regulation in the supraoptic nucleus. Prog. Brain Res. 170, 129–136.
McIlvain, V.A., Robertson, D.R., Maimone, M.M., McCasland, J.S., 2003. Abnormal thalamocortical pathfinding and terminal arbors lead to enlarged barrels in neonatal GAP-43 heterozygous mice. J. Comp. Neurol. 462, 252–264.
Mehl, A., Schanke, T.M., Torvik, A., Fonnum, F., 2007. The effect of trichlorfon and methylazoxymethanol on the development of Guinea pig cerebellum. Toxicol. Appl. Pharmacol. 219, 128–135.
Mendoza, C.E., 1976. Toxicity and effects of malathion on esterases of suckling albino rats. Toxicol. Appl. Pharmacol. 35, 229–238.

Meyer, A., Seidler, F.J., Cousins, M.M., Slotkin, T.A., 2003. Developmental neurotoxicity elicited by gestational exposure to chlorpyrifos: when is adenylyl cyclase a target? Environ. Health Perspect. 111, 1871–1876.
Meyer, A., Seidler, F.J., Aldridge, J.E., Tate, C.A., Cousins, M.M., Slotkin, T.A., 2004. Critical periods for chlorpyrifos-induced developmental neurotoxicity: alterations in adenylyl cyclase signaling in adult rat brain regions after gestational or neonatal exposure. Environ. Health Perspect. 112, 295–301.
Middlemore-Risher, M.-L., Adam, B.-L., Lambert, N.A., Terry, A.V., 2011. Effects of chlorpyrifos and chlorpyrifos-oxon on the dynamics and movement of mitochondria in rat cortical neurons. J. Pharmacol. Exp. Ther. 339, 341–349.
Mink, P.J., Kimmel, C.A., Li, A.A., 2012. Potential effects of chlorpyrifos on fetal growth outcomes: implications for risk assessment. J. Toxicol. Environ. Health B Crit. Rev. 15, 281–316.
Moreira, E.G., Yu, X., Robinson, J.F., Griffith, W., Hong, S.W., Beyer, R.P., Bammler, T.K., Faustman, E.M., 2010. Toxicogenomic profiling in maternal and fetal rodent brains following gestational exposure to chlorpyrifos. Toxicol. Appl. Pharmacol. 245, 310–325.
Mortensen, S.R., Chanda, S.M., Hooper, M.J., Padilla, S., 1996. Maturational differences in chlorpyrifos-oxonase activity may contribute to age-related sensitivity to chlorpyrifos. J. Biochem. Toxicol. 11, 279–287.
Moser, V.C., 1999. Comparison of aldicarb and methamidophos neurotoxicity at different ages in the rat: behavioral and biochemical parameters. Toxicol. Appl. Pharmacol. 157, 94–106.
Moser, V.C., Padilla, S., 1998. Age- and gender-related differences in the time course of behavioral and biochemical effects produced by oral chlorpyrifos in rats. Toxicol. Appl. Pharmacol. 149, 107–119.
Moser, V.C., Padilla, S., 2011. Esterase metabolism of cholinesterase inhibitors using rat liver *in vitro*. Toxicology 281, 56–62.
Moser, V.C., Padilla, S., 2016. Esterase detoxication of acetylcholinesterase inhibitors using human liver samples *in vitro*. Toxicology 353–354, 11–20.
Mullen, B.R., Khialeeva, E., Hoffman, D.B., Ghiani, C.A., Carpenter, E.M., 2013. Decreased reelin expression and organophosphate pesticide exposure alters mouse behaviour and brain morphology. ASN Neuro 5, e00106.
National Research Council, 1993. Pesticides in the Diets of Infants and Children. National Academy Press, Washington, DC.
Nomura, D.K., Casida, J.E., 2011. Activity-based protein profiling of organophosphorus and thiocarbamate pesticides reveals multiple serine hydrolase targets in mouse brain. J. Agric. Food Chem. 59, 2808–2815.
Nomura, D.K., Blankman, J.L., Simon, G.M., Fujioka, K., Issa, R.S., Ward, A.M., Cravatt, B.F., Casida, J.E., 2008. Activation of the endocannabinoid system by organophosphorus nerve agents. Nat. Chem. Biol. 4, 373–378.
Nurulain, S.M., Szegi, P., Tekes, K., NH Naqvi, S., 2013. Antioxidants in organophosphorus compounds poisoning. Arch. Ind. Hyg. Toxicol. 64, 169–177.
Ohnuma, S., Harris, W.A., 2003. Neurogenesis and the cell cycle. Neuron 40, 199–208.
Padilla, S., Buzzard, J., Moser, V.C., 2000. Comparison of the role of esterases in the differential age-related sensitivity to chlorpyrifos and methamidophos. Neurotoxicology 21, 49–56.
Padilla, S., Sung, H.-J., Moser, V.C., 2004. Further assessment of an *in vitro* screen that may help identify organophosphorus pesticides that are more acutely toxic to the young. J. Toxicol. Environ. Health A 67, 1477–1489.
Parran, D.K., Mundy, W.R., Barone, S., 2001. Effects of methylmercury and mercuric chloride on differentiation and cell viability in PC12 cells. Toxicol. Sci. 59, 278–290.

Petroianu, G.A., 2010. Toxicity of phosphor esters: Willy Lange (1900-1976) and Gerda von Krueger (1907-after 1970). Pharmazie 65, 776–780.

Pizzurro, D.M., Dao, K., Costa, L.G., 2014. Astrocytes protect against diazinon- and diazoxon-induced inhibition of neurite outgrowth by regulating neuronal glutathione. Toxicology 318, 59–68.

Polańska, K., Jurewicz, J., Hanke, W., 2013. Review of current evidence on the impact of pesticides, polychlorinated biphenyls and selected metals on attention deficit/hyperactivity disorder in children. Int. J. Occup. Med. Environ. Health 26, 16–38.

Pope, C.N., Liu, J., 1997. Age-related differences in sensitivity to organophosphorus pesticides. Environ. Toxicol. Pharmacol. 4, 309–314.

Pope, C.N., Chakraborti, T.K., Chapman, M.L., Farrar, J.D., Arthun, D., 1991. Comparison of *in vivo* cholinesterase inhibition in neonatal and adult rats by three organophosphorothioate insecticides. Toxicology 68, 51–61.

Popova, D., Karlsson, J., Jacobsson, S.O.P., 2017. Comparison of neurons derived from mouse P19, rat PC12 and human SH-SY5Y cells in the assessment of chemical- and toxin-induced neurotoxicity. BMC Pharmacol. Toxicol. 18, 42.

Powers, C.M., Badireddy, A.R., Ryde, I.T., Seidler, F.J., Slotkin, T.A., 2010. Silver nanoparticles compromise neurodevelopment in PC12 cells: critical contributions of silver ion, particle size, coating, and composition. Environ. Health Perspect. 119, 37–44.

Prendergast, M.A., Self, R.L., Smith, K.J., Ghayoumi, L., Mullins, M.M., Butler, T.R., Buccafusco, J.J., Gearhart, D.A., Terry, A.V., 2007. Microtubule-associated targets in chlorpyrifos oxon hippocampal neurotoxicity. Neuroscience 146, 330–339.

Qiao, D., Seidler, F.J., Slotkin, T.A., 2001. Developmental neurotoxicity of chlorpyrifos modeled *in vitro*: comparative effects of metabolites and other cholinesterase inhibitors on DNA synthesis in PC12 and C6 cells. Environ. Health Perspect. 109, 909–913.

Qiao, D., Seidler, F.J., Padilla, S., Slotkin, T.A., 2002. Developmental neurotoxicity of chlorpyrifos: what is the vulnerable period? Environ. Health Perspect. 110, 1097–1103.

Qiao, D., Seidler, F.J., Tate, C.A., Cousins, M.M., Slotkin, T.A., 2003. Fetal chlorpyrifos exposure: adverse effects on brain cell development and cholinergic biomarkers emerge postnatally and continue into adolescence and adulthood. Environ. Health Perspect. 111, 536–544.

Qiao, D., Seidler, F.J., Abreu-Villaça, Y., Tate, C.A., Cousins, M.M., Slotkin, T.A., 2004. Chlorpyrifos exposure during neurulation: cholinergic synaptic dysfunction and cellular alterations in brain regions at adolescence and adulthood. Dev. Brain Res. 148, 43–52.

Qiao, D., Seidler, F.J., Slotkin, T.A., 2005. Oxidative mechanisms contributing to the developmental neurotoxicity of nicotine and chlorpyrifos. Toxicol. Appl. Pharmacol. 206, 17–26.

Quistad, G.B., Sparks, S.E., Casida, J.E., 2001. Fatty acid amide hydrolase inhibition by neurotoxic organophosphorus pesticides. Toxicol. Appl. Pharmacol. 173, 48–55.

Quistad, G.B., Nomura, D.K., Sparks, S.E., Segall, Y., Casida, J.E., 2002a. Cannabinoid CB1 receptor as a target for chlorpyrifos oxon and other organophosphorus pesticides. Toxicol. Lett. 135, 89–93.

Quistad, G.B., Sparks, S.E., Segall, Y., Nomura, D.K., Casida, J.E., 2002b. Selective inhibitors of fatty acid amide hydrolase relative to neuropathy target esterase and acetylcholinesterase: toxicological implications. Toxicol. Appl. Pharmacol. 179, 57–63.

Quistad, G.B., Klintenberg, R., Caboni, P., Liang, S.N., Casida, J.E., 2006. Monoacylglycerol lipase inhibition by organophosphorus compounds leads to elevation of brain 2-arachidonoylglycerol and the associated hypomotility in mice. Toxicol. Appl. Pharmacol. 211, 78–83.

Raines, K.W., Seidler, F.J., Slotkin, T.A., 2001. Alterations in serotonin transporter expression in brain regions of rats exposed neonatally to chlorpyrifos. Dev. Brain Res. 130, 65–72.

Rakic, P., 1990. Principles of neural cell migration. Experientia 46, 882–891.
Rauh, V.A., Garfinkel, R., Perera, F.P., Andrews, H.F., Hoepner, L., Barr, D.B., Whitehead, R., Tang, D., Whyatt, R.W., 2006. Impact of prenatal chlorpyrifos exposure on neurodevelopment in the first 3 years of life among inner-city children. Pediatrics 118, e1845–e1859.
Ray, A., Liu, J., Ayoubi, P., Pope, C.N., 2010. Dose-related gene expression changes in forebrain following acute, low-level chlorpyrifos exposure in neonatal rats. Toxicol. Appl. Pharmacol. 248, 144–155.
Reiss, R., Chang, E.T., Richardson, R.J., Goodman, M., 2015. A review of epidemiologic studies of low-level exposures to organophosphorus insecticides in non-occupational populations. Crit. Rev. Toxicol. 45, 531–641.
Rhodes, M.C., Seidler, F.J., Qiao, D., Tate, C.A., Cousins, M.M., Slotkin, T.A., 2004. Does pharmacotherapy for preterm labor sensitize the developing brain to environmental neurotoxicants? Cellular and synaptic effects of sequential exposure to terbutaline and chlorpyrifos in neonatal rats. Toxicol. Appl. Pharmacol. 195, 203–217.
Ricceri, L., Markina, N., Valanzano, A., Fortuna, S., Cometa, M.F., Meneguz, A., Calamandrei, G., 2003. Developmental exposure to chlorpyrifos alters reactivity to environmental and social cues in adolescent mice. Toxicol. Appl. Pharmacol. 191, 189–201.
Ricceri, L., Venerosi, A., Capone, F., Cometa, M.F., Lorenzini, P., Fortuna, S., Calamandrei, G., 2006. Developmental neurotoxicity of organophosphorous pesticides: fetal and neonatal exposure to chlorpyrifos alters sex-specific behaviors at adulthood in mice. Toxicol. Sci. 93, 105–113.
Richardson, J.R., Chambers, J.E., 2003. Neurochemical effects of repeated gestational exposure to chlorpyrifos in developing rats. Toxicol. Sci. 77, 83–90.
Richardson, J.R., Chambers, J.E., 2005. Effects of repeated oral postnatal exposure to chlorpyrifos on cholinergic neurochemistry in developing rats. Toxicol. Sci. 84, 352–359.
Rosecrans, J.A., Johnson, J.H., Tilson, H.A., Hong, J.S., 1984. Hypothalamic-pituitary adrenal (HPAA) axis function in adult Fischer-344 rats exposed during development to neurotoxic chemicals perinatally. Neurobehav. Toxicol. Teratol. 6, 281–288.
Rush, T., Liu, X.Q., Hjelmhaug, J., Lobner, D., 2010. Mechanisms of chlorpyrifos and diazinon induced neurotoxicity in cortical culture. Neuroscience 166, 899–906.
Sachana, M., Flaskos, J., Alexaki, E., Glynn, P., Hargreaves, A.J., 2001. The toxicity of chlorpyrifos towards differentiating mouse N2a neuroblastoma cells. Toxicol. In Vitro 15, 369–372.
Sachana, M., Flaskos, J., Alexaki, E., Hargreaves, A.J., 2003. Inhibition of neurite outgrowth in N2a cells by leptophos and carbaryl: effects on neurofilament heavy chain, GAP-43 and HSP-70. Toxicol. In Vitro 17, 115–120.
Sachana, M., Flaskos, J., Sidiropoulou, E., Yavari, C.A., Hargreaves, A.J., 2008. Inhibition of extension outgrowth in differentiating rat C6 glioma cells by chlorpyrifos and chlorpyrifos oxon: effects on microtubule proteins. Toxicol. In Vitro 22, 1387–1391.
Sackerman, J., Donegan, J.J., Cunningham, C.S., Nguyen, N.N., Lawless, K., Long, A., Benno, R.H., Gould, G.G., 2010. Zebrafish behavior in novel environments: effects of acute exposure to anxiolytic compounds and choice of Danio rerio line. Int. J. Comp. Psychol. 23, 43–61.
Salama, M., Lotfy, A., Fathy, K., Makar, M., El-emam, M., El-gamal, A., El-gamal, M., Badawy, A., Mohamed, W.M.Y., Sobh, M., 2015. Developmental neurotoxic effects of malathion on 3D neurosphere system. Appl. Transl. Genomics 7, 13–18.
Salyha, Y.T., 2013. Chlorpyrifos leads to oxidative stress-induced death of hippocampal cells *in vitro*. Neurophysiology 45, 193–199.
Santos, H., Cintra, W.M., Aracava, Y., Maciel, C.M., Castro, N.G., Albuquerque, E.X., 2004. Spine density and dendritic branching pattern of hippocampal CA1 pyramidal

neurons in neonatal rats chronically exposed to the organophosphate paraoxon. Neurotoxicology 25, 481–494.

Saulsbury, M.D., Heyliger, S.O., Wang, K., Johnson, D.J., 2009. Chlorpyrifos induces oxidative stress in oligodendrocyte progenitor cells. Toxicology 259, 1–9.

Schoenfeld, T.A., McKerracher, L., Obar, R., Vallee, R.B., 1989. MAP 1A and MAP 1B are structurally related microtubule associated proteins with distinct developmental patterns in the CNS. J. Neurosci. 9, 1712–1730.

Selmi, S., El-Fazaa, S., Gharbi, N., 2012. Oxidative stress and cholinesterase inhibition in plasma, erythrocyte and brain of rats' pups following lactational exposure to malathion. Environ. Toxicol. Pharmacol. 34, 753–760.

Sidiropoulou, E., Sachana, M., Flaskos, J., Harris, W., Hargreaves, A.J., Woldehiwet, Z., 2009a. Diazinon oxon interferes with differentiation of rat C6 glioma cells. Toxicol. In Vitro 23, 1548–1552.

Sidiropoulou, E., Sachana, M., Flaskos, J., Harris, W., Hargreaves, A.J., Woldehiwet, Z., 2009b. Diazinon oxon affects the differentiation of mouse N2a neuroblastoma cells. Arch. Toxicol. 83, 373–380.

Sierra, V., Uphouse, L., 1986. Long-term consequences of neonatal exposure to chlordecone. Neurotoxicology 7, 609–621.

Singh, V., Panwar, R., 2014. *In vivo* antioxidative and neuroprotective effect of 4-allyl-2-methoxyphenol against chlorpyrifos-induced neurotoxicity in rat brain. Mol. Cell. Biochem. 388, 61–74.

Slotkin, T.A., Seidler, F.J., 2005. The alterations in CNS serotonergic mechanisms caused by neonatal chlorpyrifos exposure are permanent. Dev. Brain Res. 158, 115–119.

Slotkin, T.A., Seidler, F.J., 2007a. Comparative developmental neurotoxicity of organophosphates *in vivo*: transcriptional responses of pathways for brain cell development, cell signaling, cytotoxicity and neurotransmitter systems. Brain Res. Bull. 72, 232–274.

Slotkin, T.A., Seidler, F.J., 2007b. Prenatal chlorpyrifos exposure elicits presynaptic serotonergic and dopaminergic hyperactivity at adolescence: critical periods for regional and sex-selective effects. Reprod. Toxicol. 23, 421–427.

Slotkin, T.A., Seidler, F.J., 2010. Oxidative stress from diverse developmental neurotoxicants: antioxidants protect against lipid peroxidation without preventing cell loss. Neurotoxicol. Teratol. 32, 124–131.

Slotkin, T.A., Seidler, F.J., 2015. Prenatal nicotine alters the developmental neurotoxicity of postnatal chlorpyrifos directed toward cholinergic systems: better, worse, or just "different?" Brain Res. Bull. 110, 54–67.

Slotkin, T.A., Cousins, M.M., Tate, C.A., Seidler, F.J., 2001. Persistent cholinergic presynaptic deficits after neonatal chlorpyrifos exposure. Brain Res. 902, 229–243.

Slotkin, T.A., Tate, C.A., Cousins, M.M., Seidler, F.J., 2002. Functional alterations in CNS catecholamine systems in adolescence and adulthood after neonatal chlorpyrifos exposure. Dev. Brain Res. 133, 163–173.

Slotkin, T.A., Southard, M.C., Adam, S.J., Cousins, M.M., Seidler, F.J., 2004. α7 nicotinic acetylcholine receptors targeted by cholinergic developmental neurotoxicants: nicotine and chlorpyrifos. Brain Res. Bull. 64, 227–235.

Slotkin, T.A., Oliver, C.A., Seidler, F.J., 2005. Critical periods for the role of oxidative stress in the developmental neurotoxicity of chlorpyrifos and terbutaline, alone or in combination. Dev. Brain Res. 157, 172–180.

Slotkin, T.A., Levin, E.D., Seidler, F.J., 2006a. Comparative developmental neurotoxicity of organophosphate insecticides: effects on brain development are separable from systemic toxicity. Environ. Health Perspect. 114, 746–751.

Slotkin, T.A., Tate, C.A., Ryde, I.T., Levin, E.D., Seidler, F.J., 2006b. Organophosphate insecticides target the serotonergic system in developing rat brain regions: disparate effects of diazinon and parathion at doses spanning the threshold for cholinesterase inhibition. Environ. Health Perspect. 114, 1542–1546.

Slotkin, T.A., Bodwell, B.E., Levin, E.D., Seidler, F.J., 2007a. Neonatal exposure to low doses of diazinon: long-term effects on neural cell development and acetylcholine systems. Environ. Health Perspect. 116, 340–348.

Slotkin, T.A., MacKillop, E.A., Ryde, I.T., Seidler, F.J., 2007b. Ameliorating the developmental neurotoxicity of chlorpyrifos: a mechanisms-based approach in PC12 cells. Environ. Health Perspect. 115, 1306–1313.

Slotkin, T.A., Bodwell, B.E., Ryde, I.T., Levin, E.D., Seidler, F.J., 2008a. Exposure of neonatal rats to parathion elicits sex-selective impairment of acetylcholine systems in brain regions during adolescence and adulthood. Environ. Health Perspect. 116, 1308–1314.

Slotkin, T.A., Ryde, I.T., Levin, E.D., Seidler, F.J., 2008b. Developmental neurotoxicity of low dose diazinon exposure of neonatal rats: effects on serotonin systems in adolescence and adulthood. Brain Res. Bull. 75, 640–647.

Slotkin, T.A., Lassiter, T.L., Ryde, I.T., Wrench, N., Levin, E.D., Seidler, F.J., 2009a. Consumption of a high-fat diet in adulthood ameliorates the effects of neonatal parathion exposure on acetylcholine systems in rat brain regions. Environ. Health Perspect. 117, 916–922.

Slotkin, T.A., Levin, E.D., Seidler, F.J., 2009b. Developmental neurotoxicity of parathion: progressive effects on serotonergic systems in adolescence and adulthood. Neurotoxicol. Teratol. 31, 11–17.

Slotkin, T.A., Wrench, N., Ryde, I.T., Lassiter, T.L., Levin, E.D., Seidler, F.J., 2009c. Neonatal parathion exposure disrupts serotonin and dopamine synaptic function in rat brain regions: modulation by a high-fat diet in adulthood. Neurotoxicol. Teratol. 31, 390–399.

Slotkin, T.A., Skavicus, S., Card, J., Levin, E.D., Seidler, F.J., 2015a. Amelioration strategies fail to prevent tobacco smoke effects on neurodifferentiation: nicotinic receptor blockade, antioxidants, methyl donors. Toxicology 333, 63–75.

Slotkin, T.A., Skavicus, S., Levin, E.D., Seidler, F.J., 2015b. Prenatal nicotine changes the response to postnatal chlorpyrifos: interactions targeting serotonergic synaptic function and cognition. Brain Res. Bull. 111, 84–96.

Slotkin, T.A., Skavicus, S., Seidler, F.J., 2017. Diazinon and parathion diverge in their effects on development of noradrenergic systems. Brain Res. Bull. 130, 268–273.

Smith, S.L., Fishwick, J., McLean, W.G., Edwards, G., Ward, S.A., 1997. Enhanced *in vitro* neurotoxicity of artemisinin derivatives in the presence of haemin. Biochem. Pharmacol. 53, 5–10.

Smith, S.L., Sadler, C.J., Dodd, C.C., Edwards, G., Ward, S.A., Park, B.K., McLean, W.G., 2001. The role of glutathione in the neurotoxicity of artemisinin derivatives *in vitro*. Biochem. Pharmacol. 61, 409–416.

Smith, J.N., Campbell, J.A., Busby-Hjerpe, A.L., Lee, S., Poet, T.S., Barr, D.B., Timchalk, C., 2009. Comparative chlorpyrifos pharmacokinetics via multiple routes of exposure and vehicles of administration in the adult rat. Toxicology 261, 47–58.

Song, X., Seidler, F.J., Saleh, J.L., Zhang, J., Padilla, S., Slotkin, T.A., 1997. Cellular mechanisms for developmental toxicity of chlorpyrifos: targeting the adenylyl cyclase signaling cascade. Toxicol. Appl. Pharmacol. 145, 158–174.

Song, X., Violin, J., Seidler, F.J., Slotkin, T.A., 1998. Modeling the developmental neurotoxicity of chlorpyrifos *in vitro*: macromolecule synthesis in PC12 cells. Toxicol. Appl. Pharmacol. 151, 182–191.

Spyker, J.M., 1975. Assessing the impact of low level chemicals on development: behavioral and latent effects. Fed. Proc. 34, 1835–1844.

Spyker, J.M., Avery, D.L., 1977. Neurobehavioral effects of prenatal exposure to the organophosphate diazinon in mice. J. Toxicol. Environ. Health 3, 989–1002.

Stamper, C.R., Balduini, W., Murphy, S.D., Costa, L.G., 1988. Behavioral and biochemical effects of postnatal parathion exposure in the rat. Neurotoxicol. Teratol. 10, 261–266.

Stanton, M.E., Mundy, W.R., Ward, T., Dulchinos, V., Barry, C.C., 1994. Time-dependent effects of acute chlorpyrifos administration on spatial delayed alternation and cholinergic neurochemistry in weanling rats. Neurotoxicology 15, 201–208.

Strittmatter, S.M., Fankhauser, C., Huang, P.L., Mashimo, H., Fishman, M.C., 1995. Neuronal pathfinding is abnormal in mice lacking the neuronal growth cone protein GAP-43. Cell 80, 445–452.

Sun, T., Ma, T., Ho, I.K., 2003. Differential modulation of muscarinic receptors in the rat brain by repeated exposure to methyl parathion. J. Toxicol. Sci. 28, 427–438.

Tait, S., Ricceri, L., Venerosi, A., Maranghi, F., Mantovani, A., Calamandrei, G., 2009. Long-term effects on hypothalamic neuropeptides after developmental exposure to chlorpyrifos in mice. Environ. Health Perspect. 117, 112–116.

Tang, J., Carr, R.L., Chambers, J.E., 1999. Changes in rat brain cholinesterase activity and muscarinic receptor density during and after repeated oral exposure to chlorpyrifos in early postnatal development. Toxicol. Sci. 51, 265–272.

Tang, J., Carr, R.L., Chambers, J.E., 2003. The effects of repeated oral exposures to methyl parathion on rat brain cholinesterase and muscarinic receptors during postnatal development. Toxicol. Sci. 76, 400–406.

Terry, A.V., Stone, J.D., Buccafusco, J.J., Sickles, D.W., Sood, A., Prendergast, M.A., 2003. Repeated exposures to subthreshold doses of chlorpyrifos in rats: hippocampal damage, impaired axonal transport, and deficits in spatial learning. J. Pharmacol. Exp. Ther. 305, 375–384.

Tian, J., Dai, H., Deng, Y., Zhang, J., Li, Y., Zhou, J., Zhao, M., Zhao, M., Zhang, C., Zhang, Y., Wang, P., Bing, G., Zhao, L., 2015. The effect of HMGB1 on sub-toxic chlorpyrifos exposure-induced neuroinflammation in amygdala of neonatal rats. Toxicology 338, 95–103.

Timofeeva, O.A., Roegge, C.S., Seidler, F.J., Slotkin, T.A., Levin, E.D., 2008a. Persistent cognitive alterations in rats after early postnatal exposure to low doses of the organophosphate pesticide, diazinon. Neurotoxicol. Teratol. 30, 38–45.

Timofeeva, O.A., Sanders, D., Seemann, K., Yang, L., Hermanson, D., Regenbogen, S., Agoos, S., Kallepalli, A., Rastogi, A., Braddy, D., Wells, C., Perraut, C., Seidler, F.J., Slotkin, T.A., Levin, E.D., 2008b. Persistent behavioral alterations in rats neonatally exposed to low doses of the organophosphate pesticide, parathion. Brain Res. Bull. 77, 404–411.

Tortosa, E., Montenegro-Venegas, C., Benoist, M., Härtel, S., González-Billault, C., Esteban, J.A., Avila, J., 2011. Microtubule-associated protein 1B (MAP1B) is required for dendritic spine development and synaptic maturation. J. Biol. Chem. 286, 40638–40648.

Tuler, S.M., Bowen, J.M., 1989. Toxic effects of organophosphates on nerve cell growth and ultrastructure in culture. J. Toxicol. Environ. Health 27, 209–223.

Turner, C., Sawle, A., Fenske, M., Cossins, A., 2012. Implications of the solvent vehicles dimethylformamide and dimethylsulfoxide for establishing transcriptomic endpoints in the zebrafish embryo toxicity test. Environ. Toxicol. Chem. 31, 593–604.

Turu, G., Hunyady, L., 2010. Signal transduction of the CB1 cannabinoid receptor. J. Mol. Endocrinol. 44, 75–85.

Tyl, R.W., 1992. Development and reproductive toxicity of anticholinesterases. In: Ballantyne, B., Marrs, T.C. (Eds.), Clinical and Experimental Toxicology of Organophosphates and Carbamates. Elsevier–Butterworth-Heinemann Ltd, pp. 241–257.

Uphouse, L., Mason, G., Bondy, S., 1982. Comments concerning the use of dimethylsulfoxide as a solvent for studies of chlordecone neurotoxicity. Neurotoxicology 3, 149–154.

Venerosi, A., Calamandrei, G., Ricceri, L., 2006. A social recognition test for female mice reveals behavioral effects of developmental chlorpyrifos exposure. Neurotoxicol. Teratol. 28, 466–471.

Venerosi, A., Cutuli, D., Colonnello, V., Cardona, D., Ricceri, L., Calamandrei, G., 2008. Neonatal exposure to chlorpyrifos affects maternal responses and maternal aggression of female mice in adulthood. Neurotoxicol. Teratol. 30, 468–474.

Venerosi, A., Ricceri, L., Scattoni, M.L., Calamandrei, G., 2009. Prenatal chlorpyrifos exposure alters motor behavior and ultrasonic vocalization in CD-1 mouse pups. Environ. Health 8, 12.

Venerosi, A., Ricceri, L., Rungi, A., Sanghez, V., Calamandrei, G., 2010. Gestational exposure to the organophosphate chlorpyrifos alters social-emotional behaviour and impairs responsiveness to the serotonin transporter inhibitor fluvoxamine in mice. Psychopharmacology 208, 99–107.

Venerosi, A., Tait, S., Stecca, L., Chiarotti, F., De Felice, A., Cometa, M.F., Volpe, M.T., Calamandrei, G., Ricceri, L., 2015. Effects of maternal chlorpyrifos diet on social investigation and brain neuroendocrine markers in the offspring—a mouse study. Environ. Health 14, 32.

Veronesi, B., Pope, C.N., 1990. The neurotoxicity of parathion-induced acetylcholinesterase inhibition in neonatal rats. Neurotoxicology 11, 609–626.

Wei, D., Allsop, S., Tye, K., Piomelli, D., 2017. Endocannabinoid signaling in the control of social behavior. Trends Neurosci. 40, 385–396.

Whitney, K.D., Seidler, F.J., Slotkin, T.A., 1995. Developmental neurotoxicity of chlorpyrifos: cellular mechanisms. Toxicol. Appl. Pharmacol. 134, 53–62.

Wilhelm, J., Vytášek, R., Uhlík, J., Vajner, L., 2016. Oxidative stress in the developing rat brain due to production of reactive oxygen and nitrogen species. Oxidative Med. Cell. Longev. 5057610.

Williams, A.L., DeSesso, J.M., 2014. Gestational/perinatal chlorpyrifos exposure is not associated with autistic-like behaviors in rodents. Crit. Rev. Toxicol. 44, 523–534.

Wilson, C., Muñoz-Palma, E., González-Billault, C., 2017. From birth to death: a role for reactive oxygen species in neuronal development. Semin. Cell Dev. Biol. (in press).

Wu, X., Yang, X., Majumder, A., Swetenburg, R., Goodfellow, F.T., Bartlett, M.G., Stice, S.L., 2017. Astrocytes are protective against chlorpyrifos developmental neurotoxicity in human pluripotent stem cell-derived astrocyte-neuron cocultures. Toxicol. Sci. 157, 410–420.

Yang, D., Howard, A.S., Bruun, D., Ajua-Alemanj, M., Pickart, C., Lein, P.J., 2008. Chlorpyrifos and chlorpyrifos-oxon inhibit axonal growth by interfering with the morphogenic activity of acetylcholinesterase. Toxicol. Appl. Pharmacol. 228, 32–41.

Yilmaz, N., Yilmaz, M., Altuntas, I., 2012. Diazinon-induced brain toxicity and protection by vitamins E plus C. Toxicol. Ind. Health 28, 51–57.

Yuan, A., Rao, M.V., Veeranna, Nixon, R.A., 2012. Neurofilaments at a glance. J. Cell Sci. 125, 3257–3263.

Zhang, H., Liu, J., Pope, C.N., 2002. Age-related effects of chlorpyrifos on muscarinic receptor-mediated signaling in rat cortex. Arch. Toxicol. 75, 676–684.

Zhang, J., Dai, H., Deng, Y., Tian, J., Zhang, C., Hu, Z., Bing, G., Zhao, L., 2015. Neonatal chlorpyrifos exposure induces loss of dopaminergic neurons in young adult rats. Toxicology 336, 17–25.

Zheng, Q., Olivier, K., Won, Y.K., Pope, C.N., 2000. Comparative cholinergic neurotoxicity of oral chlorpyrifos exposures in preweanling and adult rats. Toxicol. Sci. 55, 124–132.

Zurich, M.-G., Honegger, P., Schilter, B., Costa, L.G., Monnet-Tschudi, F., 2004. Involvement of glial cells in the neurotoxicity of parathion and chlorpyrifos. Toxicol. Appl. Pharmacol. 201, 97–104.

CHAPTER SIX

Chronic Nicotine Effects on Memory and Nicotine Self-Administration: Age of Exposure is Key

Edward D. Levin[1]
Department of Psychiatry and Behavioral Sciences, Duke University Medical Center, Box 104790 DUMC, Durham, NC 27710, USA
[1]Corresponding author: e-mail address: edlevin@duke.edu

Contents

1. INTRODUCTION

Nicotine is a simple molecule with a wide variety of behavioral effects. Nicotine acts as an agonist and a desensitizing agent on nicotinic receptors, one of the principal subtypes of cholinergic receptors. Nicotinic receptors are found in many parts of the brain and have a variety of subtypes which partially underlies the variety of nicotine's behavioral effects. However, another key factor in determining the neuropharmacologic and neurotoxic effects of nicotine is the age of the organism when nicotine exposure occurs.

Nicotine is a well-documented neuroteratogen with early life exposure (Levin and Abreu-Villaça, 2018; Slotkin, 2008). Adverse behavioral effects after gestational exposure have been found to persist into adulthood. Adolescent response to nicotine is key in understanding tobacco addiction since

Advances in Neurotoxicology, Volume 2
ISSN 2468-7480
https://doi.org/10.1016/bs.ant.2018.03.003

the great majority of people addicted to tobacco begin use during adolescence. During adulthood nicotine has been demonstrated to produce cognitive enhancement. This may be therapeutically useful but could also potentiate addiction to tobacco. During aging, nicotine has potential therapeutic effects reducing cognitive deficits with Alzheimer's disease and related disorders. But the efficacy of nicotine in aging may be diminished by neural damage to the hippocampus as well as other brain areas. Thus, nicotine produces differential neurotoxic and even therapeutic effects depending on the age at which the exposure occurs.

Nicotine acts on the same nicotinic receptors during prenatal development as in adolescents, adults or during aging, but as the interactions of those receptors with the rest of physiology changes across the lifespan, the toxicity of nicotine changes too. Adverse effects seen after prenatal nicotine exposure, differ from effects seen after exposure later in life during adolescence, adulthood and senescence. This article covers the series of studies we have conducted over the past two decades concerning effects of chronic nicotine exposure on cognitive function and nicotine self-administration over the lifespan of the rat.

2. PRENATAL NICOTINE EXPOSURE

Exposure to nicotine or tobacco during pregnancy is well-characterized to cause persisting perturbations of neurotransmitter systems and behavioral dysfunction (Levin and Abreu-Villaça, 2018; Slotkin, 2008). Recently, we compared the persisting behavioral impacts of gestational exposure of low-dose nicotine exposure modeling exposure to second hand tobacco smoke to the effects of gestational exposure to the complex mixture of chemicals including nicotine at the same dose. Gestational exposure to 0.2 mg/kg/day of nicotine together with the variety of other compounds in tobacco smoke extract which models second hand tobacco smoke exposure causes a significant deficit in novel object recognition (Figs. 1 and 2) (Hall et al., 2016). This dose of nicotine alone also causes a similar impairment on this task.

Information concerning the likely neural systems that are involved in the cognitive effects of gestational nicotine exposure was gathered with studies of drug effects on memory function in rats exposed to nicotine during gestation and their controls. Gestational nicotine exposure significantly alters reactions to the nicotinic antagonist mecamylamine and the noradrenergic

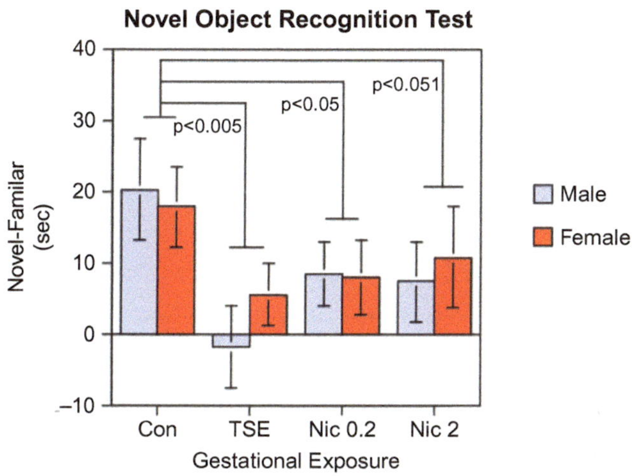

Fig. 1 Prenatal nicotine and tobacco smoke extract effects on novel object recognition (mean ± sem) (Hall et al., 2016).

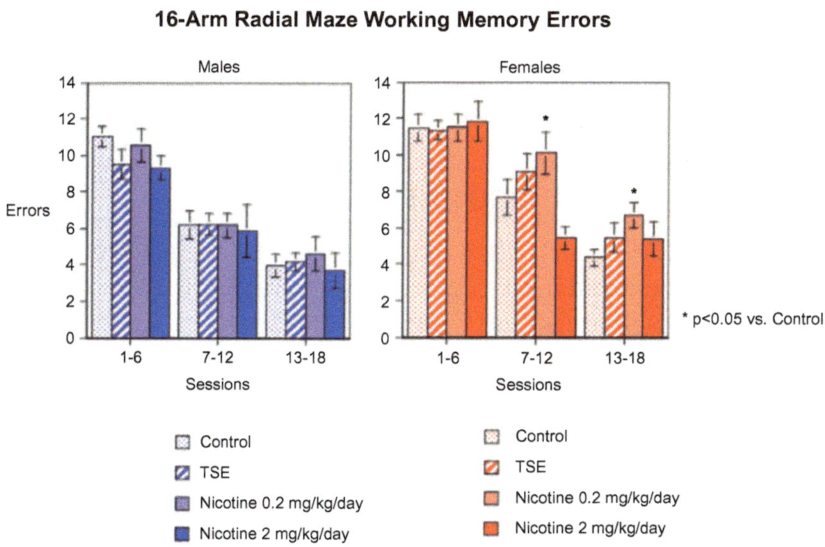

Fig. 2 Prenatal nicotine effects on the 16-arm radial maze choice accuracy (mean ± sem) (Hall et al., 2016).

releasing agent phenylpropanolamine on memory. There are significant sex differences in response to these drug challenges (Levin et al., 1993, 1996).

Prenatal nicotine exposure increases the persistence of nicotine self-administration when the female offspring were adults (Levin et al., 2006a).

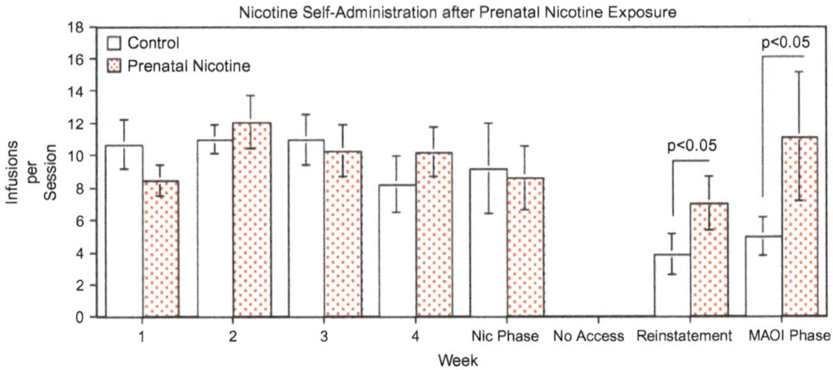

Fig. 3 Prenatal nicotine effects on nicotine self-administration in adult offspring (mean ± sem) (Levin et al., 2006a).

The rats with a history of nicotine exposure during prenatal development had their onset and maintenance of nicotine self-administration that was not different from controls, but when the rats were forced to abstain from nicotine and then given access once again the rats with nicotine exposure during prenatal development maintained a significantly higher level of nicotine self-administration (Fig. 3).

3. ADOLESCENT NICOTINE EXPOSURE

Brain development continues throughout the adolescent period. Reaction to nicotine and regulation of nicotinic receptors is different in adolescents vs. adults (Rezvani and Levin, 2004; Slotkin et al., 2008; Trauth et al., 2000). Adolescent rats have higher levels of self-administration than rats that begin nicotine self-administration in young adulthood. We have shown (Fig. 4) that access of nicotine for self-administration starting at 4–6 weeks of age, adolescence in the rat, results in greater rates of self-administration when compared with young adult rats 8 weeks of age (Levin et al., 2011). This effect is seen in both males and females, but there is a sex difference in response. Males show a greater elevation of nicotine self-administration than females. The persisting effect of adolescent-onset nicotine also differs between males and females. Both sexes decrease their self-administration as they mature from adolescence into adulthood, but female rats more than males show greater persistence of elevated nicotine self-administration into adulthood (Levin et al., 2011).

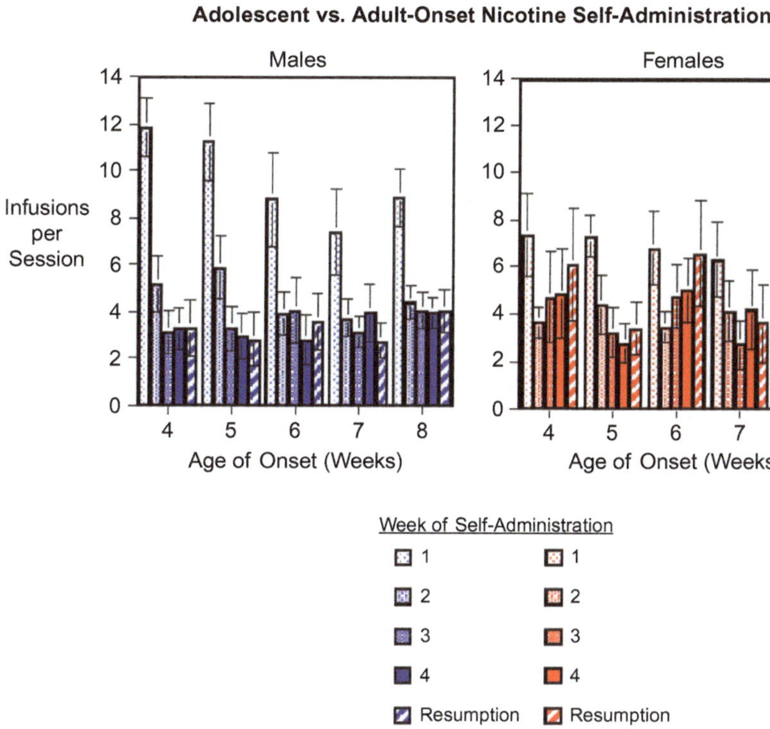

Fig. 4 Nicotine access for self-administration in adolescent and adult rats (mean ± sem) (Levin et al., 2011).

4. ADULT NICOTINE EXPOSURE

Nicotine is essential for the maintenance of tobacco addiction. As such, nicotine in adulthood has clear toxic effects by driving exposure to so many other toxic chemicals. Nicotine, when given alone, does have demonstrable pharmacological effects that improve cognitive function and help attenuate cognitive impairments. For example, people with cognitive impairment such as with attention deficit hyperactivity disorder or schizophrenia smoke at much greater rates and nicotine has been found to improve the cognitive deficits in these groups (Levin et al., 2006b). They may be self-medicating, albeit in a particularly deadly manner. We have and others have demonstrated that adult rats show improvements in cognitive function after nicotine administration in adulthood. As shown in Fig. 5, young adult rats given chronic nicotine infusions have significantly improved working memory

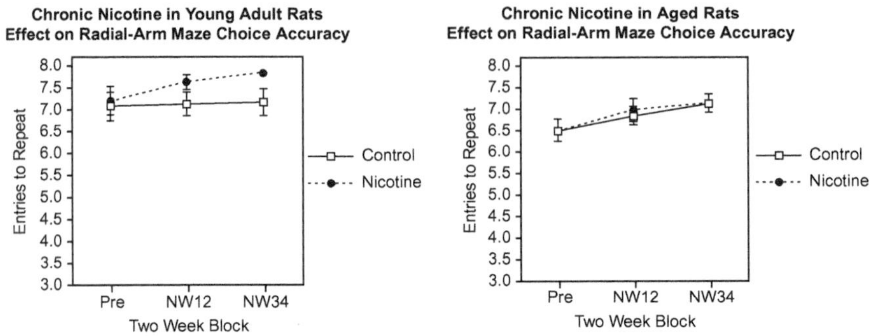

Fig. 5 Chronic nicotine effects on radial-arm maze choice accuracy in young adult and aged rats (mean ± sem) (Levin and Torry, 1996).

accuracy on the 8-arm radial maze relative to vehicle administered controls (Levin and Torry, 1996).

However, enhanced cognitive function could potentiate tobacco addiction in two ways. Some people who have cognitive impairments may have greater tobacco use to provide self-medication, albeit in a deadly package of toxins of tobacco. Additionally, nicotine could potentiate the learning of secondary cue-based reinforcement which is an important driver of nicotine and tobacco self-administration. That is, nicotine may facilitate its own addiction.

5. NICOTINE EXPOSURE DURING AGING

In aged rats the same dose of chronic nicotine infusion does not cause memory improvement (Fig. 5) (Levin and Torry, 1996). It is not impossible to improve cognitive performance in aged rats. It is just more difficult. Other manners of nicotine administration such as higher peak levels with acute injection does improve memory performance in aged rats (Levin and Torry, 1996). This resistance to nicotine induced cognitive performance during aging may be due to neural damage with aging, particularly in the hippocampus, where we have found that neural damage blocks the memory improving effects of nicotine (Levin et al., 1999). We and others have found that nicotine skin patch treatment causes significant improvements in attention and memory in humans as well (Newhouse et al., 2012; White and Levin, 1999, 2004).

6. CONCLUSIONS

Age of exposure is key to the neurotoxic effects of nicotine. Nicotine exposure during gestation causes persisting neurobehavioral impairment. Nicotine availability to self-administration during adolescence causes greater self-administration than starting in adulthood. This increase is greater in males than females. The persistence of this increased nicotine self-administration is more persistent in females. In adults, nicotine has been shown in a variety of studies to significantly improve cognitive function. Therapeutic efficacy of nicotine for aging appears to be more limited than in young adulthood, possibly due to neural damage to the hippocampus and other regions. However, during aging, nicotine has still shown cognitive benefits and may be a useful avenue for developing new therapeutic treatments.

As with all other chemicals that cause toxicity, the dose of exposure is critical to the nature and extent of toxicity. Also key in determining the extent and nature of nicotine toxicity is the age of exposure. This is true not only for nicotine but also for other toxins and toxicants as well. This is a property of the essential fact that toxicity is determined by the biology of the organism which is exposed to the chemical under consideration. When the biology changes as with the process of development and aging, the actions of chemical perturbations of that organism change as well.

ACKNOWLEDGMENTS

This research has been funded by NIDA (DA011943, DA015756, DA027840), NIMH (MH064494), NIEHS (ES02283), NIA (AG022462), US-EPA (RD83543701) and the March of Dimes.

REFERENCES

Hall, B.J., et al., 2016. Cognitive and behavioral impairments evoked by low level exposure to tobacco smoke constituents: comparison with nicotine alone. Toxicol. Sci. 15, 236–244.

Levin, E.D., Abreu-Villaça, Y., 2018. Developmental neurotoxicity of nicotine and tobacco. In: Wang, C., Paule, M., Slikker Jr., W. (Eds.), Handbook of Developmental Neurotoxicology. Academic Press, San Diego, pp. 439–452.

Levin, E.D., Torry, D., 1996. Acute and chronic nicotine effects on working memory in aged rats. Psychopharmacology 123, 88–97.

Levin, E.D., et al., 1993. Prenatal nicotine exposure and cognitive performance in rats. Neurotoxicol. Teratol. 15, 251–260.

Levin, E.D., et al., 1996. Prenatal nicotine effects on memory in rats: pharmacological and behavioral challenges. Dev. Brain Res. 97, 207–215.
Levin, E.D., et al., 1999. Ventral hippocampal ibotenic acid lesions block chronic nicotine-induced spatial working memory improvement in rats. Cogn. Brain Res. 7, 405–410.
Levin, E.D., et al., 2006a. Increased nicotine self-administration following prenatal exposure in female rats. Pharmacol. Biochem. Behav. 85, 669–674.
Levin, E.D., McClernon, F.J., Rezvani, A.H., 2006b. Nicotinic effects on cognitive function: behavioral characterization, pharmacological specification and anatomic localization. Psychopharmacology 184, 523–539.
Levin, E.D., et al., 2011. Threshold of adulthood for the onset of nicotine self-administration in male and female rats. Behav. Brain Res. 225, 473–481.
Newhouse, P.A., et al., 2012. Transdermal nicotine treatment of mild cognitive impairment: a six-month double-blind pilot clinical trial. Neurology 78, 91–101.
Rezvani, A.H., Levin, E.D., 2004. Adolescent and adult rats respond differently to nicotine and alcohol: motor activity and body temperature. Int. J. Dev. Neurosci. 22, 349–354.
Slotkin, T.A., 2008. If nicotine is a developmental neurotoxicant in animal studies, dare we recommend nicotine replacement therapy in pregnant women and adolescents? Neurotoxicol. Teratol. 30, 1–19.
Slotkin, T.A., et al., 2008. Adolescent nicotine treatment changes the response of acetylcholine systems to subsequent nicotine administration in adulthood. Brain Res. Bull. 76, 152–165.
Trauth, J.A., et al., 2000. Modeling adolescent nicotine exposure: effects on cholinergic systems in rat brain regions. Brain Res. 873, 18–25.
White, H.K., Levin, E.D., 1999. Four-week nicotine skin patch treatment effects on cognitive performance in Alzheimer's disease. Psychopharmacology 143, 158–165.
White, H.K., Levin, E.D., 2004. Chronic transdermal nicotine patch treatment effects on cognitive performance in age-associated memory impairment. Psychopharmacology 171, 465–471.

CHAPTER SEVEN

Developmental Neurotoxicity of Endocrine Disruptor Chemicals: A Challenge for Behavioral Toxicology

Gemma Calamandrei[1], Laura Ricceri
Center for Behavioral Sciences and Mental Health, Istituto Superiore di Sanità, Viale Regina Elena 299, I-00161 Rome, Italy
[1]Corresponding author: e-mail address: gemma.calamandrei@iss.it

Contents

1. THE DEVELOPING BRAIN: PLASTICITY AND VULNERABILITY

At variance from other organs, the mammalian brain develops over a very long period extending from the embryonic period through puberty.

Advances in Neurotoxicology, Volume 2
ISSN 2468-7480
https://doi.org/10.1016/bs.ant.2018.03.008

Brain development is made up of several sequential, precisely-timed and interactive events that start as soon as 2 weeks after conception. The formation of the neural tube from the neural plate implies rapid cell proliferation and migration in the proper position. In the human fetus, cell migration is nearly complete in the neocortex and in most of the brain by the sixth month of gestation (Gupta et al., 2005). The remainder of intrauterine development is dedicated to differentiation and formation of connections among cells. The period of rapid synaptogenesis or the so-called brain growth spurt (in humans occurring between the last 3 months of pregnancy and the second postnatal year) is considered one of the most important processes that take place during brain development (Garner et al., 2002). During this period substantially more connections are formed than those that will be eventually retained, as many of them will be gradually discarded by apoptotic mechanisms. This process is crucial not only in neurodevelopment but also for synaptic plasticity, learning, memory, and adaptation throughout life. Without this process, no complex brain network can be established, as synapse is the fundamental unit of connectivity and communication between neurons (Tau and Peterson, 2010). The process of synapse formation and elimination is a normal part of development and both endogenous (temporally-determined expression of genes increasing the levels of neurotrophic factors or hormones) and environmental factors (i.e., nutrients or sensorial inputs) concur to influence the fine matchup between pre- and postsynaptic neurons and ensure the maturation of proper connections between brain areas.

Thus the remarkable plasticity of the brain during the fetal and neonatal periods is a double-edged sword as it renders this organ extremely responsive to environmental factors. On one side, early environmental inputs can interact with genetic factors to promote neural circuit and functions development (Sale et al., 2014); on the other hand, early negative conditions (i.e., stress, toxic insults) can act in the opposite direction, increasing the chances of neurodevelopmental disorders (NDDs) or diseases in adulthood (Kundakovic and Jaric, 2017).

A detailed description of the molecular events underpinning the development of connections in the human central nervous system is beyond the scope of the present chapter (Stiles and Jernigan, 2010). Variation in neuronal activity, hormonal milieu, or environmental cues can permanently change the formation of specific circuits and the phenotype of individual neurons during developmental critical periods. Paradigmatic examples of the complex interplay between the different molecular actors at the boundary between plasticity and vulnerability are the functional specialization of the visual cortex (Berardi et al., 2003) and the development of human language (Neville and Bavelier, 2002).

The budding field of epigenetics offered a new perspective for investigation of gene × environment interaction in development (Fagiolini et al., 2009). Epigenetic mechanisms are related to relevant modifications in the genome that influence gene expression without changes in nucleotide sequence. These modifications alter how the histone proteins interact with DNA to form the chromatin, and may be stable across generations (Bohacek and Mansuy, 2015; Tordjman et al., 2014). They include several processes, such as DNA methylation, histone modifications, different mechanisms of chromatin organizations orchestrated also by noncoding RNAs (Chuang and Jones, 2007), which have the potential to mediate the effects of environmental exposure by switching on and off gene expression.

The role of epigenetic modification in sustaining the effects of environmental experience on neurobehavioral development has been shown in the context of early mother–infant interactions. Individual variations in maternal care in the postpartum period in rats are associated with changes in offspring hypothalamic–pituitary–adrenal (HPA) activity, neuroendocrine systems and hippocampal plasticity (Meaney, 2001). Specifically, high levels of maternal care are associated in offspring with decreased promoter methylation within both the hippocampal glucocorticoid receptor (GR) and hypothalamic estrogen receptor (ERα) genes and this in turn increases gene expression (Champagne et al., 2006). These epigenetic changes are likely mediated by increased binding of nerve growth factor-inducible protein A (NGFI-A) to the GR exon promoter among offspring who receive high levels of care in infancy (Weaver et al., 2007). These effects on GR receptors are long lasting and might underlay the behavioral expression of fearfulness observed in adult rats exposed to different levels of maternal care (Caldji et al., 1998, 2011).

Another example of the continuous interplay between plasticity and vulnerability comes from the effects of stressful/traumatic experiences during early development and early life, which have been related to the risk of developing psychopathologies, cognitive, anxiety and mood disorders. These early life periods profoundly affect development of the brain limbic structures and more importantly the functioning of the HPA axis implicated in stress and coping responses. The expression of brain derived neurotrophic factor (BDNF) gene, a protein involved in neurodevelopment and neuroplasticity has been associated to early-life adversity: rodent models for early-life adversity and depression showed that the epigenetic processes controlling BDNF transcription were sensitive to stress. The BDNF promoter region was hypermethylated with ensuing decreased expression levels in stressed animals (Roth et al., 2009). In this way, adverse psychosocial experience molds synaptic connections and possibly modifies the lifelong individual trajectory.

All this considered, it is clear that when the proper environmental input (i.e., visual stimuli, maternal care, and nutritional factors) reaches the developing brain at the appropriate time it drives harmonic brain and behavior development, whereas diverse environmental stressors—chemical pollutants, drugs, maternal infection, stress, deprivation—may interfere with typical brain developmental trajectories. In the framework of the Developmental Origins of Health and Disease (DOHaD) paradigm proposed originally by Barker et al. (1993), in utero or early-life conditions program lifelong health trajectories. The DOHaD paradigm focuses on organisms' biological plasticity to adjust their phenotype to their environment over the short and long term through epigenetic processes such as DNA methylation. In order to cope with insults perturbing prenatal environment, the fetus develops compensatory strategies that could be adaptive or disruptive. In this perspective mismatches between the pre-/postnatally anticipated and the actual mature environment predisposes organisms to disease (Leenen et al., 2016). Environmental exposures during critical periods of development can permanently reprogram normal physiological responses, thereby increasing susceptibility to disease later in life—a process known as developmental reprogramming (Fig. 1).

2. IMPACT OF ENDOGENOUS HORMONES IN BRAIN AND BEHAVIOR DEVELOPMENT

Among the molecular actors that play a leading role in brain and behavior development, steroid hormones act as powerful regulators of neuronal function also outside the context of reproduction. The fetal brain is exposed to endogenous hormones from both the fetus' own developing reproductive system as well as that of its mother. Hormones play important roles not only in brain development but also in sexual dimorphism in the brain, with changes persisting into adolescence and adulthood during which sexually dimorphic behaviors are expressed (McCarthy, 2016). The discovery of binding sites and receptors for estrogens in the hypothalamus and their influence on reproductive behaviors dates back to the early 1970s, but the turning point

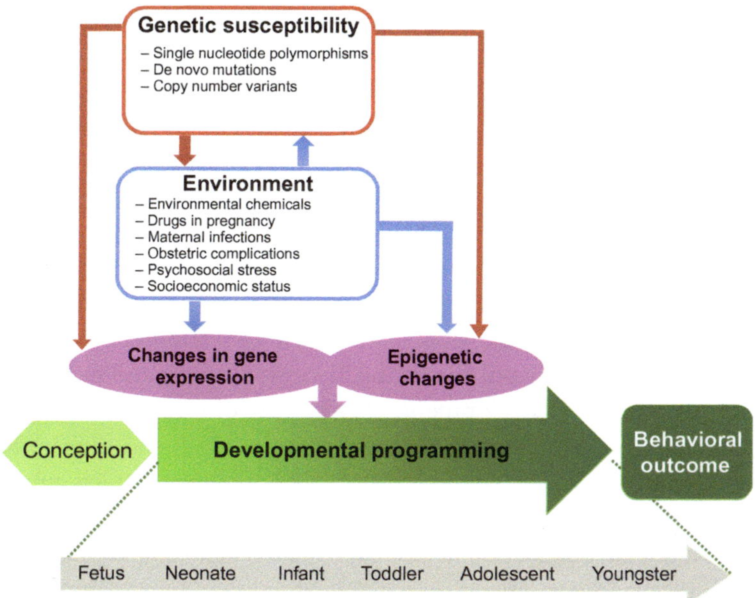

Fig. 1 The complex interaction between genetic susceptibility and diverse environmental stressors interferes with ongoing developmental programming from the fetal stage to young age. Delayed effects on behavior may be mediated by genetic modification, epigenetic events and direct changes in the internal organism environment.

was the finding that steroid hormones receptors were also in the hippocampus (Gerlach and McEwen, 1972), a brain region implicated in learning, memory and stress response. Since then much research in rodent models has demonstrated the wide range of action of steroid hormones in the brain well beyond sex and reproduction, and paved the way for subsequent discovery on the modes and mechanisms of action of hormone in the central nervous system. Today is clear that not all steroid hormones' action in the brain occur via nuclear receptors but rather they operate through cell receptors and other signaling pathways (McEwen et al., 2012).

Mccarthy et al. (2008) have beautifully shown that estradiol modulates neuron and astrocyte growth and differentiation in the developing brain by different mechanisms that involve GABA neurotransmitter, levels of inflammatory mediators and glutamate receptors. In particular, they also showed that the activation of estrogen receptors by estradiol in a nongenomic fashion enhanced glutamate release from presynaptic neurons, subsequent activation of ionotropic glutamate receptors and finally induction of dendritic spine formation. These results revealed a trans-neuronal mechanism by which estradiol acts during a sensitive period to establish a profound and lasting sex difference in hypothalamic synaptic patterning. To date, the programming action of steroids in the brain involves NMDA receptors, glutamate release, cholinergic inhibitory or disinhibitory action on cortical neurons; to further complicate the picture, the different classes of steroids interact with endogenous opioids, hypothalamic neuropeptides, HPA axis and dopaminergic pathways (for extensive review see McEwen and Milner, 2017).

The developmental effects of steroids are typically irreversible and are referred to as "organizational" (McCarthy, 2010). We know now that not only sex steroids but other hormones such as thyroid and hypothalamic peptides are implicated in synaptic formation, dendritic length, and neuronal connectivity throughout development. Severe thyroid hormone deficiency produces severe brain damage (Chen and Hetzel, 2010) and their moderate or even transient insufficiency can cause developmental defects in rodents (Goodman and Gilbert, 2007) and humans (Kooistra, 2006). Gilbert and Sui found that a 28% reduction in circulating levels of T4 in pregnant rats produced significant adverse effects on synaptic function of hippocampal neurons of their adult offspring despite no detected change in serum T4 levels in the pups after birth (Gilbert and Sui, 2008). Thus, even at very small concentrations hormones can induce or participate to permanent sexual dimorphisms in certain brain regions and exert marked effects on the development of the neuroendocrine system and related behaviors.

Sexual dimorphism in brain organization is reflected at the behavioral level: rodent studies have largely shown that there are specific behavioral differences between males and females and they are not limited to reproductive behaviors but include differences in infant play, aggression, learning, exploration, activity level, food intake and preference, and even cognitive capabilities (McCarthy, 2016).

The human brain is also structurally and functionally sexually dimorphic (Cahill, 2006), as sex differences in reproductive and parental behaviors have been selected by the evolutionary process as critical for species survival (Dulac and Kimchi, 2007). In this framework even subtle changes in brain organization and activation from chemical exposures that disrupt this delicate balance can have profound effects on physiology and behaviors. In the last two decades, many chemicals have been classified as endocrine disrupting chemicals (EDCs). They produce their effects by mimicking, antagonizing or altering endogenous hormone levels via changing rates of their synthesis or metabolism and/or expression or action at receptor targets. There is growing concern that disruption of neuroendocrine pathways might be a plausible mechanism by which environmental chemical exposures increase the risk of NDDs in children (Bellanger et al., 2015; Engel and Wolff, 2013).

3. DEVELOPMENTAL NEUROTOXICITY OF LOW-DOSE CHEMICALS

There is vast literature on the developmental neurotoxicity of environmental chemicals that indicates an inverse association between chemical exposure and child neurodevelopment starting by the early 1970s (Bellinger et al., 2016). However it was not until the early 1990s that significant evidence emerged showing that low-dose exposures to metals such as methylmercury and lead, though not producing teratogenic effects evident at birth could result in later behavioral dysfunctions in children (Banks et al., 1997; Gilbert and Grant-Webster, 1995). Similar results were later identified for the perinatal exposure to other environmental contaminants such as pesticides (Eskenazi et al., 2007; Perera et al., 2009), polychlorobiphenyls (Winneke, 2011), polybrominated diphenyl ethers (Herbstman et al., 2010), and phthalates (Whyatt et al., 2012). These observations were consistent with the emergent concept of developmental origin of health and disease, and shifted the emphasis of developmental toxicology from the study of teratogenic effects to functional changes occurring during development and becoming manifest only after a latent period (Bellinger, 2009). Mechanistic studies carried out in animals and in in vitro models point to multiple pathways and targets of toxicity (including hormonal pathways) for several established neurotoxicants, depending on the dosage or time window of exposure (Neal et al., 2011; Roy et al., 2004; Slotkin and Seidler, 2009).

Toxic disruption of early maturational events are more likely to result in neural tube defects or major malformation (i.e., anencephaly, spina bifida), while toxic interference with later developmental events results more commonly in cytoarchitectural and molecular alterations that might be expressed as behavioral dysfunction. Specifically, damage or destruction of neurons by chemical compounds when they are in the process of synapses formation, integration and formation of neural networks will derange the organization and function of these networks, thereby setting the stage for subsequent behavioral dysfunction (Kovacs et al., 2014). Many of the chemical compounds found to affect neurobehavioral development in epidemiological and experimental studies are also endowed with endocrine disrupting activity: they are industrial, agricultural chemicals or therapeutical drugs. For some of them (i.e., atrazine, bisphenol A (BPA), genistein, phthalates, organochlorines, polychlorinated biphenyls) specific interaction with hormone receptors and metabolism has been clearly demonstrated. For others, already established as developmental neurotoxicants (i.e., heavy metals, organophosphates) increasing evidence suggests that interference with neuroendocrine pathways might be a relevant component of their neurobehavioral toxicity (Gore et al., 2014). Over the last two decades, there has been burgeoning scientific evidence based on research in wildlife species, epidemiological data and laboratory studies using cell cultures and animal models that provided insight into how EDCs may cause biological changes in the developing organism (Gore et al., 2015; WHO, 2012). However, as described in the following paragraph, the definition of EDC among scientists and regulators is still a matter of controversy.

4. FIFTY YEARS OF EDC RESEARCH: FROM AVIAN EGGSHELL THINNING TO CHILD NDDs

The scientific understanding of what is an EDC has undergone a remarkable evolution since the 1970s when physicians and researchers began to link the presence of certain chemicals in the environment with rare cancers and reproductive effects in humans and wildlife. During the Rochester Conference held in the United States in 1969 the scientific community debated about the potential health effects over wildlife of pesticides with hormone-like activity. At that time researchers proposed for the first time that reproductive anomalies among avian populations around the Great Lakes could be due to exposure to pesticides like DDT with estrogenic properties (reviewed in Schecter et al., 1994). In the 1970s the discovery of the health effects of prenatal exposure to diethylstilbestrol in humans highlighted the potential for chemicals to cause not only physical anomalies at birth but also more subtle health effects that emerge many years later (McLachlan, 1993). The growing consciousness of harmful environmental chemicals attracted the attention of the scientific community on hormone-like chemicals. Scientists from a variety of fields convened in 1991 at the Wingspread Conference and it was there that the terms "endocrine disruption" and "endocrine disruptors" were firstly coined (Colborn and Clement, 1992), although we should wait until 1999 for a first scientific definition of EDCs by the National Academy of Sciences (Zoeller et al., 2014). In 1992 the publication of the study by Carlsen et al. (1992) that associated falling sperm counts with environmental exposures in Scandinavian men over a 50-year period suggested a growing impact of estrogenic chemicals on male gonadal function and enlightened the alarming perspective of reduced male fertility due to environmental influences. This prompted much research on the effects of hormone-like chemicals on human health: as a consequence during the 1990s and into the early 2000s, the list of EDCs has increased rapidly from a few pesticides (such as DDT, chlordecone, and methoxychlor) and industrial chemicals (such as PCB congeners), as new compounds with estrogenic activity were found among plastics, disinfectants, and personal care products. Studies began to suggest that EDCs could cause health effects even at extremely low doses and show non-monotonic dose response (NMDR) curves. In parallel, epidemiological studies clearly showed that higher levels of PCBs in cord blood were associated with lower neuropsychological scores in children (Engel and Wolff, 2013) extending to neurodevelopment the pervasive health effects of EDCs.

EDCs can interfere with biological processes by mimicking hormones, activating or blocking the body's hormone receptors, disrupting the synthesis of hormones, or altering their degradation. Connections are emerging between EDC and the epigenetic field (Walker and Gore, 2017). The potential implication of EDC chemicals in the pathogenesis of child NDDs has become one of the critical issues in EDC research, in the light of the alarming increase of NDDs with sex-biased prevalence rates worldwide (Colborn, 2004).

4.1 Key Uncertainties and Hot Issues in the Debate on EDCs

Extensive evidence to date shows that EDCs alter developmental programming by interfering with the action of steroid and thyroid hormones. However, hardly has an issue raised so much debate among scientists and regulators as the criteria to identify chemicals that act as EDCs. The definition of EDC has changed over years, in parallel with the increasing evidence on the associated possible risk for human health defined by an Endocrine Society statement of principles: an EDC is "an exogenous chemical, or mixture of chemicals, that interferes with any aspect of hormone action." The Endocrine Society's 2015 scientific statement, EDC-2, further qualified this definition by emphasizing the particular importance of developmental exposures to EDCs as most problematic (Gore et al., 2015).

To date the major controversy is about the endocrinological, pharmacological and toxicological principles that should support scientific criteria for the identification of a chemical as endocrine disruptor. Prominent in this debate were the definition of endocrine function and relative adverse effects, the questionable existence of thresholds for EDCs, and the significance of non-monotonic dose–response relationships. In 2013 two positions have faced up, representing, the community of toxicologists and that of endocrinologists, respectively (Bergman et al., 2015; Dietrich et al., 2013). First of all, the debate has involved the definition of endocrine system, an aspect particularly important for the clarification of the adverse effects of these chemicals. The endocrine system has a fundamental homeostatic role for organisms exposed to a varying environment. Thus, seen in such homeostatic perspective any variation in the endocrine activity may be viewed as an adaptive rather than an adverse effect as supported by Dietrich et al. (2013). However, as stated by Zoeller et al. (2014) the endocrine system also plays essential roles in brain development and sexual differentiation as well as in establishing the set-point for metabolism or stress responses later in life. Within the DOHaD framework, any alteration of the programming role of hormones during prenatal and postnatal development has to be considered as adverse in that it might permanently change the way the organism will respond to environmental challenges later in life.

A second hot issue in the debate concerns the concept of threshold for toxicity. A large body of evidence indicates that EDCs can have effects at doses much lower than what is usually used in toxicological tests, often several orders of magnitude below current reference doses. The toxicology approach assumes a direct relationship between dose and toxicity, namely there will be no effect of a chemical below a given "threshold" of exposure, but there will be effects at doses above. But such a relationship is not the rule in endocrinology: endocrine-signaling pathways that mediate responses to hormones have rather evolved to act as powerful amplifiers (vom Saal and Hughes, 2005; Zoeller et al., 2012). Thus, both endogenous hormones and EDCs can trigger response in the target tissues by binding to a small fraction of available receptors and recruiting co-activators able to amplify the response.

A third issue refers to NMDR shown by many EDCs. For example, they may exhibit an inverted U-shaped dose–response curve, with toxic effects maximal at very low and very high exposure levels (Vandenberg et al., 2012; Villar-Pazos et al., 2017). The existence of NMDR implies that it cannot be always assumed that EDCs which are toxic at high doses are much less risky at lower, environmentally relevant levels. As a result, common concepts in regulatory toxicology cannot be easily applied to EDCs.

As several developmental neurotoxicants, EDCs do not express at environmentally relevant doses their effects in the form of tissue pathology, malformation or clinical disorders but they rather shift the organism's typical patterns of responses to environmental challenges or produce subclinical alterations that may be amplified by later environmental hits. Finally, in common with low-dose exposure to developmental neurotoxicants, the time span between early exposures and the manifestation of disease later in life makes it challenging to trace morbidity due to EDC exposure.

EDC research generated considerable controversies in the scientific, medical, and regulatory policy communities, specifically where basic research intersected with toxicology and risk assessment. EDCs can act simultaneously at the level of the receptor, hormone synthesis, and hormone degradation. An EDC can have estrogenic or antiandrogenic effects, sometimes creating integrated estrogenic signals not predicted by studying each action alone. Further complicating the picture, compounds that alter thyroid signaling can affect the action of other hormones.

Although this level of complexity is familiar in both endocrinology and neuroscience, it does not easily fit into the traditional frameworks of risk assessment and hazard management, which largely rely on calculating a threshold exposure level below which an exposure can be considered "safe." If EDCs interact like hormones, the most sensitive endpoint can change, depending on other endocrine active compounds present in the body, included endogenous hormones.

The problems related to the definition of an EDC for risk assessment purposes are further complicated by the discovery of the hormone-like activity of many classical neurotoxicants, whose safe levels were originally defined on the basis of their acute toxicity in the adult organism. For example, low-dose exposure to metals and organophosphate pesticides during development have been found to produce sex-dependent behavioral effects in humans and animals, and interferences with thyroid hormone levels as well as with key enzymes in steroid metabolism (i.e., aromatase) and hypothalamic neuropeptides. We will show in the following paragraphs that the behavioral and physiological endpoints affected by low-dose exposure to two chemical compounds with very different structure and activity, namely the estrogen mimic BPA and the organophosphate insecticide chlorpyrifos (CPF), partially overlap, possibly due to their convergence on neuroendocrine pathways modulating brain and behavioral development.

5. THE DEVELOPING BRAIN AS A TARGET OF BPA

Perhaps no EDC has received more attention than BPA, a synthetic chemical used in the production of plastics since the 1950s. The estrogenic properties of BPA were first identified in the 1930s but it never found use as a drug. The commercial production of the first epoxy resins containing BPA, used for production of protective coatings on metal equipment (e.g., interior of food cans) began in the early 1950s. From the 1960s, BPA has been used to produce a hard plastic called polycarbonate, and it is now used worldwide in the production of plastics found in numerous commercial products, including laptops, cell phones, baby bottles, water main pipes, laboratory and hospital equipment, and food containers (Vogel, 2009).

The ubiquity of BPA containing products means that there are several potential sources of exposure to this synthetic estrogen. And yet, although BPA's estrogen-like properties were never completely forgotten, its safety was defined by its commercial use in plastics.

In 1993 Krishnan and coworkers (Krishnan et al., 1993) found that BPA leaching from plastic flasks was capable of activating estrogen-responsive breast cancer cells. In those same years, other researchers began investigating synthetic hormones in developing organisms. vom Saal and coworkers showed for the first time that fetal exposure to environmentally relevant parts-per-billion doses of BPA, in the range currently being consumed by people, can alter the adult reproductive system in mice increasing prostate weight (Welshons et al.,1999). Low-dose effects of BPA were then reported on mammary tissue (Markey et al., 2001) and reproductive system (Markey et al., 2005).

The subsequent finding that environmentally relevant doses of BPA given in the perinatal phase altered the brain sexual differentiation in mouse offspring was not surprising, given the crucial role of estrogens in shaping brain synaptogenesis and connectivity (Rubin et al., 2006). In 2007 Richter revised laboratory rodent studies as for the in vivo effects of BPA [at doses well below the tolerable daily intake (TDI) of 50 μg/kg bw/day established by EPA at that time] and pointed to long lasting, organizational effects in response to developmental exposure to BPA specifically in the brain, the male reproductive system, and metabolic processes (Richter et al., 2007).

This research fueled the discussion about low-dose effects of EDCs, and BPA became a paradigmatic example in EDC research, also considering that the 2003–04 National Health and Nutrition Examination Survey reported that 93% of people in the United States had measurable amounts of BPA in their urine, and a similar proportion was found in Europe. The European Food Safety Agency in 2015 released a scientific opinion that lowered the TDI from 50 to 4 μg/kg bw/day (EFSA, 2015). The TDI was made temporary and EFSA committed to re-evaluate BPA toxicity again on the basis of the results of a 2-year study by the US National Toxicology Program. Thus EFSAs endorsed a scientific protocol for the upcoming re-evaluation of BPA hazards, scheduled to start in 2018.

5.1 Neurobehavioral Effects of BPA in Children

In the last decade, studies in human cohorts in Europe and the United States have started to analyze the potential repercussion of developmental exposure to BPA on children's behavior in the age range 3–8 years (Braun et al., 2009, 2011; Casas et al., 2015; Evans et al., 2014; Harley et al., 2013; Miodovnik et al., 2011; Perera et al., 2012; Roen et al., 2015). The results of these studies did not evidence a strong association between BPA maternal urinary levels during pregnancy and child neuropsychological outcome; inconsistency is likely attributable to the modest sample size, the presence of major confounders (such as socioeconomic status and maternal education) and the episodic nature of BPA exposure that, together with its short biological half-life (<6 h), generates high within-person variation of urinary BPA concentrations. However, it is worth noting that the behavioral effects reported by these studies identified subclinical scores in several domains, mainly internalizing and externalizing behaviors, withdrawn/depressed behavior, somatic problems, and oppositional/defiant behaviors. Furthermore, the association between prenatal BPA and some behaviors were modulated by child sex, with boys more affected than girls in most studies (Casas et al., 2015; Evans et al., 2014; Harley et al., 2013; Perera et al., 2012). Two independent studies reported no association between prenatal BPA exposure and social impairment as measured in children age 7–9 years (Braun et al., 2014; Miodovnik et al., 2011). Altogether, there are indications from prospective studies that BPA exposure during pregnancy may be associated with child behavior in a sex-dependent manner. Disruption of maternal thyroid or gonadal hormones critical to proper brain development may explain the effects of BPA. For example, prenatal BPA exposure is associated with decreased maternal serum free thyroxine (T4) and decreased serum thyroid stimulating hormone in male neonates (Chevrier et al., 2013). The evidence coming from in vivo animal studies indicates that effects not directly related to hormone levels in the developing brain have to be taken into account (see below).

5.2 Behavioral Toxicology of BPA in Rodent Models

A great body of animal data on BPA effects on neurodevelopment has been produced since the beginning of this century, mainly aimed at supporting the risk represented by this EDC for child neurobehavioral disorders. Description of all the studies performed so far it is out of the scope of this contribution but it might be useful to go in depth of some of the behavioral effects evidenced in relation to the potential molecular mechanisms to appreciate the complexity of EDC research (refer to Palanza et al., 2016).

Fetal/neonatal exposure has been the major focus of research assessing the impact of BPA on neurodevelopment at doses below the 50 μg/kg/day (recommended as reference safe daily limit by US EPA). Animal studies suggest possible mechanisms by which prenatal BPA exposure may directly alter fetal neurodevelopment (Richter et al., 2007; Wolstenholme et al., 2011a,b). Many of these changes are sexually dimorphic and associated with aberrant behavior in different domains, namely parental/social behaviors, anxiety and learning/memory functions.

Prenatal and early postnatal exposure to BPA affects socio-sexual interactions, play behavior and maternal care in rodents, and non-human primates (reviewed in Palanza et al., 2016). More in detail Palanza et al. (2002) described changes in maternal behavior in female mice prenatally exposed to very low doses of BPA, reporting reduced time spent in nursing and huddling over the pups and more time engaged in exploring and nest-building behaviors. Alteration of the normal pattern of maternal care was also reported in the rat (Boudalia et al., 2014) and in macaques (Nakagami et al., 2009). The long-term effects of BPA on social behaviors were analyzed in different experimental contexts, e.g., home-cage observations, social interaction and preference tests. Wolstenholme et al. in 2011 showed that gestational exposure to BPA increased play behavior and enhanced social investigation in juvenile mice without impairing social recognition of a familiar conspecific (Wolstenholme et al., 2011a,b). The alterations in these behavioral domains were transmitted up to the third generation (Wolstenholme et al., 2012, 2013). Behavioral changes were paralled by lower gene expression of several estrogen receptors, and decreased expression of vasopressin and oxytocin genes in embryonic brains of the F1-generation and the F4-generation with sex-dependent effects.

Kundakovic et al. (2013) described low-dose BPA effects on home-cage social behavior in adolescents offspring of both sexes with moderate effects on aggression and social dominance at adulthood. This study also examined markers for gene expressions in offspring showing dose-dependent, and brain region-specific changes in expression of genes encoding estrogen receptors (ERs, ERα, ERβ, ERγ). The effects measured in this study appeared to be of different extent and direction depending on BPA dose, suggesting that low doses can be more effective in disturbing social behavior than the higher ones, in agreement with NMDR.

Despite some inconsistencies due to differences in strain, species and behavioral tasks applied, several experimental evidence also pointed to anxiogenic properties of prenatal BPA exposure in rodent models. All of the studies included either an open field test procedure or an Elevated Plus Maze (EPM) test for evaluation of anxiety-like behavior. Several of the above studies report an increased anxiety-like behavior after BPA exposure, and some report no effect on such behavior. Patisaul et al. (2012) showed that BPA-exposed juveniles displayed increased anxiety-like behavior in the light/dark box and the EPM, and BPA-exposed adult offspring showed disappearance of the normal sexual dimorphism in exploratory behavior. Gioiosa et al. (2013) found that gestational BPA consistently eliminated the sex dimorphism in anxiety-like/exploratory response in mice. Jašarević et al. (2013) used outbred deer mice (*Peromyscus maniculatus* bairdii), whose dams were fed a diet supplemented with either ethinyl estradiol or different doses of BPA, which showed increased anxiety-like behavior in the EPM. Kundakovic et al. (2013) found that BPA exposure induced persistent, largely sex-specific effects on anxiety-like behavior, leading to disruption of sexually dimorphic behaviors in adult mice.

Of all of the observed effects regarding the toxicity of BPA to the central nervous one of the critical effect involves the impairment of memory and learning, concurrent with a decrease in the expression of various subunits of glutamate NMDA (*N*-methyl-D-aspartate) receptors, which are particularly involved in synaptic and neuronal plasticity and in memory and learning processes.

The study by Xu et al. (2013) showed a link between changes in synaptic and neuronal plasticity mechanisms in specific cerebral regions (hippocampus) and functional behavioral impairment (spatial learning and conditioning) following BPA exposure during gestation.

The same authors reported effects of perinatal exposure to BPA at very low doses on number of pyramidal cells in hippocampal region and on expression levels of synaptic markers in developing and young adult male. More specifically, BPA significantly reduced synaptic density and altered the synaptic morphology by down-regulating NMDA receptor and AMPA receptor subunits during development till young adulthood. These results suggest that perinatal exposure to low level of BPA inhibits synaptogenesis and affects synaptic structural modification after birth.

5.3 BPA Effects on Brain and Behavior: Are They Mediated by its Estrogen-Like Activity?

There is enough evidence to support a role for BPA at environmentally relevant doses in modulation of several sex dimorphic behavioral patterns. Traditionally BPA was classified as selective estrogen receptor modulator that binds both classic estrogen receptors with significantly lower potency than the endogenous hormone estradiol. However the in vivo findings showed that the effects of BPA can be detected even for exposure at very low doses, and a growing body of data point to several mechanisms implicated, including both genomic and nongenomic actions of estrogens, interaction with the ERγ estrogen receptors, with androgen receptors, disruption of thyroid hormone signaling and inhibition of key enzymes involved in steroidogenesis (reviewed in Gore et al., 2015). In the brain BPA is able to modulate rapidly synaptic plasticity in cultured hippocampal neurons and spinogenesis in hippocampal slices (Tanabe et al., 2012). BPA acts as agonist of the membrane associated estrogen receptors and activate the extracellular signal regulated kinase (ERK) pathway (Chen et al., 2017). BPA can trigger methylation of ERα gene codifying for Esr1 in prenatally exposed offspring (Kundakovic et al., 2013). In juvenile rats, perinatal exposure to low-dose BPA alters significantly the transcription genes encoding rate-limiting enzymes for DA and 5-HT synthesis (Castro et al., 2015). Other recent studies have reported that GABA neurotransmission could be affected in different brain regions after early-life exposure to BPA; Ogi et al. have recently shown that prenatal and lactational exposure to BPA leads to increased levels of GABA and glutamate in almost all brain regions and of norepinephrine in the cortex, hypothalamus and thalamus, likely by modulation of gene expression (Ogi et al., 2015; Yeo et al., 2013; Zalko et al., 2016).

It is clear that, when an EDC like BPA interferes with an ubiquitous neurotransmitter like GABA other disruptive effects can occur everywhere GABA plays a role, i.e., on neural connectivity in the early developmental periods (see discussion of this aspect in the recent review by Pinson et al., 2017).

Kundakovic et al. found that prenatal BPA induces lasting DNA methylation changes in the transcriptionally relevant region of the BDNF gene in the hippocampus and blood of mice. These changes seem consistent with BDNF changes in the cord blood of humans exposed to high maternal BPA levels in utero (Kundakovic et al., 2015). As previously described, epigenetic modification of BDNF expression is implicated in brain plasticity and associated with early-life adversity. These findings open new perspectives on the potential convergence of hormonal, psychosocial and chemical factors on crucial molecular pathways orchestrating brain and behavior development.

6. CHLORPYRIFOS AS A NEUROENDOCRINE DISRUPTOR

6.1 Neurobehavioral Effects of CPF in Children

An increasing body of data collected in the past decades supports the developmental neurotoxicity of organophosphate insecticides (OPs), and in particular of CPF (Burke et al., 2017; Saunders et al., 2012). Although non-agricultural uses of CPF (e.g., residential control of insect pests) were phased out [in the United States in 2001 (EPA, 2002) and in EU in 2005 (EC, 1451/2007)], CPF still remains the most used OP insecticide in agriculture in these areas. Exposure to CPF in the general population is ubiquitous, primarily occurring through diet; pregnant women, infants, and children represent vulnerable populations (Eaton et al., 2008). As a consequence, several epidemiological studies on developmental OP exposure and neuropsychological maturation carried out in the last decade included CPF or its metabolites (Bouchard et al., 2011; Engel et al., 2011; Eskenazi et al., 2007; Lovasi et al., 2011; Marks et al., 2010; Rauh et al., 2006). Recent data on 6/9-year-old children living nearby banana plantations with extensive pesticide use revealed that urinary concentrations of TCPy (3,5,6-trichloro-2-pyridinol) were associated with impaired neurobehavioral outcomes (e.g., poorer working memory in boys, poorer visual motor coordination, increased prevalence of cognitive problems and inattention) (van Wendel de Joode et al., 2016). This is a piece of evidence in line with urban birth cohort studies, in which in utero CPF exposure from household use appear to delay cognitive and psychomotor development (Lovasi et al., 2011; Rauh et al., 2006). Low concentrations of CPF in cord blood were associated with 1.4% decrease of child's IQ and 2.8% decrement of working memory (Rauh et al., 2011), an effect more evident in males (Horton et al., 2012). Significant associations between prenatal CPF exposure and structural changes in the developing human brain were also observed: high-exposed children showed frontal and parietal cortical thinning with disappearance or even reversal of the typical sex dimorphism in several cortical areas. In addition, a positive association between autism spectrum disorders and prenatal residential proximity to overall organophosphates (third trimester) and CPF (second trimester) has been evidenced (Shelton et al., 2014).

6.2 Neurobehavioral Toxicology of CPF in Rodent Models

Studies from different laboratories have shown in rodent models that developmental exposure to CPF doses below the threshold of acetylcholinesterase (AChE) inhibition induce long-term behavioral changes in male and female mice. These changes are mainly related to exploration and response to novelty, levels of arousal/activity and social responses. In particular, CPF influences different items of the social repertoire of laboratory mice, such as maternal behavior induction in virgin females, social recognition and achievement of social rank in both sexes (reviewed in Venerosi et al., 2012). We hypothesized that CPF exposure might lead to inappropriate processing of the salience of those same social cues, possibly by affecting the maturation of sex-dimorphic brain areas implicated in social investigation and discrimination (Venerosi et al., 2015). Mechanisms of CPF action to explain the effects reported for low-dose developmental exposures have not been fully clarified but several lines of evidence indicate that CPF can directly interact with non-AChE targets. Upregulation of serotonin (5-HT) receptors and/or transporters has been observed in adult rodents prenatally exposed to CPF (Aldridge et al., 2004; Venerosi et al., 2010); prenatal CPF may alter axonal transport and outgrowth mediated by tubulin polymerization (Middlemore-Risher et al., 2011) with potential detrimental impact on axonal outgrowth, neuronal connections and brain mitochondrial function. CPF-oxon can directly interact with muscarinic M2 and cannabinoid CB1 receptors that both can have several effects on brain development as well as on peripheral tissues (Busquets Garcia et al., 2016). Developmental CPF exposure also decreased growth factor levels, another effect potentially contributing to inhibition of axonal growth (Betancourt and Carr, 2004). A detailed toxicogenomic profiling confirmed that low doses (e.g., 4mg/kg) can exert profound alterations in gene expression in the fetal brain across diverse signaling pathways, ranging from chromatin remodeling to long term potentiation (Moreira et al., 2010). These changes in gene expression could be indicative of epigenetic changes: indeed DNA methylation and histone modifications by CPF (and methyl CPF) were observed in different experimental models, including rodent brain regions and human neuronal progenitor cells (Kim et al., 2016; Shin et al., 2015) and are likely to occur during brain development, according to sexual dimorphic patterns. Microarray technology has shown that besides direct modifications of DNA and histones, CPF can target ncRNAs [upregulation of miRNA-132 and -212 in the hippocampus of adult rats treated with CPF (10mg/kg) for 3 weeks] (Lee et al., 2016). Finally, an additional level of complexity, so far neglected in developmental toxicity is represented by gut microbiota alterations [with wide consequences on brain and behavioral functions (Dinan and Cryan, 2015)] that have been recently pointed out for CPF (Fang et al., 2018). It is more than conceivable that altered microbiota composition (with associated reduced serum insulin, abundance of opportunistic pathogens and short fatty acid producing bacteria) is a challenging factor to be considered for risk assessment of CPF developmental toxicity.

6.3 CPF Effects on Brain and Behavior are Also Mediated by Hormones?

Although CPF is not classified as an endocrine disruptor, the wide range of pathways targeted by its developmental exposure do suggest that the boundaries of the functional impairments can easily extend beyond the nervous system to endocrine and neuroendocrine functions as well. At least in animals, following prenatal/perinatal exposure to subtoxic CPF, not only adult thyroid function is affected (De Angelis et al., 2009; Holzer et al., 2017; Slotkin et al., 2013) but also oxytocin and vasopressin brain circuitries are targeted (Tait et al., 2009; Venerosi et al., 2015).

Three features of these CPF neuroendocrine effects are worth of interest: (i) they can be amplified if occurring under conditions of oxidative stress (De Felice et al., 2016); (ii) in the mouse species the long term consequences of prenatal CPF are based in two brain regions, hypothalamus and amygdala, which have a key role in neuroendocrine control of social responses, in agreement with behavioral data (Ricceri et al., 2006; Venerosi et al., 2006, 2010); (iii) finally, these effects are often sexually dimorphic, with greater vulnerability of the male sex [in males CPF increased expression of ERβ in hypothalamus and decreased oxytocin expression in amygdala; CPF increased vasopressin 1a receptor expression in amygdala in both sexes (Venerosi et al., 2012, 2015)]. Importantly, the finding that CPF produces sex-dimorphic changes in social/investigative behaviors and masculinize the females' responses fits well with the altered expression of hypothalamic and amygdaloid neuropeptides. The very recent report carried out within the BPA Clarity Consortium evidences a similar scenario for BPA, as developmental BPA exposure can interfere with estrogen, oxytocin, and vasopressin signaling pathways in the amygdala and alter signaling pathways critical for synaptic organization and transmission (Arambula et al., 2018).

7. CONCLUSIONS

Several publications in the past few years have reflected the intense debate among scientific community and regulators on the criteria to identify EDCs and their potential health effects (Zoeller et al., 2014). Notwithstanding the different points of view, there is large consensus on the fact that the traditional toxicology approach—based on the definition of dose threshold and dose–response relationship to derive the adverse effects of chemicals—is no longer sufficient to address the EDC topic (Weiss, 2012). In actual fact from the beginning of the 1990s developmental neurotoxicology has deeply challenged the traditional toxicological paradigm, by showing that delayed functional effects may be induced by chemical exposures at levels below the Not Observed Effect Adverse Level (NOAEL) (Slikker, 1994). In this chapter we have attempted to describe the complexity of brain and behavior during development, when hormones may behave as neurotransmitters or neurotrophic factors and vice versa. We have also shown that CPF, an environmental chemical known to act by AChE inhibition, interfere with brain developmental programming by previously unsuspected hormone-like activity.

The definition of EDCs for risk assessment purposes is undoubtedly a challenging task.

Schug et al. in their recent review posed a series of questions very important for EDC research, which are extremely relevant for developmental neurotoxicology as well (Schug et al., 2016). The main question concerns the multiple mechanisms of action of recognized EDCs such as BPA that are mediated by the classical nuclear hormone receptors, membrane-bound estrogen receptors, interaction with targets in the cytosol, changes in metabolism of endogenous hormones, cross-talk between genomic and nongenomic pathways, interference with HPA feedback regulation and changes in DNA methylation or histone modifications. The same multiplicity of mechanisms is also common to classical neurotoxicants when exposure at low doses occurs during development, resulting in outcomes that are not necessarily predicted by effects reported at higher toxic doses, as in the case of the OP insecticide CPF.

The variety of mechanisms of both EDCs' and neurotoxicants' action on brain development makes extremely complicated to select behavioral endpoints representing the whole spectrum of potential effects of these chemicals in human populations. We need predictive biomarkers of exposure and effects that would allow the tracing of EDC health effects in a shorter period of time than in epidemiological studies. The search for biomarkers reflecting different levels of the organism's response to both endogenous and external environment is pursued by the integrated exposome approach and it appears as a promising strategy to address the EDC issue in the DOHaD perspective (Robinson and Vrijheid, 2015).

This strategy implies the establishment of reliable in vitro and in vivo models to mimic plausible human exposure scenarios to either EDCs or neurotoxicants. In vivo studies are pivotal as they allow one to anchor biomarkers to behavioral changes: they require the establishment of prospective animal cohorts and the use of developmental exposures with multilevel analysis of the embryos, fetus and pups, as well as of later life stages to trace dysfunction/disease. The key periods when exposure to EDCs might cause the most damage and how exposures during these windows contribute to the health effects are still undefined. Laboratory research must use environmentally relevant doses, different windows of exposure, multidose designs for identification of NMDR and the examination of the relationship between the molecular mechanism proposed and the behavioral phenotype, always including the two sexes. This would also entail the use of the new technologies of gene expression profiling and epigenetics including "omic" signatures that may predict a specific disease or dysfunction later in life.

In close relationship with the more updated in vitro and in silico models, these targeted in vivo studies will possibly empower the human exposome with a set of peripheral accessible markers of susceptibility, vulnerability and effects, mirroring the effects in brain and able to predict the adverse behavioral outcomes of developmental EDC exposure. The approach of behavioral toxicology may be instrumental to identify candidate biomarkers and pinpoint sex vulnerability or life stages to be translated to large prospective studies (De Felice et al., 2015).

As highlighted in this chapter, EDCs act across multiple levels of biological organization; in agreement, the EDC topic cuts across multiple disciplines. The development of scientific criteria for the identification of EDCs and their role in disease etiology could greatly benefit from the research background of developmental neurotoxicology and behavioral sciences.

ACKNOWLEDGMENTS

Supported by FP7 HEALS Grant No. 603946 "Health and Environment-wide Associations based on Large Population Surveys," and by H2020-MSCA-ITN-2017 NEUROSOME EC 766251 "Exploring the neurological exposome."

REFERENCES

Aldridge, J.E., Seidler, F.J., Slotkin, T.A., 2004. Developmental exposure to chlorpyrifos elicits sex-selective alterations of serotonergic synaptic function in adulthood: critical periods and regional selectivity for effects on the serotonin transporter, receptor subtypes, and cell signaling. Environ. Health Perspect. 112 (2), 148–155.

Arambula, S.E., Jima, D., Patisaul, H.B., 2018. Prenatal bisphenol A (BPA) exposure alters the transcriptome of the neonate rat amygdala in a sex-specific manner: a CLARITY-BPA consortium study. Neurotoxicology 65, 207–220.

Banks, E.C., Ferretti, L.E., Shucard, D.W., 1997. Effects of low level lead exposure on cognitive function in children: a review of behavioral, neuropsychological and biological evidence. Neurotoxicology 18 (1), 237–281.

Barker, D.J., Gluckman, P.D., Godfrey, K.M., Harding, J.E., Owens, J.A., Robinson, J.S., 1993. Fetal nutrition and cardiovascular disease in adult life. Lancet 341 (8850), 938–941.

Bellanger, M., Demeneix, B., Grandjean, P., Zoeller, R.T., Trasande, L., 2015. Neurobehavioral deficits, diseases, and associated costs of exposure to endocrine-disrupting chemicals in the European Union. J. Clin. Endocrinol. Metab. 100 (4), 1256–1266.

Bellinger, D.C., 2009. Interpreting epidemiologic studies of developmental neurotoxicity: conceptual and analytic issues. Neurotoxicol. Teratol. 31 (5), 267–274.

Bellinger, D.C., Matthews-Bellinger, J.A., Kordas, K., 2016. A developmental perspective on early-life exposure to neurotoxicants. Environ. Int. 94, 103–112.

Berardi, N., Pizzorusso, T., Ratto, G.M., Maffei, L., 2003. Molecular basis of plasticity in the visual cortex. Trends Neurosci. 26 (7), 369–378.

Bergman, Å., Becher, G., Blumberg, B., Bjerregaard, P., Bornman, R., Brandt, I., Casey, S.C., Frouin, H., Giudice, L.C., Heindel, J.J., Iguchi, T., Jobling, S., Kidd, K.A., Kortenkamp, A., Lind, P.M., Muir, D., Ochieng, R., Ropstad, E., Ross, P.S., Skakkebaek, N.E., Toppari, J., Vandenberg, L.N., Woodruff, T.J., Zoeller, R.T., 2015. Manufacturing doubt about endocrine disrupter science—a rebuttal of industry-sponsored critical comments on the UNEP/WHO report 'State of the Science of Endocrine Disrupting Chemicals 2012'. Regul. Toxicol. Pharmacol. 73 (3), 1007–1017.

Betancourt, A.M., Carr, R.L., 2004. The effect of chlorpyrifos and chlorpyrifos-oxon on brain, cholinesterase, muscarinic receptor binding, and neurotrophin levels in rats following early postnatal exposure. Toxicol. Sci. 77 (1), 63–71.

Bohacek, J., Mansuy, I.M., 2015. Molecular insights into transgenerational non-genetic inheritance of acquired behaviours. Nat. Rev. Genet. 16 (11), 641–652.

Bouchard, M.F., Chevrier, J., Harley, K.G., Kogut, K., Vedar, M., Calderon, N., Trujillo, C., Johnson, C., Bradman, A., Barr, D.B., Eskenazi, B., 2011. Prenatal exposure to organophosphate pesticides and IQ in 7-year-old children. Environ. Health Perspect. 119 (8), 1189–1195.

Boudalia, S., Berges, R., Chabanet, C., Folia, M., Decocq, L., Pasquis, B., Abdennebi-Najar, L., Canivenc-Lavier, M.C., 2014. A multi-generational study on low-dose BPA exposure in Wistar rats: effects on maternal behavior, flavor intake and development. Neurotoxicol. Teratol. 41, 16–26.

Braun, J.M., Yolton, K., Dietrich, K.N., Hornung, R., Ye, X., Calafat, A.M., Lanphear, B.P., 2009. Prenatal bisphenol A exposure and early childhood behavior. Environ. Health Perspect. 117 (12), 1945–1952.

Braun, J.M., Kalkbrenner, A.E., Calafat, A.M., Bernert, J.T., Ye, X., Silva, M.J., Barr, D.B., Sathyanarayana, S., Lanphear, B.P., 2011. Variability and predictors of urinary bisphenol a concentrations during pregnancy. Environ. Health Perspect. 119 (1), 131–137.

Braun, J.M., Lanphear, B.P., Calafat, A.M., Deria, S., Khoury, J., Howe, C.J., Venners, S.A., 2014. Early-life bisphenol a exposure and child body mass index: a prospective cohort study. Environ. Health Perspect. 122 (11), 1239–1245.

Burke, R.D., Todd, S.W., Lumsden, E., Mullins, R.J., Mamczarz, J., Fawcett, W.P., Gullapalli, R.P., Randall, W.R., Pereira, E.F.R., Albuquerque, E.X., 2017. Developmental neurotoxicity of the organophosphorus insecticide chlorpyrifos: from clinical findings to preclinical models and potential mechanisms. J. Neurochem. 142 (S2), 162–177.

Busquets Garcia, A., Soria-Gomez, E., Bellocchio, L., Marsicano, G., 2016. Cannabinoid receptor type-1: breaking the dogmas. F1000Res. 5, 990.

Cahill, L., 2006. Why sex matters for neuroscience. Nat. Neurosci. Rev. 7, 477–484.

Caldji, C., Tannenbaum, B., Sharma, S., Francis, D., Plotsky, P.M., Meaney, M.J., 1998. Maternal care during infancy regulates the development of neural. Proc. Natl. Acad. Sci. U.S.A. 95 (9), 5335–5340.

Caldji, C., Hellstrom, I.C., Zhang, T.-Y., Diorio, J., Meaney, M.J., 2011. Environmental regulation of the neural epigenome. FEBS Lett. 585 (13), 2049–2058.

Carlsen, E., Giwercman, A., Keiding, N., Skakkebaek, N.E., 1992. Evidence for decreasing quality of semen during past 50 years. BMJ 305, 609–613.

Casas, M., Forns, J., Martínez, D., Avella-García, C., Valvi, D., Ballesteros-Gómez, A., Luque, N., Rubio, S., Julvez, J., Sunyer, J., Vrijheid, M., 2015. Exposure to bisphenol A during pregnancy and child neuropsychological development in the INMA-Sabadell cohort. Environ. Res. 142, 671–679.

Castro, B., Sánchez, P., Miranda, M.T., Torres, J.M., Ortega, E., 2015. Identification of dopamine- and serotonin-related genes modulated by bisphenol A in the prefrontal cortex of male rats. Chemosphere 139, 235–239.

Champagne, F.A., Weaver, I.C.G., Diorio, J., Dymov, S., Szyf, M., Meaney, M.J., 2006. Maternal care associated with methylation of the estrogen receptor-alpha1b promoter and estrogen receptor-alpha expression in the medial preoptic area of female offspring. Endocrinology 147 (6), 2909–2915.

Chen, Z.P., Hetzel, B.S., 2010. Cretinism revisited. Best Pract. Res. Clin. Endocrinol. Metab. 24 (1), 39–50.

Chen, X., Wang, Y., Xu, F., Wei, X., Zhang, J., Wang, C., Wei, H., Xu, S., Yan, P., Zhou, W., Mody, I., Xu, X., Wang, Q., 2017. The rapid effect of bisphenol-A on long-term potentiation in hippocampus involves estrogen receptors and ERK activation. Neural Plast. 2017.

Chevrier, J., Gunier, R.B., Bradman, A., Holland, N.T., Calafat, A.M., Eskenazi, B., Harley, K.G., 2013. Maternal urinary bisphenol a during pregnancy and maternal and neonatal thyroid function in the CHAMACOS study. Environ. Health Perspect. 121 (1), 138–144.

Chuang, J.C., Jones, P.A., 2007. Epigenetics and microRNAs. Pediatr. Res. 61 (5 Pt. 2 Suppl), 24R–29R.

Colborn, T., 2004. Neurodevelopment and endocrine disruption. Environ. Health Perspect. 112 (9), 944–949.

Colborn, T., Clement, C., 1992. Chemically-induced alterations in sexual and functional development: the wildlife/human connection. Adv. Mod. Environ. Toxicol. 21, 403.

De Angelis, S., Tassinari, R., Maranghi, F., Eusepi, A., Di Virgilio, A., Chiarotti, F., Ricceri, L., Venerosi Pesciolini, A., Gilardi, E., Moracci, G., Calamandrei, G., Olivieri, A., Mantovani, A., 2009. Developmental exposure to chlorpyrifos induces alterations in thyroid and thyroid hormone levels without other toxicity signs in CD-1 mice. Toxicol. Sci. 108 (2), 311–319.

De Felice, A., Ricceri, L., Venerosi, A., Chiarotti, F., Calamandrei, G., 2015. Multifactorial origin of neurodevelopmental disorders: approaches to understanding complex etiologies. Toxics 3, 89–129.

De Felice, A., Greco, A., Calamandrei, G., Minghetti, L., 2016. Prenatal exposure to the organophosphate insecticide chlorpyrifos enhances brain oxidative stress and prostaglandin E2 synthesis in a mouse model of idiopathic autism. J. Neuroinflammation 13 (1), 149.

Dietrich, D.R., von Aulock, S., Marquardt, H., Blaauboer, B., Dekant, W., Kehrer, J., Hengstler, J., Collier, A., Gori, G.B.a., Pelkonen, O., Lang, F., Nijkamp, F.P., Stemmer, K., Li, A., Savolainen, K., Hayes, A.W., Gooderham, N., Harvey, A., 2013. Scientifically unfounded precaution drives European Commission's recommendations on EDC regulation, while defying common sense, well-established science and risk assessment principles. Food Chem. Toxicol. 62, A1–A4.

Dinan, T.G., Cryan, J.F., 2015. The impact of gut microbiota on brain and behaviour: implications for psychiatry. Curr. Opin. Clin. Nutr. Metab. Care 18 (6), 552–558.

Dulac, C., Kimchi, T., 2007. Neural mechanisms underlying sex-specific behaviors in vertebrates. Curr. Opin. Neurobiol. 17 (6), 675–683.

Eaton, D.L., Daroff, R.B., Autrup, H., Bridges, J., Buffler, P., Costa, L.G., Coyle, J., McKhann, G., Mobley, W.C., Nadel, L., Neubert, D., Schulte-Hermann, R., Spencer, P.S., 2008. Review of the toxicology of chlorpyrifos with an emphasis on human exposure and neurodevelopment. Crit. Rev. Toxicol. 38 (Suppl. 2), 1–125.
EC, 1451/2007. Official Journal of the European Union L 325/3.
EFSA, 2015. Scientific opinion on the risks to public health related to the presence of bisphenol A (BPA) in foodstuffs. EFSA J. 13 (1), 3978. https://doi.org/10.2903/j.efsa.2015.3978.
Engel, S.M., Wolff, M.S., 2013. Causal inference considerations for endocrine disruptor research in children's health. Annu. Rev. Public Health 34 (1), 139–158.
Engel, S.M., Wetmur, J., Chen, J., Zhu, C., Barr, D.B., Canfield, R.L., Wolff, M.S., 2011. Prenatal exposure to organophosphates, paraoxonase 1, and cognitive development in childhood. Environ. Health Perspect. 119 (8), 1182–1188.
EPA, US Chlopyrifos Facts, 2002, http://www2.epa.gov/ingredients-used-pesticide-products/chlorpyrifos (accessed April 2018).
Eskenazi, B., Marks, A.R., Bradman, A., Harley, K., Barr, D.B., Johnson, C., Morga, N., Jewell, N.P., 2007. Organophosphate pesticide exposure and neurodevelopment in young Mexican-American children. Environ. Health Perspect. 115 (5), 792–798.
Evans, S.F., Kobrosly, R.W., Barrett, E.S., Thurston, S.W., Calafat, A.M., Weiss, B., Stahlhut, R., Yolton, K., Swan, S.H., 2014. Prenatal bisphenol A exposure and maternally reported behavior in boys and girls. Neurotoxicology 45, 91–99.
Fagiolini, M., Jensen, C.L., Champagne, F.A., 2009. Epigenetic influences on brain development and plasticity. Curr. Opin. Neurobiol. 19 (2), 207–212.
Fang, B., Li, J.W., Zhang, M., Ren, F.Z., Pang, G.F., 2018. Chronic chlorpyrifos exposure elicits diet-specific effects on metabolism and the gut microbiome in rats. Food Chem. Toxicol. 111, 144–152.
Garner, C.C., Zhai, R.G., Gundelfinger, E.D., Ziv, N.E., 2002. Molecular mechanisms of CNS synaptogenesis. Trends Neurosci. 25 (5), 243–250.
Gerlach, J.L., McEwen, B.S., 1972. Rat brain binds adrenal steroid hormone: radioautography of hippocampus with corticosterone. Science (New York, N.Y.) 175 (26), 1133–1136.
Gilbert, S.G., Grant-Webster, K.S., 1995. Neurobehavioral effects of developmental methylmercury exposure. Environ. Health Perspect. 103 (6), 135–142.
Gilbert, M.E., Sui, L., 2008. Developmental exposure to perchlorate alters synaptic transmission in hippocampus of the adult rat. Environ. Health Perspect. 116 (6), 752–760.
Gioiosa, L., Parmigiani, S., vom Saal, F.S., Palanza, P., 2013. The effects of bisphenol A on emotional behavior depend upon the timing of exposure, age and gender in mice. Horm. Behav. 63 (4), 598–605.
Goodman, J.H., Gilbert, M.E., 2007. Modest thyroid hormone insufficiency during development induces a cellular malformation in the corpus callosum: a model of cortical dysplasia. Endocrinology 148 (6), 2593–2597.
Gore, A.C., Martien, K.M., Gagnidze, K., Pfaff, D., 2014. Implications of prenatal steroid perturbations for neurodevelopment, behavior, and autism. Endocr. Rev. 35 (6), 961–991.
Gore, A.C., Chappell, V.A., Fenton, S.E., Flaws, J.A., Nadal, A., Prins, G.S., Toppari, J., Zoeller, R.T., 2015. Executive summary to EDC-2: the Endocrine Society's second Scientific Statement on endocrine-disrupting chemicals. Endocr. Rev. 36 (6), 593–602.
Gupta, R.K., Hasan, K.M., Trivedi, R., Pradhan, M., Das, V., Parikh, N.A., Narayana, P.A., 2005. Diffusion tensor imaging of the developing human cerebrum. J. Neurosci. Res. 81 (2), 172–178.
Harley, K.G., Gunier, R.B., Kogut, K., Johnson, C., Bradman, A., Calafat, A.M., Eskenazi, B., 2013. Prenatal and early childhood bisphenol A concentrations and behavior in school-aged children. Environ. Res. 126, 43–50.

Herbstman, J.B., Sjödin, A., Kurzon, M., Lederman, S.A., Jones, R.S., Rauh, V., Needham, L.L., Tang, D., Niedzwiecki, M., Wang, R.Y., Perera, F., 2010. Prenatal exposure to PBDEs and neurodevelopment. Environ. Health Perspect. 118 (5), 712–719.

Holzer, G., Besson, M., Lambert, A., François, L., Barth, P., Gillet, B., Hughes, S., Piganeau, G., Leulier, F., Viriot, L., Lecchini, D., Laudet, V., 2017. Fish larval recruitment to reefs is a thyroid hormone-mediated metamorphosis sensitive to the pesticide chlorpyrifos. eLife 6, e27595.

Horton, M.K., Kahn, L.G., Perera, F., Barr, D.B., Rauh, V., 2012. Does the home environment and the sex of the child modify the adverse effects of prenatal exposure to chlorpyrifos on child working memory? Neurotoxicol. Teratol. 34 (5), 534–541.

Jašarević, E., Williams, S.A., Vandas, G.M., Ellersieck, M.R., Liao, C., Kannan, K., Roberts, R.M., Geary, D.C., Rosenfeld, C.S., 2013. Sex and dose-dependent effects of developmental exposure to bisphenol A on anxiety and spatial learning in deer mice (Peromyscus maniculatus bairdii) offspring. Horm. Behav. 63 (1), 180–189.

Kim, H.Y., Wegner, S.H., Van Ness, K.P., Park, J.J., Pacheco, S.E., Workman, T., Hong, S., Griffith, W., Faustman, E.M., 2016. Differential epigenetic effects of chlorpyrifos and arsenic in proliferating and differentiating human neural progenitor cells. Reprod. Toxicol. 65, 212–223.

Kooistra, L., 2006. Neonatal effects of maternal hypothyroxinemia during early pregnancy. Pediatrics 117 (1), 161–167.

Kovacs, G.G., Adle-Biassette, H., Milenkovic, I., Cipriani, S., van Scheppingen, J., Aronica, E., 2014. Linking pathways in the developing and aging brain with neurodegeneration. Neuroscience 269, 152–172.

Krishnan, A.V., Stathis, P., Permuth, S.F., Tokes, L., Feldman, D., 1993. Bisphenol-A: an estrogenic substance is released from polycarbonate flasks during autoclaving. Endocrinology 132 (6), 2279–2286.

Kundakovic, M., Jaric, I., 2017. The epigenetic link between prenatal adverse environments and neurodevelopmental disorders. Genes 8 (3), 104.

Kundakovic, M., Gudsnuk, K., Franks, B., Madrid, J., Miller, R.L., Perera, F.P., Champagne, F.A., 2013. Sex-specific epigenetic disruption and behavioral changes following low-dose in utero bisphenol A exposure. Proc. Natl. Acad. Sci. U.S.A. 110 (24), 9956–9961.

Kundakovic, M., Gudsnuk, K., Herbstman, J.B., Tang, D., Perera, F.P., Champagne, F.A., 2015. DNA methylation of BDNF as a biomarker of early-life adversity. Proc. Natl. Acad. Sci. U.S.A. 112 (22), 6807–6813.

Lee, Y.S., Lewis, J.A., Ippolito, D.L., Hussainzada, N., Lein, P.J., Jackson, D.A., Stallings, J.D., 2016. Repeated exposure to neurotoxic levels of chlorpyrifos alters hippocampal expression of neurotrophins and neuropeptides. Toxicology 340, 53–62.

Leenen, F.A.D., Muller, C.P., Turner, J.D., 2016. DNA methylation: conducting the orchestra from exposure to phenotype? Clin. Epigenetics 8 (1), 92.

Lovasi, G.S., Quinn, J.W., Rauh, V.A., Perera, F.P., Andrews, H.F., Garfinkel, R., Hoepner, L., Whyatt, R., Rundle, A., 2011. Chlorpyrifos exposure and urban residential environment characteristics as determinants of early childhood neurodevelopment. Am. J. Public Health 101 (1), 63–70.

Markey, C.M., Luque, E.H., Munoz De Toro, M., Sonnenschein, C., Soto, A.M., 2001. In utero exposure to bisphenol A alters the development and tissue organization of the mouse mammary gland. Biol. Reprod. 65 (4), 1215–1223.

Markey, C.M., Wadia, P.R., Rubin, B.S., Sonnenschein, C., Soto, A.M., 2005. Long-term effects of fetal exposure to low doses of the xenoestrogen bisphenol-A in the female mouse genital tract. Biol. Reprod. 72 (6), 1344–1351.

Marks, A.R., Harley, K., Bradman, A., Kogut, K., Barr, D.B., Johnson, C., Calderon, N., Eskenazi, B., 2010. Organophosphate pesticide exposure and attention in young Mexican-American children: the CHAMACOS study. Environ. Health Perspect. 118 (12), 1768–1774.

McCarthy, M.M., 2010. How it's made: organisational effects of hormones on the developing brain. J. Neuroendocrinol. 22 (7), 736–742.
McCarthy, M.M., 2016. Sex differences in the developing brain as a source of inherent risk. Dialogues Clin. Neurosci. 18 (4), 361–372.
Mccarthy, M.M., Schwarz, J.M., Wright, C.L., Dean, S.L., 2008. Mechanisms mediating oestradiol modulation of the developing brain. J. Neuroendocrinol. 20 (6), 777–783.
McEwen, B.S., Milner, T.A., 2017. Understanding the broad influence of sex hormones and sex differences in the brain. J. Neurosci. Res. 95 (1–2), 24–39.
McEwen, B.S., Akama, K.T., Spencer-Segal, J.L., Milner, T.A., Waters, E.M., 2012. Estrogen effects on the brain: actions beyond the hypothalamus via novel mechanisms. Behav. Neurosci. 126 (1), 4–16.
McLachlan, J.A., 1993. Functional toxicology: a new approach to detect biologically active xenobiotics. Environ. Health Perspect. 101 (5), 386–387.
Meaney, M., 2001. Maternal care, gene expression, and the transmission of individual differences in stress reactivity across generations. Annu. Rev. Neurosci. 24, 1161–1192.
Middlemore-Risher, M.-L., Adam, B.-L., Lambert, N.A., Terry, A.V., 2011. Effects of chlorpyrifos and chlorpyrifos-oxon on the dynamics and movement of mitochondria in rat cortical neurons. J. Pharmacol. Exp. Ther. 339 (2), 341–349.
Miodovnik, A., Engel, S.M., Zhu, C., Ye, X., Soorya, L.V., Silva, M.J., Calafat, A.M., Wolff, M.S., 2011. Endocrine disruptors and childhood social impairment. Neurotoxicology 32 (2), 261–267.
Moreira, E.G., Yu, X., Robinson, J.F., Griffith, W., Hong, S.W., Beyer, R.P., Bammler, T.K., Faustman, E.M., 2010. Toxicogenomic profiling in maternal and fetal rodent brains following gestational exposure to chlorpyrifos. Toxicol. Appl. Pharmacol. 245 (3), 310–325.
Nakagami, A., Negishi, T., Kawasaki, K., Imai, N., Nishida, Y., Ihara, T., Kuroda, Y., Yoshikawa, Y., Koyama, T., 2009. Alterations in male infant behaviors towards its mother by prenatal exposure to bisphenol A in cynomolgus monkeys (Macaca fascicularis) during early suckling period. Psychoneuroendocrinology 34 (8), 1189–1197.
Neal, A.P., Worley, P.F., Guilarte, T.R., 2011. Lead exposure during synaptogenesis alters NMDA receptor targeting via NMDA receptor inhibition. Neurotoxicology 32 (2), 281–289.
Neville, H., Bavelier, D., 2002. Human brain plasticity: evidence from sensory deprivation and altered language experience. Prog. Brain Res. 138, 177–188.
Ogi, H., Itoh, K., Ikegaya, H., Fushiki, S., 2015. Alterations of neurotransmitter norepinephrine and gamma-aminobutyric acid correlate with murine behavioral perturbations related to bisphenol A exposure. Brain Dev. 37 (8), 739–746.
Palanza, P., Howdeshell, K.L., Parmigiani, S., vom Saal, F.S., 2002. Exposure to a low dose of bisphenol A during fetal life or in adulthood alters maternal behavior in mice. Environ. Health Perspect. 110 (Suppl. 3), 415–422.
Palanza, P., Nagel, S.C., Parmigiani, S., vom Saal, F.S., 2016. Perinatal exposure to endocrine disruptors: sex, timing and behavioral endpoints. Curr. Opin. Behav. Sci. 7, 69–75.
Patisaul, H.B., Sullivan, A.W., Radford, M.E., Walker, D.M., Adewale, H.B., Winnik, B., Coughlin, J.L., Buckley, B., Gore, A.C., 2012. Anxiogenic effects of developmental bisphenol a exposure are associated with gene expression changes in the juvenile rat amygdala and mitigated by soy. PLoS One 7 (9), e43890.
Perera, F.P., Li, Z., Whyatt, R., Hoepner, L., Wang, S., Camann, D., Rauh, V., 2009. Prenatal airborne polycyclic aromatic hydrocarbon exposure and child IQ at age 5 years. Pediatrics 124 (2), e195–e202.
Perera, F., Vishnevetsky, J., Herbstman, J.B., Calafat, A.M., Xiong, W., Rauh, V., Wang, S., 2012. Prenatal bisphenol A exposure and child behavior in an inner-city cohort. Environ. Health Perspect. 120 (8), 1190–1194.

Pinson, A., Franssen, D., Gérard, A., Parent, A.-S., Bourguignon, J.-P., 2017. Neuroendocrine disruption without direct endocrine mode of action: polychloro-biphenyls (PCBs) and bisphenol A (BPA) as case studies. C. R. Biol. 340, 432–438.
Rauh, V.A., Garfinkel, R., Perera, F.P., Andrews, H.F., Hoepner, L., Barr, D.B., Whitehead, R., Tang, D., Whyatt, R.W., 2006. Impact of prenatal chlorpyrifos exposure on neurodevelopment in the first 3 years of life among inner-city children. Pediatrics 118 (6), e1845–e1859.
Rauh, V., Arunajadai, S., Horton, M., Perera, F., Hoepner, L., Barr, D.B., Whyatt, R., 2011. Seven-year neurodevelopmental scores and prenatal exposure to chlorpyrifos, a common agricultural pesticide. Environ. Health Perspect. 119 (8), 1196–1201.
Ricceri, L., Venerosi, A., Capone, F., Cometa, M.F., Lorenzini, P., Fortuna, S., Calamandrei, G., 2006. Developmental neurotoxicity of organophosphorous pesticides: fetal and neonatal exposure to chlorpyrifos alters sex-specific behaviors at adulthood in mice. Toxicol. Sci. 93 (1), 105–113.
Richter, C.A., Birnbaum, L.S., Farabollini, F., Newbold, R.R., Rubin, B.S., Talsness, C.E., Vandenbergh, J.G., Walser-Kuntz, D.R., vom Saal, F.S., 2007. In vivo effects of bisphenol A in laboratory rodent studies. Reprod. Toxicol. 24 (2), 199–224.
Robinson, O., Vrijheid, M., 2015. The pregnancy exposome. Curr. Environ. Health Rep 2 (2), 204–213.
Roen, E.L., Wang, Y., Calafat, A.M., Wang, S., Margolis, A., Herbstman, J., Hoepner, L.A., Rauh, V., Perera, F.P., 2015. Bisphenol A exposure and behavioral problems among inner city children at 7-9 years of age. Environ. Res. 142, 739–745.
Roth, T.L., Lubin, F.D., Funk, A.J., Sweatt, J.D., 2009. Lasting epigenetic influence of early-life adversity on the BDNF gene. Biol. Psychiatry 65 (9), 760–769.
Roy, T.S., Seidler, F.J., Slotkin, T.A., 2004. Morphologic effects of subtoxic neonatal chlorpyrifos exposure in developing rat brain: regionally selective alterations in neurons and glia. Dev. Brain Res. 148 (2), 197–206.
Rubin, B.S., Lenkowski, J.R., Schaeberle, C.M., Vandenberg, L.N., Ronsheim, P.M., Soto, A.M., 2006. Evidence of altered brain sexual differentiation in mice exposed perinatally to low, environmentally relevant levels of bisphenol A. Endocrinology 147 (8), 3681–3691.
Sale, A., Berardi, N., Maffei, L., 2014. Environment and brain plasticity: towards an endogenous pharmacotherapy. Physiol. Rev. 94 (1), 189–234.
Saunders, M., Magnanti, B.L., Correia Carreira, S., Yang, A., Alamo-Hernández, U., Riojas-Rodriguez, H., Calamandrei, G., Koppe, J.G., Krayer Von Krauss, M., Keune, H., Bartonova, A., 2012. Chlorpyrifos and neurodevelopmental effects: a literature review and expert elicitation on research and policy. Environ. Health 11 (Suppl. 1), S5.
Schecter, A., Startin, J., Wright, C., Kelly, M., Papke, O., Lis, A., Ball, M., Olson, J.R., 1994. Congener-specific levels of dioxins and dibenzofurans in U.S. food and estimated daily dioxin toxic equivalent intake. Environ. Health Perspect. 102 (11), 962–966.
Schug, T.T., Johnson, A.F., Birnbaum, L.S., Colborn, T., Guillette, L.J., Crews, D.P., Collins, T., Soto, A.M., vom Saal, F.S., McLachlan, J.A., Sonnenschein, C., Heindel, J.J., 2016. Minireview: endocrine disruptors: past lessons and future directions. Mol. Endocrinol. 30 (8), 833–847.
Shelton, J.F., Geraghty, E.M., Tancredi, D.J., Delwiche, L.D., Schmidt, R.J., Ritz, B., Hansen, R.L., Hertz-Picciotto, I., 2014. Neurodevelopmental disorders and prenatal residential proximity to agricultural pesticides: the charge study. Environ. Health Perspect. 122 (10), 1103–1109.
Shin, H.S., Seo, J.H., Jeong, S.H., Park, S.W., Park, Y.I., Son, S.W., Kang, H.G., Kim, J.S., 2015. Effect on the H19 gene methylation of sperm and organs of offspring after chlorpyrifos-methyl exposure during organogenesis period. Environ. Toxicol. 30 (12), 1355–1363.

Slikker, W., 1994. Principles of developmental neurotoxicology. Neurotoxicology 15 (1), 11–16.
Slotkin, T.A., Seidler, F.J., 2009. Oxidative and excitatory mechanisms of developmental neurotoxicity: transcriptional profiles for chlorpyrifos, diazinon, dieldrin, and divalent nickel in PC12 cells. Environ. Health Perspect. 117 (4), 587–596.
Slotkin, T.A., Cooper, E.M., Stapleton, H.M., Seidler, F.J., 2013. Does thyroid disruption contribute to the developmental neurotoxicity of chlorpyrifos? Environ. Toxicol. Pharmacol. 36 (2), 284–287.
Stiles, J., Jernigan, T.L., 2010. The basics of brain development. Neuropsychol. Rev. 20 (4), 327–348.
Tait, S., Ricceri, L., Venerosi, A., Maranghi, F., Mantovani, A., Calamandrei, G., 2009. Long-term effects on hypothalamic neuropeptides after developmental exposure to chlorpyrifos in mice. Environ. Health Perspect. 117 (1), 112–116.
Tanabe, N., Yoshino, H., Kimoto, T., Hojo, Y., Ogiue-Ikeda, M., Shimohigashi, Y., Kawato, S., 2012. Nanomolar dose of bisphenol A rapidly modulates spinogenesis in adult hippocampal neurons. Mol. Cell. Endocrinol. 351 (2), 317–325.
Tau, G.Z., Peterson, B.S., 2010. Normal development of brain circuits. Neuropsychopharmacol. Rev. 35115 (10), 147–168.
Tordjman, S., Somogyi, E., Coulon, N., Kermarrec, S., Cohen, D., Bronsard, G., Bonnot, O., Weismann-Arcache, C., Botbol, M., Lauth, B., Ginchat, V., Roubertoux, P., Barburoth, M., Kovess, V., Geoffray, M.M., Xavier, J., 2014. Gene × environment interactions in autism spectrum disorders: role of epigenetic mechanisms. Front. Psychol. 5, 53.
van Wendel de Joode, B., Mora, A.M., Lindh, C.H., Hernández-Bonilla, D., Córdoba, L., Wesseling, C., Hoppin, J.A., Mergler, D., 2016. Pesticide exposure and neurodevelopment in children aged 6–9 years from Talamanca, Costa Rica. Cortex 85, 137–150.
Vandenberg, L.N., Colborn, T., Hayes, T.B., Heindel, J.J., Jacobs, D.R., Lee, D.H., Shioda, T., Soto, A.M., vom Saal, F.S., Welshons, W.V., Zoeller, R.T., Myers, J.P., 2012. Hormones and endocrine-disrupting chemicals: low-dose effects and nonmonotonic dose responses. Endocr. Rev. 33 (3), 378–455.
Venerosi, A., Calamandrei, G., Ricceri, L., 2006. A social recognition test for female mice reveals behavioral effects of developmental chlorpyrifos exposure. Neurotoxicol. Teratol. 28 (4), 466–471.
Venerosi, A., Ricceri, L., Rungi, A., Sanghez, V., Calamandrei, G., 2010. Gestational exposure to the organophosphate chlorpyrifos alters social-emotional behaviour and impairs responsiveness to the serotonin transporter inhibitor fluvoxamine in mice. Psychopharmacology 208 (1), 99–107.
Venerosi, A., Ricceri, L., Tait, S., Calamandrei, G., 2012. Sex dimorphic behaviors as markers of neuroendocrine disruption by environmental chemicals: the case of chlorpyrifos. Neurotoxicology 33 (6), 1420–1426.
Venerosi, A., Tait, S., Stecca, L., Chiarotti, F., De Felice, A., Cometa, M.F., Volpe, M.T., Calamandrei, G. and Ricceri, L., Effects of maternal chlorpyrifos diet on social investigation and brain neuroendocrine markers in the offspring—a mouse study. Environ. Health, 14(1), 32, 2015.
Villar-Pazos, S., Martinez-Pinna, J., Castellano-Muñoz, M., Alonso-Magdalena, P., Marroqui, L., Quesada, I., Gustafsson, J.A., Nadal, A., 2017. Molecular mechanisms involved in the non-monotonic effect of bisphenol-A on Ca^{2+} entry in mouse pancreatic β-cells. Sci. Rep. 7 (1), 11770.
Vogel, S.A., 2009. The politics of plastics: the making and unmaking of bisphenol a "safety" Am. J. Public Health 99 (Suppl. 3), s559–s566.
vom Saal, F.S., Hughes, C., 2005. An extensive new literature concerning low-dose effects of bisphenol A shows the need for a new risk assessment. Environ. Health Perspect. 113 (8), 926–933.

Walker, D.M., Gore, A.C., 2017. Epigenetic impacts of endocrine disruptors in the brain. Front. Neuroendocrinol. 44, 1–26.

Weaver, I.C.G., D'Alessio, A.C., Brown, S.E., Hellstrom, I.C., Dymov, S., Sharma, S., Szyf, M., Meaney, M.J., 2007. The transcription factor nerve growth factor-inducible protein a mediates epigenetic programming: altering epigenetic marks by immediate-early genes. J. Neurosci. 27 (7), 1756–1768.

Weiss, B., 2012. The intersection of neurotoxicology and endocrine disruption. Neurotoxicology 33, 1410–1419.

Welshons, W.V., Nagel, S.C., Thayer, K.A., Judy, B.M., Vom Saal, F.S., 1999. Low-dose bioactivityof xenoestrogens in animals: fetal exposure to low doses of methoxychlor andother xenoestrogens increases adult prostate size in mice. Toxicol. Ind. Health 15 (1-2), 12–25.

WHO, 2012. State of the science of endocrine disrupting chemicals. In: Bergman, Å., Heindel, J.J., Jobling, S., Kidd, K.A., Zoeller, R.T. (Eds.), United Nations Environment Programme and the World Health Organization. WHO/UNEP. http://www.who.int/ceh/publications/endocrine/en/. 2013 ISBN: 978-92-807-3274-0 (UNEP) and 978 92 4 150503 1 (WHO).

Whyatt, R.M., Liu, X., Rauh, V.A., Calafat, A.M., Just, A.C., Hoepner, L., Diaz, D., Quinn, J., Adibi, J., Perera, F.P., Factor-Litvak, P., 2012. Maternal prenatal urinary phthalate metabolite concentrations and child mental, psychomotor, and behavioral development at 3 years of age. Environ. Health Perspect. 120 (2), 290–295.

Winneke, G., 2011. Developmental aspects of environmental neurotoxicology: lessons from lead and polychlorinated biphenyls. J. Neurol. Sci. 308 (1–2), 9–15.

Wolstenholme, J.T., Rissman, E.F., Connelly, J.J., 2011a. The role of bisphenol A in shaping the brain, epigenome and behavior. Horm. Behav. 59 (3), 296–305.

Wolstenholme, J.T., Taylor, J.A., Shetty, S.R.J., Edwards, M., Connelly, J.J., Rissman, E.F., 2011b. Gestational exposure to low dose bisphenol A alters social behavior in juvenile mice. PLoS One 6 (9), e25448.

Wolstenholme, J.T., Edwards, M., Shetty, S.R.J., Gatewood, J.D., Taylor, J.A., Rissman, E.F., Connelly, J.J., 2012. Gestational exposure to bisphenol A produces transgenerational changes in behaviors and gene expression. Endocrinology 153 (8), 3828–3838.

Wolstenholme, J.T., Goldsby, J.A., Rissman, E.F., 2013. Transgenerational effects of prenatal bisphenol A on social recognition. Horm. Behav. 64 (5), 833–839.

Xu, X., Xie, L., Hong, X., Ruan, Q., Lu, H., Zhang, Q., Zhang, G., Liu, X., 2013. Perinatal exposure to bisphenol-A inhibits synaptogenesis and affects the synaptic morphological development in offspring male mice. Chemosphere 91 (8), 1073–1081.

Yeo, M., Berglund, K., Hanna, M., Guo, J.U., Kittur, J., Torres, M.D., Abramowitz, J., Busciglio, J., Gao, Y., Birnbaumer, L., Liedtke, W.B., 2013. Bisphenol A delays the perinatal chloride shift in cortical neurons by epigenetic effects on the Kcc2 promoter. Proc. Natl. Acad. Sci. U.S.A. 110 (11), 4315–4320.

Zalko, D., Soto, A.M., Canlet, C., Tremblay-Franco, M., Jourdan, F., Cabaton, N.J., 2016. Bisphenol a exposure disrupts neurotransmitters through modulation of transaminase activity in the brain of rodents. Endocrinology 157 (5), 1736–1739.

Zoeller, R.T., Brown, T.R., Doan, L.L., Gore, A.C., Skakkebaek, N.E., Soto, A.M., Woodruff, T.J., Vom Saal, F.S., 2012. Endocrine-disrupting chemicals and public health protection: a statement of principles from the Endocrine Society. Endocrinology 153 (9), 4097–4110.

Zoeller, R.T., Bergman, A., Becher, G., Bjerregaard, P., Bornman, R., Brandt, I., Iguchi, T., Jobling, S., Kidd, K.A., Kortenkamp, A., Skakkebaek, N.E., Toppari, J., Vandenberg, L.N., 2014. A path forward in the debate over health impacts of endocrine disrupting chemicals. Environ. Health 13, 118.

CHAPTER EIGHT

Neurodevelopment: Inflammation Matters

Randall L. Davis[1]
Department of Pharmacology and Physiology, Center for Health Sciences, Oklahoma State University, Tulsa, OK, United States
[1]Corresponding author: e-mail address: randall.davis@okstate.edu

Contents

1. INTRODUCTION

There are numerous neurodevelopmental disorders that impact individuals, families and society. Neurodevelopment can be adversely affected by genetic and environmental factors. Inflammation is one environmental factor that can be detrimental to brain development and will be the focus of this chapter. Among neurodevelopmental disorders, evidence for the etiological involvement of inflammation is likely the strongest for cerebral palsy (CP), autism spectrum disorder (ASD), and schizophrenia (SZ). Exposure to

Advances in Neurotoxicology, Volume 2
ISSN 2468-7480
https://doi.org/10.1016/bs.ant.2018.03.002

inflammatory insult is commonly associated with maternal infection, perinatal infection in fetus/neonate, fetal hypoxia, and preterm birth. Extraordinary advances have been made in terms of developing animal models of perinatal infection/inflammation that recapitulate the brain damage and behavioral consequences observed in CP, ASD, and SZ. Maternal inflammation or maternal immune activation (MIA) is typically achieved by intraperitoneal (i.p.) or intrauterine injection of a bacterial or viral mimetic [bacterial lipopolysaccharide, LPS; polyriboinosinic–polyribocytidylic acid, poly(I:C), respectively] during a critical period of gestation. Perinatal infection can also be modeled by administering the inflammatory mediator directly into the neonate. Hypoxia–ischemia is commonly induced through blood vessel ligation and/or use of hypoxic chambers; and may be combined with LPS or poly(I:C) injection. The details for these methods will be further defined in later sections, in the context of specific studies.

Neurodevelopment of course involves dynamic changes in cell number and function of multiple glial and neuronal cell types, as well as interactions among these cells. Therefore, this chapter will begin with a brief discussion of the key stages in neurodevelopment as well as the primary roles of these cell types in neurodevelopment and responses to inflammatory insult. The remainder of the chapter will include a discussion of our understanding of the cellular and molecular events in neurodevelopment that are affected by inflammation and the extent to which these events may contribute to the etiology of neurodevelopmental disorders. Lastly, will be a discussion of potential therapeutic strategies targeting inflammatory pathways as a means of preventing neurodevelopmental disorders.

2. KEY STAGES OF NEURODEVELOPMENT

Human brain development is a remarkably complex process that is controlled through interactions among intracellular signaling, secreted signaling molecules, and cell surface receptors (Andersen, 2003; Reemst et al., 2016; Semple et al., 2013). These interactions are impacted by genetic, epigenetic, and environmental factors. The stages of brain development in both humans and rodents have been very well characterized through gestation into the postnatal period (Andersen, 2003; Semple et al., 2013). The timeline for the major neurodevelopmental processes are summarized in Table 1. This comparison of the neurodevelopmental process between humans and rodents is

Table 1 Temporal Comparison of Neurodevelopmental Processes in Humans and Rodents.

Developmental Process	Humans	Rodents
Neurogenesis	GW3 to 1 year	GD9.5 to P15
Gliogenesis	GW18 to 1 year	GD11 to P30
Synaptogenesis	GW18 to 15 years	P1 to P25
Myelination	GW30 to 20 years	P10 to P120
Synaptic pruning	1 to 20 years	P40 to P100

Gestational weeks (GW), gestational day (GD), postnatal day (P).
Information derived from Andersen (2003), Watson et al. (2006), Semple et al. (2013), and Cordeiro et al. (2015).

crucial given that preclinical studies using rodent models are instrumental in obtaining mechanistic insights into the impact of inflammation during critical periods on brain development and neurodevelopmental disorders.

3. GLIA: ROLES IN NEURODEVELOPMENT AND NEUROIMMUNE FUNCTION

3.1 Microglia

Microglia originate in the yolk sac from erythromyeloid precursor cells during the first trimester, which is prior to embryonic day (E)7.5 in mice, and are recognized as $CD45^{+}$ and $CD11b^{+}$ macrophage (Ginhoux et al., 2010; Gomez Perdiguero et al., 2015). These immature macrophage then mature and microglia are detectable in the brain by E9.5 in mice. Microglia continue to proliferate and colonize the brain in a highly orchestrated manner throughout gestation and into the postnatal period (Pierre et al., 2017). Microglia are very well characterized as the primary immune cell of the brain and thus instrumental in response to injury and infection through immune surveillance and phagocytosis (Gomez-Nicola and Perry, 2015). Microglia are also essential to neurodevelopmental processes including, neurogenesis, synaptic pruning, myelination, and angiogenesis (Gomez-Nicola and Perry, 2015; Reemst et al., 2016). Interestingly, Matcovitch-Natan et al. (2016) recently used genome-wide chromatin and expression profiling, in conjunction with single-cell transcriptome analysis in mouse microglia, to describe how microglia proceed through a stepwise developmental program. Essentially they discovered that the pattern of global gene expression changed in three distinct phases from E10.5 to adulthood and these phases are linked with differences in chromatin landscape (Matcovitch-Natan et al., 2016). The authors suggested that these phases in microglia development may help explain the diverse, age-dependent functions of microglia (e.g., synaptic pruning to immune surveillance and inflammatory signaling) from embryonic development into adulthood (Matcovitch-Natan et al., 2016).

3.1.1 Role of Microglia in Neurogenesis

Microglia are essential to neurogenesis and the development of neuronal circuitry (Pierre et al., 2017; Shigemoto-Mogami et al., 2014). Neuronal precursor cells (NPC) release factors that recruit microglia; in turn, microglia release trophic factors that promote neuronal survival (Arcuri et al., 2017). Microglia also have a critical role in removing excess neurons (approximately 50% of all neurons are removed) in the developing brain (Andersen, 2003; Arcuri et al., 2017). More specifically, microglia release cytokines, such as tumor necrosis factor-alpha (TNF-α), that induce neuronal apoptosis, but also, they phagocytize NPC and neuronal debris as a means of regulating neurogenesis (Arcuri et al., 2017; Cunningham et al., 2013; Mazaheri et al., 2014).

3.1.2 Role of Microglia in Synaptic Pruning

Normal synaptogenesis is dependent upon appropriate synaptic pruning, which involves removal of unnecessary synapses and promotion of appropriate synapses (Pierre et al., 2017). Synapses are targeted for removal by microglia when opsonized with complement factor 1q (C1q) or complement factor 3 (C3) (Bialas and Stevens, 2013; Schafer et al., 2012). Microglia also express the chemokine receptor, CX3CR1, which recognizes fractalkine, and is instrumental in synaptic pruning in postnatal neurodevelopment (Paolicelli et al., 2011).

3.1.3 Role of Microglia in Myelination

Oligodendrocytes myelinate neurons in the central nervous system (CNS). Differentiation and maturation of OLs is dependent upon microglia-derived cytokines including interleukin-1β (IL-1β) and interleukin-6 (IL-6) (Shigemoto-Mogami et al., 2014). Microglia also signal oligodendrocytes to produce key components of the myelin sheath, including myelin basic protein (MBP) and proteolipid protein (Hamilton and Rome, 1994).

3.1.4 Role of Microglia in Angiogenesis

The critical importance of microglia in brain vascularization was highlighted by demonstrating that angiogenesis is impaired in mice that are deficient in microglia (Fantin et al., 2010). Indeed, increasing evidence suggests that microglia have an integral role in angiogenesis within the developing brain (Arnold and Betsholtz, 2013; Reemst et al., 2016). While the signaling molecules and mechanistic aspects remain to be fully elucidated, microglia-derived TNF-α may be involved (Li et al., 2014).

3.2 Astrocytes

Astrocytes are essential to normal neurodevelopment with roles in neuronal survival, axonal guidance, synaptic pruning, synaptogenesis, and angiogenesis (Clarke and Barres, 2013). In humans, astrogenesis occurs predominantly from mid-to-late gestation until about 15 months of age; and in rodents, astrogenesis generally peaks within the first 2 weeks after birth (Reemst et al., 2016; Semple et al., 2013). In adults, astrocytes are estimated to account for up to 40% of the total number of cells in the brain and are integral to the blood–brain barrier, trophic/metabolic support of neurons, neurotransmitter recycling, synaptic transmission, and are involved in innate and adaptive immune responses (Ben Haim and Rowitch, 2017; Colombo and Farina, 2016; Herculano-Houzel, 2014; Khakh and Sofroniew, 2015).

3.2.1 Role of Astrocytes in Neuronal Survival

Neurons require trophic support to survive and integrate into appropriate neural circuits, and astrocytes are instrumental in this process (Clarke and Barres, 2013; Deshmukh and Johnson, 1997; Krieglstein et al., 2002). Astrocytes actively regulate the extracellular environment in the proximity of neurons, through potassium ion buffering, neurotransmitter recycling and regulation of energy metabolism (Belanger et al., 2011; Kimelberg and Nedergaard, 2010). For example, astrocytes utilize excitatory amino acid transporters to remove excess extracellular glutamate (Bjornsen et al., 2014). In addition to these neuroprotective actions, astrocytes are a major source of cytokines and neurotrophic factors (Choi et al., 2014; Park et al., 2006; Reemst et al., 2016).

3.2.2 Role of Astrocytes in Axonal Guidance

Astrocytes guide axonal outgrowth by establishing boundaries through release of extracellular matrix glycoproteins, laminin and fibronectin (Liesi and Silver, 1988; Spohr et al., 2014; Tonge et al., 2012). There are in fact a host of synaptogenic molecules that are released by astrocytes during neurodevelopment that modulate axon dynamics (Clarke and Barres, 2013). For example, ephrins (e.g., ephrin A3, A4) are also secreted by astrocytes and signal through ephrin receptor tyrosine kinase (e.g., EPHA4) to induce dendritic spine retraction (Carmona et al., 2009).

3.2.3 Role of Astrocytes in Synaptic Pruning

While much attention has been afforded microglia in terms of synaptic pruning (Paolicelli et al., 2011; Pierre et al., 2017; Schafer et al., 2012), astrocytes are also instrumental in this process of neurodevelopment (Mottahedin et al., 2017; Reemst et al., 2016; Stevens et al., 2007). As discussed above, microglia phagocytose C1q-tagged synapses, but, emerging evidence suggests that C1q expression in developing synapses is dependent upon astrocytes. For example, astrocyte-derived transforming growth factor-β (TGF-β) induces neuronal C1q expression (Bialas and Stevens, 2013). Microglia are also a major source of C1q (Fonseca et al., 2017) and presumably, astrocyte-derived factors also affect C1q expression in microglia during brain development, but to my knowledge this has yet to be demonstrated.

Another pathway through which developing astrocytes impact synaptic pruning involves the phagocytic receptors, multiple epidermal growth factor (EGF)-like domains 10 (MEGF10) and MER tyrosine kinase (MERTK) (Chung et al., 2013). It has been suggested that these astrocytic receptors recognize and remove synapses tagged with phosphatidylserine (Hochreiter-Hufford and Ravichandran, 2013).

3.2.4 Role of Astrocytes in Synaptogenesis

There has been an increasing appreciation for the role of astrocytes in synapse formation and maintenance during neurodevelopment (Clarke and Barres, 2013; Reemst et al., 2016). Astrocytes secrete a variety of synaptogenic molecules including, brain-derived neurotrophic factor (BDNF), ephrins, glypicans, hevin, thrombospondins, and TGF-β (Bosworth and Allen, 2017; Clarke and Barres, 2013). Astrocytes also express fatty acid binding protein 7 (FABP7) which facilitates fatty acid uptake and intracellular mobilization; and FABP7 is required for normal synaptogenesis (Ebrahimi et al., 2016; Kagawa et al., 2015). Additionally, astrocytes synthesize cholesterol that is essential to synaptogenesis/synapse maturation; this synthesis is controlled in part by transcription factors called sterol regulatory element binding proteins or SREBPs (van Deijk et al., 2017). It is also important to note that the synaptogenic properties of astrocytes differ among brain regions, as indicated by differing expression levels of synaptogenic factors (Buosi et al., 2018).

3.2.5 Role of Astrocytes in Angiogenesis

Astrocytes are essential to normal angiogenesis in the developing brain, as inhibition of astrogenesis is detrimental to cortical blood vessel branching (Ma et al., 2012). However, while the importance of astrocyte-derived angiogenic factors (e.g., vascular endothelial growth factor, angiopoietin-1, and angiopoietin-2) have been described for adult angiogenesis, much less is known about the specific role of these factors in perinatal neurodevelopment (Reemst et al., 2016).

3.3 Oligodendrocytes

Oligodendrocyte development occurs predominately toward the end of gestation, and continues into the postnatal period (Forbes and Gallo, 2017; Semple et al., 2013). Oligodendrocyte precursor cells (OPCs) appear in select brain regions during mid-to-late gestation, then migrate throughout the brain, with many ultimately populating the white matter (Forbes and Gallo, 2017). OPCs proliferate and differentiate into mature, myelinating oligodendrocytes throughout the latter stages of gestation, postnatally, and into adulthood (Yeung et al., 2014). Myelination is a highly orchestrated process that is regulated by many factors (Emery and Lu, 2015; Mitew et al., 2014) and can be impacted by environmental factors, including infection/inflammation (Hagberg et al., 2015; Knuesel et al., 2014; Makinodan et al., 2008). White matter injury is common in preterm infants and infants exposed to maternal/perinatal inflammation (Back, 2017; van Tilborg et al., 2016) thus, there is much interest in understanding the effects of inflammation on oligodendrocytes during critical periods of neurodevelopment.

4. CLINICAL EVIDENCE FOR THE DETRIMENTAL IMPACT OF INFLAMMATION ON NEURODEVELOPMENT

4.1 Cerebral Palsy

Cerebral palsy is among the most common neurodevelopmental disorders with a prevalence of approximately 3 per 1000 live births (Colver et al., 2014). Non-progressive white matter damage in the developing brain, involving paraventricular leukomalacia, is central to CP neuropathology (Rumajogee et al., 2016; Zaghloul et al., 2017). CP is characterized by impaired motor function and in many cases, deficits in cognition, communication, and perception (Colver et al., 2014). Among the greatest risk factors for CP are preterm birth, intraventricular hemorrhage (IVH), chorioamnionitis, hypoxia–ischemia, intrauterine inflammation and neonatal infection (Hirvonen et al., 2014; McIntyre et al., 2013; Oskoui et al., 2013). Particularly relevant to this chapter is the plethora of evidence that perinatal infection/inflammation and very-preterm birth are associated with a higher risk of CP (Alshaikh et al., 2013; Kuban et al., 2014;

Mitha et al., 2013; Wang et al., 2014). For instance, Yanni et al. (2017) recently investigated the risk factors for structural and functional neurological disorders in extremely-preterm infants (delivered prior to 28 weeks gestation) as part of the ELGAN (Extremely Low Gestational Age Newborns) study (O'Shea et al., 2009). It was determined that exposure to two separate inflammatory insults increased the risk for spastic CP, more than a single inflammatory hit (Yanni et al., 2017). More specifically, placental inflammation, in combination with increased TNF-α or IL-8 levels, increased the risk for CP (Yanni et al., 2017). Based on these and other findings, animal models used to study CP typically involve hypoxia–ischemia, with or without LPS administration, as will be discussed later in detail.

4.2 Autism Spectrum Disorders

ASDs include several related neurodevelopmental disorders with symptom onset and diagnosis occurring in early childhood, with a prevalence of 1 out of 68 according to a 2016 report by the Center for Disease Control and Prevention (Christensen et al., 2016). Individuals with ASD display a host of behavioral features including deficits in social reciprocity and communication, restricted interests, and repetitive behaviors (American Psychiatric Association, 2013). The etiology of ASDs is thought to involve a complex interaction among genetics and environmental factors such as, pregnancy complications involving hypoxia–ischemia, gut microbiota, and maternal/perinatal infection or inflammation (Gottfried et al., 2015; Modabbernia et al., 2017; Patterson, 2011; Vuong and Hsiao, 2017). Preclinical, rodent models that recapitulate critical aspects of brain damage and neurobehavioral deficits of ASD have become instrumental in unveiling the specific effects of inflammation on neurodevelopment, with potential insights into the causes of ASD. Key findings from these preclinical investigations on the role of inflammation in neurodevelopment and ASD-like neuropathology, will be discussed later in this chapter, with a particular emphasis on the most recent reports.

4.3 Schizophrenia

An estimated 1 in 200 people will be diagnosed with SZ (Simeone et al., 2015). Schizophrenia is typically first diagnosed in early adulthood and across a lifetime, symptoms can include delusions, hallucinations, blunted affect, alogia, amotivation, anticipatory anhedonia, social withdrawal, and deficits in both cognitive function and social cognition (American Psychiatric Association, 2013; Millan et al., 2014). The etiology of SZ is not completely understood, yet increasing evidence supports the role of maternal/perinatal inflammation (Brown and Patterson, 2011; Khandaker et al., 2013). It has been suggested for many years that SZ may result from "two-hits," one during very early neurodevelopment, and a second during a later critical period of brain development (Debnath et al., 2015; Feigenson et al., 2014; Maynard et al., 2001). For example, exposure to infection/inflammation in utero, followed by a second stressor during adolescence, such as infection, trauma or hormonal changes, may be instrumental in the etiology of this neurological disorder (Debnath et al., 2015). As will be discussed below, preclinical animal models have become a useful tool in the discovery of inflammatory pathways that may impact neurodevelopment, specifically during early stages of brain development, and provide key insights into the etiology of SZ (Cassella et al., 2016; Moller et al., 2015).

5. PRECLINICAL MODELS OF INFLAMMATION AND NEURODEVELOPMENT

5.1 Maternal Inflammation

5.1.1 LPS

Much of what is known about the effects of inflammation on neurodevelopment stems from preclinical studies using rodent models of maternal infection/inflammation (Hagberg et al., 2015). There are several different models of MIA, but they typically involve administration of LPS or poly(I:C) to the pregnant mother, via either i.p. or intrauterine injection at critical stages of gestation. Inflammation, neurodevelopment, and behavior are then assessed postnatally. There is considerable evidence that this form of maternal inflammation is detrimental to neurodevelopment in the offspring, yet there remains much to learn in terms of specific cell types and cellular mechanisms involved (Hagberg et al., 2015).

Avendano et al. (2015) discovered that unopposed plasma membrane channels formed by connexins or pannexins, which are involved in gliotransmitter-mediated intercellular communication, are influenced by maternal inflammation. In this study, pregnant mice were administered a single LPS injection (i.p., 0.01 μg/g) on gestation day (GD)17; pups were then carried full term. Astrocytes were collected from the pups in order to study inflammatory and paracrine signaling. Astrocytes from pups exposed to maternal inflammation expressed more IL-1β, TNF-α, and inducible nitric oxide synthase (iNOS) than did astrocytes from untreated dams (Avendano et al., 2015). Maternal inflammation also led to increased nitric oxide release and increased spontaneous $[Ca^{++}]_i$ oscillations in offspring astrocytes. These investigators reportedly showed for the first time that maternal inflammation alters offspring astrocytes, as evidenced by increased connexin 43 (Cx43) and pannexin 1 (Panx1) unopposed channel opening (Avendano et al., 2015). Furthermore, the adenosine triphosphate that was released through these channels induced neuronal cell death (Avendano et al., 2015). The extent to which these alterations in offspring astrocyte function contribute to overall neuronal development remains to be determined.

In a similar model of MIA, Fernandez de Cossio et al. (2017) administered a single LPS injection (i.p., 100 μg/kg) on GD15 to pregnant C57BL/6J mice. MIA-induced ASD-like behaviors (reduced social interest and increased repetitive/stereotypic movements) in adult (8–9 weeks of age) male and female offspring. At an earlier time point (P15), they discovered that spine density in the granule cells of the dentate gyrus was increased in only the male offspring exposed to maternal inflammation (Fernandez de Cossio et al., 2017). Furthermore, the microglia receptor instrumental in synaptic pruning, CX3CR1, was decreased in the hippocampus of offspring of LPS-exposed dams, and in a male-specific manner (Fernandez de Cossio et al., 2017). The extent to which these apparent deficits in synaptic pruning contribute to the ASD-like behaviors requires further investigation.

The amygdala is also sensitive to MIA (i.p. LPS on GD12) as demonstrated by an increase in IL-1β on P7 followed by increased expression of cyclooxygenase-2 (COX-2), iNOS, and the chemokine CCL2 (also known as monocyte chemotactic protein-1) on P40 (O'Loughlin et al., 2017). The postnatal amygdala in LPS-exposed offspring also exhibited microglia/astroglia activation, decrease volume (central nucleus of the amygdala) and a reduction in neuronal cell number. The extent to which these developmental changes contribute to long-term behavioral deficits remain to be determined (O'Loughlin et al., 2017).

Makinson et al. (2017) studied the effects of intrauterine inflammation on subsequent inflammation in the placenta and fetal brain, and the long-term consequences on neuronal development and behavior. Pregnant CD-1 mice received a single intrauterine injection (IUI) of LPS (50 μg) on GD15. To assess the acute effects of IUI on the placenta and inflammation in the fetal brain, tissues were collected 48 h after LPS injection, on E17 (Makinson et al., 2017). Intrauterine injection of LPS was previously shown to induce an acute increase in proinflammatory cytokines in the placenta at 6 h (Elovitz et al., 2011). However, by 48 h, inflammation in the placenta was essentially resolved, with the exception of CCL2, which remained elevated for both sexes (Makinson et al., 2017). Levels of IL-1β, TNF-α, and suppressor of cytokine 3 (SOCS3) were increased in fetal brain tissue of male offspring from LPS-treated dams, but not in female offspring; and cytokine levels were not elevated in maternal blood (Makinson et al., 2017). Brain tissue was also collected at P2 to assess expression of transcripts for C1q, C3, and TGF-β, which are integral in synaptic pruning during neurodevelopment (Bialas and Stevens, 2013; Schafer et al., 2012). As noted for the proinflammatory cytokines in fetal brain, expression of both complement factors and TGF-β were elevated in whole brain of male offspring exposed to intrauterine LPS, but unchanged or reduced in females (Makinson et al., 2017). Together with the findings of Fernandez de Cossio et al. (2017), there is mounting evidence that factors involved in synaptic pruning are particularly vulnerable to MIA in a male-specific manner (Makinson et al., 2017). However, there remains much to learn in terms of the differential effects of MIA on synaptic pruning related factors (e.g., C1q, C3, TGF-β, CX3CL1, CX3CR1) across brain regions and temporally throughout neurodevelopment; and the extent to which these changes impact behavior and CNS function.

Interestingly, the stepwise development of microglia that was recently described by Matcovitch-Natan et al. (2016) is vulnerable to MIA. This group discovered that microglia from adult, germ-free mice exhibited an underdeveloped phenotype. Furthermore, MIA induced a more advanced transcriptional phenotype in the early stage of neurodevelopment, then an overall normal expression phenotype was observed in adult offspring. Thus, the timing of inflammatory insult impacts the specific long-term neurological deficits and may explain why in some cases a normal microglia phenotype is observed in later stages of neurodevelopmental diseases (Matcovitch-Natan et al., 2016; Meyer et al., 2006).

The long-term effects of intrauterine LPS have also been revealed by assessing white matter damage in adult offspring (Makinson et al., 2017). Levels of MBP were reduced in the corpus callosum of both male and female adult offspring exposed to intrauterine LPS, whereas, MBP was unaffected in the anterior commissure (Makinson et al., 2017). Further analysis revealed that intrauterine LPS resulted in increased microglia number in the brain of both male and female offspring. Interestingly, at 50 weeks of age, acute challenge with LPS (i.p., 10 μg) induced a greater cytokine response in males exposed to intrauterine LPS, as indicated by increased IL-1β, TNF-α, CXCL10, and SOCS3 in the hypothalamus, amygdala, and prefrontal cortex (Makinson et al., 2017). Whereas, the response to acute LPS was similar among females, regardless of intrauterine exposure conditions (Makinson et al., 2017). Others have found that LPS responses in adults are potentiated by neonatal infection (Bilbo and Schwarz, 2009), but this is reportedly the first demonstration that intrauterine LPS exposure potentiates LPS-induced cytokine expression in the adult brain in a male-specific manner (Makinson et al., 2017). These investigators also found that glutamate decarboxylase 1 (*GAD1*) and dopamine type 2 receptor (*DRD2*) mRNA expression were increased in the hypothalamus and amygdala of adult male mice that were exposed to intrauterine LPS (Makinson et al., 2017). *GAD1* codes for the glutaminase decarboxylase 1 enzyme (GAD67) which is critical for synthesis of the neurotransmitter, gamma-aminobutyric acid (GABA), and subsequent neuronal inhibition (Chattopadhyaya et al., 2007; Huang et al., 2007). DRD2, is expressed on GABAergic cells (De Bundel et al., 2016) and dopamine signaling through these receptors is important in numerous behaviors, including risk seeking (Zalocusky et al., 2016). Thus, modulation of GAD67 and/or DRD2 may be important mechanistically in neurodevelopment consequent to maternal inflammation, but additional investigation is needed (Makinson et al., 2017).

Another group also found evidence of a dopamine connection in terms of the impact of maternal inflammation on the development of ASD-like behaviors in juvenile male (but not female) offspring (Kirsten et al., 2015). Pregnant Wistar rats received a single i.p. injection of LPS (100 μg/kg) on GD9.5. Treatment with LPS resulted in maternal hypozincemia, possibly due to sequestration of zinc in metallothionein (Kirsten et al., 2015). These researchers did not specifically measure zinc levels in the fetuses, but suggested that future studies should be performed to confirm levels are indeed reduced for a period in the fetuses as well. Nonetheless, LPS exposure during this critical period resulted in maternal hypozincemia and subsequent disruptions in behavior (i.e., social deficits, repetitive behavior, and cognitive inflexibility) in the male offspring (Kirsten et al., 2015). Evaluation of the brains from the offspring revealed that prenatal LPS exposure reduced striatal tyrosine hydroxylase (TH) (Kirsten et al., 2015) and dopamine levels (Kirsten et al., 2012). Importantly, subcutaneous administration of zinc to the pregnant rats 1 h after the LPS injection prevented both the decrease in striatal TH and autistic-like behaviors in juvenile offspring. These findings suggest that reduced dopaminergic signaling may be critical in the development of ASD-like behaviors associated with maternal inflammation. Additionally, it remains to be determined the extent to which the male-specific effects of perinatal inflammation on neurodevelopment identified in preclinical studies translate mechanistically to the greater prevalence of ASD in males that is observed clinically (Fernandez de Cossio et al., 2017; Kirsten et al., 2012; Loomes et al., 2017; Makinson et al., 2017).

Maternal intrauterine LPS administration (GD28) has also been used to induce chorioamnionitis in New Zealand White (NZW) rabbits and CP-like pathology in the offspring (Zhang et al., 2016). In this model, the kits display a CP phenotype including microglial activation and elevated cytokine expression in the periventricular region (PVR), as well as reduced myelination and motor deficits that are positively correlated with the degree of microglial activation (Kannan et al., 2012). Also, Zhang et al. (2016) demonstrated for the first time that intrauterine LPS exposure leads to increased glutamate carboxypeptidase II (GCPII) levels in activated microglia within the PVR white matter. Astrocytes in the white matter were also impacted as indicated by soma hypertrophy, retracted processes, and altered expression of enzymes and cellular components involved glutamate homeostasis (Zhang et al., 2016). Specifically, in the PVR of offspring exposed to intrauterine inflammation, there was increased glutaminase and *N*-methyl-D-aspartate receptor expression concurrent with decreased glial L-glutamate transporter 1 expression. Additionally, TGF-β, which is generally considered neuroprotective, was reduced in offspring that were exposed to maternal inflammation (Zhang et al., 2016). Thus, these studies in NZW rabbits suggest that maternal inflammation adversely affects glutamate homeostasis in the PVR white matter of offspring, likely contributing to glutamate excitotoxicity and neuroinflammation; and may partly explain the mechanism by which neurodevelopment is disrupted by inflammation.

Chorioamnionitis involves not only inflammation, but also hypoxia (Yanowitz et al., 2002, 2006). Therefore, another model for investigating the neurodevelopmental consequences of maternal inflammation involves prenatal transient systemic hypoxia–ischemia (TSHI) plus intra-amniotic LPS administration. In pregnant Sprague Dawley rats, on GD18, uterine arteries were occluded for 1 h, blood flow was then restored, followed by injection of 4 μg LPS into each amniotic sac. The pups (P2–P28) exhibit ventriculomegaly, gliosis, reduced MBP expression, decreased expression of potassium-chloride co-transport 2 (which is instrumental in GABAergic stimulation), and motor function deficits (Jantzie et al., 2014, 2015). The consequences of prenatal TSHI + LPS are also evident in the placenta and fetal/newborn brain. For instance, prenatal TSHI + LPS resulted in placental inflammation, hemorrhage and apoptotic cell death (Maxwell et al., 2015). Pups also had systemic inflammation as indicated by increased levels of IL-1β, IL-6, TNF-α and CXCL1 in circulation. Furthermore, CXCL1 expression was increased in the brain of P0 pups exposed prenatally to TSHI + LPS, whereas, IL-1β and TNF-α were not elevated (Maxwell et al., 2015). The increase in CXCL1 is very interesting given the critical role of this cytokine in the proliferation and migration of OPCs during neurodevelopment, as well as its role in neutrophil recruitment (Robinson and Franic, 2001; Robinson et al., 1998; Tsai et al., 2002). More recently, it was determined that experimental chorioamnionitis not only affects CXCL1 expression in the placenta and offspring, but also modulates expression of CXCR2, the receptor through which CXCL1 signals (Yellowhair et al., 2017). Chorioamnionitis led to increases in CXCL1 and CXCR2 expression in the placenta and brain out to P7, along with accumulation of $CXCR2^+$ neutrophils in placenta and brain (Yellowhair et al., 2017). In addition to neutrophilia, microgliosis, and in some cases peripheral macrophage infiltration, was present in the brain of pups exposed to chorioamnionitis. These authors suggested that neuroinflammation in the affected offspring likely contributed to the white matter damage observed later at P15 (Yellowhair et al., 2017), although, additional investigation is needed to fully appreciate the extent to which aberrant CXCL1/CXCR2 signaling following chorioamnionitis contributes to neurodevelopment disorders.

5.1.2 Poly(I:C)

There is evidence that viral infection during pregnancy can negatively impact neurodevelopment and is associated with developmental neuropsychiatric disorders such SZ and ASD (Brown and Derkits, 2010; Sakamoto et al., 2015). The dsRNA, poly(I:C) is commonly used as a viral mimetic to induce maternal inflammation (Meyer, 2014; Reisinger et al., 2015). Administration of poly(I:C) during pregnancy not only leads to MIA, as indicated by increased proinflammatory cytokine expression, but also leads to inflammation in the fetal brain and behavioral deficits in adult offspring (Choi et al., 2016; Giovanoli et al., 2016a; Reisinger et al., 2015).

A deficit in prepulse inhibition (PPI) of the acoustic startle reflex is indicative of altered sensorimotor gating and is characteristic of individuals with neurodevelopmental disorders such as SZ and ASD (Hazlett et al., 2008; Perry et al., 2007; Swerdlow et al., 2017). PPI is also commonly used to measure sensorimotor gating in animal models of neurological disorders. For instance, intravenous administration of poly(I:C) to pregnant mice on GD9 results in deficits in sensorimotor gaiting in adult offspring, as determined by diminished PPI of the acoustic startle reflex (Giovanoli et al., 2016a). Further investigation in this model revealed that poly(I:C)-induced maternal inflammation also results in hippocampal synaptic deficits in adult offspring. Notably, expression of the presynaptic proteins, synaptophysin and bassoon, was reduced in adult, but not pubescent, offspring (Giovanoli et al., 2016a).

Conversely, synaptic Ras GTPase-activating protein 1 (SynGap) expression was decreased only in pubescent offspring; and postsynaptic density protein 95 expression was suppressed in both pubescent and adult offspring (Giovanoli et al., 2016a). Thus, there appear to be age-related hippocampal synaptic deficits in offspring exposed prenatally to poly(I:C)-induced maternal inflammation, such that postsynaptic deficits occur in pubescence, whereas, adults are prone to presynaptic impairments (Giovanoli et al., 2016a). These synaptic deficiencies occurred in the absence of increased microglia density and activation in the hippocampus, however, there was elevated hippocampal IL-1β expression in adult offspring that were exposed to maternal inflammation (Giovanoli et al., 2016a). The lack of microglia activation was unexpected given the substantial evidence that microglia are instrumental in synaptic pruning (Paolicelli et al., 2011; Schafer et al., 2012). The authors suggest that the hippocampal synaptic deficits are not a direct consequence of altered synaptic pruning by microglia (Giovanoli et al., 2016a). However, it may be that microglia were affected in an earlier critical period of neurodevelopment, in a manner that contributed to delayed development of synaptic deficits; yet, additional studies are needed to elucidate the role of microglia and the mechanism by which poly(I:C)-induced maternal inflammation leads to synaptic deficits in affected offspring. The presence of synaptic deficits in SZ and ASD make these findings particularly interesting (Hazlett et al., 2008, Perry et al., 2007, Swerdlow et al., 2017).

A recent study by Choi et al. (2016) found that poly(I:C) administration (20 mg/kg; i.p.) on GD12.5 to pregnant mice resulted in increased cytokine (IL-6, IL-1β, TNF-α and IFN-β) levels in maternal circulation 3 h postinjection; subsequently on GD14.5, IL-17a levels were increased. Furthermore, IL-17a receptor subunit a (IL-17Ra) mRNA was elevated in the cortex of fetal brain following MIA (Choi et al., 2016). The increased maternal IL-17a contributed to abnormal cortical development in offspring as pretreatment of pregnant dams with IL-17a blocking antibodies prevented poly(I:C)-induced disruption of cortical development. MIA-induced ASD-like behavior abnormalities in offspring, and similar to the disruptions in cortical development, the abnormal behaviors were mediated by increased maternal IL-17a (Choi et al., 2016). Additionally, maternal T helper 17 (T_H17) lymphocytes [specifically, $CD4^+$, retinoic acid receptor-related orphan nuclear receptor gamma t $(ROR\gamma t)^+$ T_H17 cells] were the primary source of the IL-17a (Choi et al., 2016). The IL-17a-mediated events were dependent upon IL-6, as poly(I:C) administration failed to increase circulating levels of IL-17a levels in pregnant IL-6 KO mice. Also, injection of IL-6 into pregnant dams resulted in increased IL-17a expression similar to that induced by poly(I:C) injection. Lastly, intraventricular administration of IL-17a in the fetus also led to ASD-like phenotypes. Altogether, these findings provide compelling evidence for the importance of maternal T_H17 cells and IL-17a in abnormal neurodevelopment and ASD-like behaviors induced by MIA (Choi et al., 2016).

Induction of MIA with a substantially lower dose of poly(I:C), 2 mg/kg, also induced behavioral deficits in adult offspring (Meyer and Feldon, 2012; Meyer et al., 2005). This same dose of poly(I:C) was used to induce maternal inflammation in CD-1 mice on GD12 (Goeden et al., 2016). Consistent with mild, transient inflammation, IL-6 levels were increased in circulation (serum) 24 h after immune activation, then returned to baseline by 48 h. Conversely, IL-6 levels in placenta and fetal brain were not affected by poly(I:C) injection. Maternal inflammation resulted in increased conversion of tryptophan to serotonin in the placenta with a subsequent increase in serotonin within the fetal brain (Goeden et al., 2016). The elevated serotonin in the brain disrupted neurodevelopment as indicated by deficits in serotonergic axon outgrowth/circuit formation (Goeden et al., 2016). While placenta tryptophan levels returned to baseline by 48 h, tryptophan hydroxylase 1 (TPH1) activity increased in the placenta leading to increased output of serotonin to the fetus and elevated concentrations of serotonin in the brain. However, the extent to which these changes in serotonin levels and altered neuronal circuitry in the fetal brain lead to behavioral deficits in adulthood requires further investigation (Goeden et al., 2016).

5.2 Postnatal Inflammation

5.2.1 LPS

It is well established that perinatal infection increases the risk of brain damage, namely white matter injury, as well as neurodevelopmental disorders (Alshaikh et al., 2013; Knuesel et al., 2014; Mitha et al., 2013; Yanni et al., 2017). Postnatal LPS administration is often used in preclinical models of perinatal infection. For instance, Pang et al. (2016) injected Sprague Dawley rat pups with LPS (i.p., 1 mg/kg) on P3. On P6, microglia activation was observed throughout the brain, but was most pronounced in the white matter tracts (Pang et al., 2016). Interestingly, the M2 phenotype predominated, rather than the proinflammatory, M1 polarization that occurs following intracerebral or intravenous LPS administration in adult rats (Banks and Robinson, 2010; Gottschall et al., 1992). These differences in microglia responses may be related to differences in developmental stages of these cells or reflect direct versus indirect LPS-mediated activation of microglia, depending on central or systemic LPS administration (Pang et al., 2016). It is also interesting to note that P3 postnatal LPS exposure induced an M2-like, hyperplastic response in the subventricular zone, and dentate gyrus of the hippocampus, rather than a proinflammatory, neurotoxic M1-like response (Pang et al., 2016). There was a reduction in developmentally appropriate apoptosis (which is critical for refining cell populations) in these brain regions of LPS-exposed pups (Pang et al., 2016). The effects on cell numbers may be related to increased levels of microglia-derived trophic factors (e.g., TGF-β) that were detected and effects of these factors on proliferation and/or apoptosis (Pang et al., 2016). The prolonged proliferation period for oligodendrocytes was not accompanied by a comparable increase in myelination (Pang et al., 2016). Behaviorally, postnatal LPS administration induced both communicative and cognitive impairments in offspring at P10 and P40, respectively. Together these finding are expected to provide insights into the mechanism by which perinatal inflammation disrupts neurodevelopment (Pang et al., 2016).

Perinatal inflammation often takes the form of a severe, more prolonged inflammatory condition, septicemia. Neonatal septicemia is of particular concern as it can be fatal and is associated with abnormal neurodevelopment including white matter injury, CP, and cognitive/affective disorders (Cardoso et al., 2015; Comim et al., 2016; Mitha et al., 2013; Schlapbach et al., 2011). Septicemia can be induced experimentally with repeated LPS injections, however, Cardoso et al. (2015) were among the first to induce neonatal septicemia in mice. These researchers injected mice with LPS (6 mg/kg; i.p.) daily from P4 to P6, resulting in pronounced neuroinflammation as evidenced by increased proinflammatory cytokine expression, glial activation, and subsequently a reduction in brain weight. More specifically, repeated LPS exposure resulted in cerebellar hypoplasia, neuronal loss, and hypomyelination (Cardoso et al., 2015). As discussed above, both astrocytes and microglia are instrumental in the development and maintenance of synapse formation, thus, the temporal dynamics of gliosis observed following LPS exposure, including an early activation of astrocytes and microglia predominantly in proximity to the vasculature, may have contributed to the neuronal atrophy (Cardoso et al., 2015). The decreased myelination was suggested to be the result of delayed maturation of OPC into myelin-producing cells (Cardoso et al., 2015).

5.2.2 Cytomegalovirus

Congenital cytomegalovirus (CMV) infection occurs in as many as 1% of live births in the United States and can result in abnormal neurodevelopment (Boppana et al., 2013). Indeed, recent reports highlight the association of congenital CMV infection and development of ASD (Garofoli et al., 2017). To model congenital CMV infection observed in humans, BALB/cJ mouse pups were infected (i.p.) with CMV within 12h of birth (consistent with second trimester in humans). Infection with CMV during this critical period leads to pronounced neuroinflammation involving monocyte infiltration, increased levels of proinflammatory cytokines/chemokines [TNF-α, IFN-γ, IFN-β, CCL2, and CCL5 (also known as regulated on activation, normal T cell expressed and secreted or RANTES)] and altered neurodevelopment on P8, most notably in the cerebellum (Seleme et al., 2017). Treatment of CMV-infected pups with daily (P2–P7) i.p. injections of neutralizing antibody to TNF-α (TNF-nAb) was neuroprotective (Seleme et al., 2017). Specifically, TNF-nAb decreased $CD45^{+}$ $Ly6C^{hi}$ $CD11b^{+}$ $CCR2^{+}$ activated mononuclear cells in the brain, which was secondary to a reduction in this cell population in circulation, as opposed to decreased activation in brain (Seleme et al., 2017). TNF-nAb also prevented the proinflammatory cytokine expression and developmental abnormalities in the cerebellum of CMV-infected mice. Yet, TNF-nAb did not significantly reduce viral replication in the brain or periphery. These findings indicated that virus-induced inflammation is instrumental in abnormal neurodevelopment associated with congenital CMV infection (Seleme et al., 2017). Also, a select population of peripheral monocytes were instrumental in mediating these insults in the developing brain (Seleme et al., 2017).

5.2.3 Hypoxia–Ischemia

Perinatal hypoxic–ischemic brain injury involves neuroinflammation and can lead to deficits in neurodevelopment and subsequent life-long neurological disorders (Li et al., 2017). HI-induced brain damage in neonatal rodents has been used to gain insights into the role of neuroinflammation in neurodevelopmental deficits (Li et al., 2017). While there is considerable data to suggest that white matter tracts are particularly vulnerable to hypoxic insults, there remains much to learn about the specific cellular and molecular events responsible for this damage. Certainly, HI adversely affects oligodendrocytes resulting in hypomyelination (Watzlawik et al., 2015; Zaghloul et al., 2017) and Deng et al. (2014) provided insights in terms of the role of astrocytes in HI-induced hypomyelination in the developing brain.

Wistar rat pups were placed in a hypoxic chamber (5% oxygen/95% nitrogen) for 2 h then allowed to recover under normoxic conditions for 7 or 28 days (Deng et al., 2014). Expression of TNF-α and IL-1β was increased in astrocytes within the paraventricular white matter (PWM) 7 days after hypoxic insult. There were fewer oligodendrocytes in the PWM of hypoxia-exposed rats, and the percentage of these oligodendrocytes undergoing apoptosis was greater than in controls (Deng et al., 2014). Importantly, the oligodendrocytes in the PWM of hypoxia-exposed rats had increased TNF-α receptor 1 (TNF-R1) and IL-1β receptor 1 (IL-1R1) expression. Further in vitro experiments showed that 3 h exposure to hypoxia resulted in increased TNF-α and IL-1β in astrocytes and increased TNF-R1 and IL-1R1 in oligodendrocytes. Together these findings suggest that astrocyte-derived cytokines are instrumental in hypomyelination and PWM damage induced by HI (Deng et al., 2014).

More recently, Lan et al. (2016) subjected Sprague Dawley rat pups (P5) to HI by ligating the right common carotid artery, followed by 90 min exposure in a hypoxic (6% oxygen/94% nitrogen) environment. Examination of behavior at P8 and P20 revealed significant deficits in sensorimotor and cognitive function in HI-exposed rats. Additionally, at P21, HI-exposed rats displayed pronounced brain injury in brain regions that are instrumental in the examined behaviors. More specifically, there were fewer mature oligodendrocytes and hypomyelination in white matter tracts; neuronal degeneration in the striatum, cortex, and hippocampus; and microglial activation in the white matter tracts, striatum, cortex, and hippocampus (Lan et al., 2016).

Another group used C57/BL6 mice [wild-type and with genetic alterations in signal transducer and activator of transcription 3 (STAT3) expression] to study the effects of HI on neurodevelopment (Hristova et al., 2016). The transcription factor STAT3 has been implicated in HI-induced brain injury (Heusch et al., 2011; Shrivastava et al., 2013), but the role of STAT3 activation in specific brain cell types in HI-induced neonatal brain development are not fully understood. Therefore, HI was induced in P7 pups via unilateral carotid artery ligation followed by 60 min exposure to a hypoxic (8% oxygen/92% nitrogen) environment (Hristova et al., 2016). Similar to other models of HI (Lan et al., 2016), this procedure leads to brain damage as evidenced by neuronal death, microglia and astrocyte activation, and neurobehavioral deficits (Hristova et al., 2016; Vannucci and Vannucci, 1997). Hristova et al. (2016) were able to demonstrate that neuronal-specific deletion of STAT3, and to a lesser extent, astrocyte-specific STAT3 deletion protected against HI-induced neuronal death and glial activation. Further investigation revealed that pharmacologic inhibition of STAT3 activation also prevented HI-induced damage in neonatal mice. Overall, these findings suggest that STAT3 activation, particularly in neurons, is instrumental in HI-induced damage in the developing brain (Hristova et al., 2016).

The impact of chronic, perinatal hypoxia (pCH) on early neurodevelopment has also been recently investigated in a preclinical mouse model (Ortega et al., 2016). Exposure of dams and pups to a hypoxic (10% oxygen) environment from P3 to P14 or P3 to P28 resulted in reduced cortical myelination, that was still present in mice that were exposed to hypoxia out to P28, followed by normoxic conditions from P28 to P56. Vestibulomotor function was also disrupted in the mice that were exposed to pCH followed by 4 weeks of normal oxygen levels. Mechanistically, there was evidence that hypomyelination was due in part to a loss of myelin-producing oligodendrocytes and an influx of MBP-specific, autoreactive CD4 T cells into the developing brain during pCH (Ortega et al., 2016).

There is substantial evidence that HI, in conjunction with bacterial infection, increases the risk of neonatal encephalopathy, including long-term neurological disorders, such as CP (Wang et al., 2014). Girard et al. (2009) was among the first to describe a rat model of CP that exhibited characteristic, developmental brain damage along with motor deficits similar to those in very-preterm human neonates. This preclinical CP model involved exposing pregnant Lewis rats to LPS in combination with HI (Girard et al., 2009). Others have used similar protocols to induce neonatal encephalopathy reminiscent of that observed in the human cases of CP, in order to identify the role of inflammation in neurodevelopment (Osredkar et al., 2015; Savard et al., 2015). To model infection in combination with HI in term neonates, P12 rats were administered a single i.p. injection of LPS; then 4 h later subjected to unilateral carotid artery ligation and hypoxia (8% oxygen/92% nitrogen) for 90 min (Savard et al., 2015). Neuroinflammation was evident 4–48 h after hypoxic insult, as IL-1β, CCL5, and CCL2 were elevated in the neocortex, white matter, and striatum early in the postinsult period; then microglia activation and polymorphonuclear cell infiltration were evident at 48 h post-LPS + HI exposure. Exposure to LPS + HI also resulted in increased matrix metalloproteinase-9 (MMP-9) expression in the brain; interestingly neurons, and not microglia or astrocytes, were identified as the source. Additionally, neuroinflammation contributed to apoptotic and necrotic cell death and subsequent motor function deficits out to P120. Finally, pharmacologic blockade of IL-1β or MMP-9 (with IL-1 receptor antagonist (IL-1Ra) or SB3CT, respectively) protected against LPS + HI-mediated brain damage and motor function deficits (Savard et al., 2015).

More recently, van Tilborg et al. (2017) used a different model of perinatal LPS + HI to assess the detrimental effects on white matter development and accompanying motor/behavioral abnormalities. MIA was induced in pregnant rats with i.p. LPS injections on GD18 and GD19; then on P4 offspring were subjected to hypoxia for 140 min. Combined MIA and postnatal hypoxia led to glial activation and deficits in myelination at P18; and impaired oligodendrocyte maturation was still present 10 weeks after birth (van Tilborg et al., 2017). The white matter injury was associated with motor function deficits and ASD-like behaviors, including asocial behavior and increased repetitive grooming (van Tilborg et al., 2017).

Epigenetic modifications, including histone acetylation/deacetylation are instrumental in gene expression, neurodevelopment, and brain function (Rudenko and Tsai, 2014). White matter injury associated with perinatal inflammation appears to be due in part to dysregulation of histone acetylation in the developing oligodendrocytes (Huang et al., 2018). Huang et al. (2018) demonstrated that LPS + HI-induced white matter injury was prevented by i.p. administration of the histone deacetylase inhibitor, sodium butyrate. In addition to maintaining normal histone acetylation, treatment with sodium butyrate prevented the reduced expression of BDNF and its receptor, tropomyosin receptor kinase B (TrkB) (Huang et al., 2018). More studies are needed to determine the specific cell types in which BDNF/TrkB expression were affected, but these findings are consistent with a critical role of BDNF-TrkB signaling in neurodevelopment (Huang et al., 2018; Lu et al., 2014).

Chorioamnionitis and neonatal sepsis are often caused by Gram-positive organisms that stimulate immune cells through activation of Toll-like receptor 2 (TLR2) (Stoll et al., 2004). Therefore, Mottahedin et al. (2017) utilized the lipopeptide TLR2 ligand, $Pam_3CysSerLys_4$ (P3C) in combination with HI to induce neonatal brain injury. Mouse pups (TLR2 +/+ and TLR2 −/−) on P8 were injected i.p. with P3C, then 14 h later subjected to HI (unilateral carotid artery ligation) followed by 50 min in a hypoxic chamber (10% oxygen). Inflammation and brain damage were then assessed 5 days later on P14. Injection of P3C plus HI resulted in significant neural tissue loss (cerebrum) and hypomyelination in TLR2 +/+ mice, whereas, TLR2 −/− were protected (Mottahedin et al., 2017). Previously it was shown that repeated P3C injections from P3 to P11 induced proinflammatory cytokine expression, microglia accumulation, and neuronal tissue loss on P12, but by adulthood there was apparently full recovery (Du et al., 2011). Thus, the more recent findings make clear that a single injection of P3C in combination with HI during brain development results in even more pronounced neonatal brain injury. The 2017 study by Mottahedin et al. is one of the first studies to specifically address the role of TLR2 in HI related neonatal brain injury, and together with previous insights, suggests that TLR2 signaling pathways are potentially instrumental in Gram-positive mediated perinatal infection and injury to the developing brain.

5.2.4 Intraventricular Hemorrhage

A common complication of preterm birth is IVH, and it is associated with neurodevelopmental impairment (Mukerji et al., 2015). IVH induces inflammation in the PWM, typically during periods of oligodendrocyte maturation (Dohare et al., 2016; Georgiadis et al., 2008; Vinukonda et al., 2010). The increased microglia/astrocyte activation and elevated levels of proinflammatory cytokines likely contribute to this white matter damage (Chua et al., 2009; Vinukonda et al., 2010). More specifically, it is the premyelinating oligodendrocytes that are thought to be most vulnerable to inflammatory insult, particularly HI and glutamate receptor mediated toxicity (Back et al., 2002; Fern and Moller, 2000; Itoh et al., 2002). Dohare et al. (2016) further addressed these aspects of IVH-induced inflammation and developmental brain damage in premature rabbits. NZW rabbit pups were delivered 2 days preterm (GD29) by C-section, then 4 h later administered i.p. glycerol to induce vessel rupture in the ganglionic eminence (Georgiadis et al., 2008). IVH resulted in inflammation of periventricular white matter as indicated by increased proinflammatory cytokine expression (TNF-α, IL-1β, and IL-6) and microglia infiltration (Dohare et al., 2016). There was also an increase in apoptotic OPCs, decreased OPC maturation, and hypomyelination (Dohare et al., 2016). The neurological insults were prevented by pharmacologic inhibition of AMPA-kainate receptors, thereby providing insight into the critical role of these receptors in IVH-induced perinatal brain damage (Dohare et al., 2016).

5.3 Therapeutic Strategies

As mentioned briefly in the previous sections, pharmacologic agents have been used to identify mechanisms involved in perinatal-induced inflammation, yet, there remains a great need for the development of neuroprotective strategies to prevent or reduce adverse effects of perinatal infection/inflammation on brain development. The neuroprotective effects of numerous pharmacologic agents have been tested across a variety of preclinical models of perinatal infection/inflammation. A previous review by Hagberg et al. (2015) summarized the extent to which different agents protected against inflammation-induced perinatal brain injury; among these agents were: corticosteroids, melatonin, minocycline, *N*-acetylcysteine (NAC), statins, COX inhibitors, IL-1Ra, and a TNF Rc blocker (Hagberg et al., 2015). All of these agents provided some degree of protection in animal models of perinatal-induced brain damage (Hagberg et al., 2015). There have since been additional studies investigating the therapeutic potential of these, and other, pharmacologic agents; the key findings from some of these recent studies will be discussed next.

5.3.1 NAC

Systemic inflammation was induced in pregnant rats with an i.p. injection of LPS (25 mg/kg) on GD14 (Chao et al., 2016). Maternal inflammation resulted in increased oxidative stress and inflammation as indicated by elevated reactive oxygen species and IL-6 in the amniotic fluid as well as increased expression of antioxidant enzymes (catalase, heme oxygenase-1, superoxide dismutase), and microglial activation in the fetal brain on GD18 (Chao et al., 2016). Furthermore, there was significant intrauterine fetal death on GD18. Concurrent treatment of the pregnant dam with NAC was neuroprotective and prevented fetal loss, highlighting the importance of oxidative stress in the LPS-induced fetal brain damage. Maternal inflammation also led to disruption of laminar organization of the developing neocortex, but NAC was neuroprotective (Chao et al., 2016). These findings add support to potential therapeutic benefits of NAC or related compounds in the prevention of perinatal inflammation and associated neurodevelopmental abnormalities (Awad et al., 2011; Chang et al., 2011; Lante et al., 2008).

5.3.2 Minocycline

The protective effects of minocycline were tested in a preclinical rodent model of SZ (Giovanoli et al., 2016b). In this two-hit model of neurodevelopmental stress, pregnant C57BL6/J mice received i.p. poly(I:C) at a dose of 1 mg/kg on GD9, then peripubertal offspring were subjected to variable, unpredictable stress between P30 and P40 (stressors included: electric foot shock, restraint, swimming, food restriction, and repeated change of home cage) (Giovanoli et al., 2016b). As adults (P70–P90), offspring were subjected behavioral analysis; neuroimmune activation was assessed in a subset of mice on P41 (1 day after the last peripubertal stressor). As expected, the prenatal inflammatory stress followed by subchronic stress during peripubertal maturation, resulted in neuroinflammation as revealed by increased microglial activation and IL-1β expression in the hippocampus and prefrontal cortex of offspring exposed to LPS/stress during neurodevelopment (Giovanoli et al., 2016b). Offspring exposed to combined prenatal immune activation followed by peripubertal stress also displayed anxiety-like behavior and deficits in PPI in adulthood. When offspring were exposed prenatally to LPS-induced inflammation then treated with minocycline (in drinking water) 24 h prior to the first peripubertal stressor, and throughout the completion of the final stressor, they were protected from neuroinflammation and sensorimotor gating remained normal; however, anxiety-like behavior remained (Giovanoli et al., 2016b). Further investigation into the neuroprotective actions of minocycline are certainly warranted, yet, these findings support a potential role for minocycline treatment in the prevention of certain neurodevelopmental disorders, such as SZ.

5.3.3 7,8-Dihydrooxyflavone

BDNF signaling through TrkB is critical for normal neurodevelopment, and decreased expression of both BDNF and TrkB have been observed in the hippocampus and prefrontal cortex of individuals with SZ (Nieto et al., 2013; Ray et al., 2014; Takahashi et al., 2000; Waterhouse and Xu, 2009). Additionally, a prospective analysis revealed that subjects exposed to fetal hypoxia, then later diagnosed with SZ, also had reduced BDNF levels in cord blood obtained at birth (Cannon et al., 2008). In humans, there is a lack of information on the effects of perinatal infection/inflammation on BDNF levels in the fetal brain; however, it is known that cord BDNF levels are reduced in fetal hypoxia (Cannon et al., 2008) and prematurity (Dhobale et al., 2012). There is interest in elucidating the impact of perinatal inflammation on BDNF signaling during neurodevelopment and the potential for pharmacologic intervention. Han et al. (2016) recently used ddY mice to investigate the effects of MIA [poly(I:C), i.p., daily from GD12–GD17] on BDNF-TrkB signaling in the brain and SZ-like behaviors in adult offspring. Prenatal poly(I:C) exposure resulted in decreased BDNF-TrkB signaling in the prefrontal cortex, but not the nucleus accumbens or hippocampus, in juvenile (4–5 weeks of age) offspring; as well as cognitive deficits. Adult (10–12 weeks of age) offspring exposed prenatally to poly(I:C) had reduced BDNF-TrkB signaling in the prefrontal cortex and CA1 region of the hippocampus along with deficits in cognition and PPI (Han et al., 2016). Supplementation with the TrkB agonist, 7,8-dihydroxyflavone, in the drinking water during juvenile and adolescent stages prevented MIA-induced deficits in BDNF-TrkB signaling and behavior in adult offspring (Han et al., 2016).

More recently, this same model of MIA was used to demonstrate that treatment of the pregnant dam with 7,8-dihydroxyflavone also provided neuroprotection in adult offspring (Han et al., 2017). More specifically, poly(I:C)-injected, pregnant dams were provided 7,8-dihydroxyflavone in drinking water from GD12 to P21 (weaning of offspring), which was sufficient to prevent the reduction in BDNF-TrkB signaling (in PFC) and cognitive deficits in adult offspring. These novel findings suggest that prophylactic treatment with a TrkB agonist may protect against inflammation-mediated brain injury during neurodevelopment, thus providing a potential therapeutic strategy to prevent SZ, ASD, or other related disorders (Han et al., 2017).

5.3.4 Melatonin

The pineal gland is a major site of melatonin synthesis, but numerous other tissues can also produce melatonin (Hardeland et al., 2011). Melatonin is well characterized in terms of its physiological functions, including circadian regulation, immune function, antioxidant properties, anti-apoptotic actions, and mitochondrial function (Hardeland et al., 2011). Administration of melatonin can be neuroprotective in the context of inflammation-induced preterm birth and perinatal brain injury (Dominguez Rubio et al., 2014; Wong et al., 2014; Xu et al., 2007). Recently, Dominguez Rubio et al. (2017) demonstrated that maternal melatonin administration protected the fetal brain from MIA. In this study, two separate i.p. injections of LPS to pregnant BALB/C mice on GD15 resulted in 100% preterm birth/pup death and neuroinflammation in the fetus as indicated by microglia activation and iNOS expression (2 h after last LPS injection). Subcutaneous implantation of a 25 mg melatonin pellet in pregnant dams 1 day prior to LPS injections resulted in 50% preterm birth/pup survival, and reduced neuroinflammation. Behavioral measures of exploratory and anxiety-like behaviors on P90 revealed no differences among the offspring from the three groups of mothers with surviving offspring, control-, melatonin-, or melatonin + LPS-treated (Dominguez Rubio et al., 2017). Further analysis of the neuroprotective effects of melatonin using additional models of maternal inflammation are warranted.

5.3.5 Paliperidone

The atypical antipsychotic, paliperidone was recently evaluated in terms of its immunomodulatory and neuroprotective effects in a murine model of SZ (MacDowell et al., 2017). Pregnant C57BL/6J mice were injected with poly(I:C) (i.p., 5 mg/kg) on GD9.5; then on P60, young-adult offspring were treated daily with paliperidone or saline (i.p.) until P80. MIA resulted in neuroinflammation and oxidative stress in the frontal cortex as indicated by increased expression of inflammatory mediators/proinflammatory cytokines, and lipid peroxidation. Prenatal exposure to maternal inflammation also caused deficits in working memory and cognitive flexibility in adult offspring. However, paliperidone treatment enhanced endogenous antioxidants and anti-inflammatory systems, namely nuclear factor E2-related factor 2 (NRF2)- and peroxisome proliferator-activated receptor gamma (PPAR-γ)-mediated pathways, and induced M2 polarization of microglia in the frontal cortex. Paliperidone also protected against behavioral deficits in adult offspring exposed prenatally to maternal inflammation (MacDowell et al., 2017).

5.3.6 Hyaluronidase

IVH in premature infants results in inflammation and white matter damage (Mukerji et al., 2015). Deficits in OPC maturation are instrumental in the hypomyelination present in white matter injury (Back, 2017; Buser et al., 2012). Hyaluronan (HA), is a glycosaminoglycan polymer and a major component of the extracellular matrix in the brain and has been implicated in the etiology of white matter injury, particularly following IVH (Buser et al., 2012). Vinukonda et al. (2016) used a preclinical model of IVH in preterm rabbits to induce neuroinflammation, hypomyelination, and motor function deficits; and tested the protective effects of intracerebroventricular hyaluronidase treatment (Vinukonda et al., 2016). The levels of HA in the brain were not affected by IVH, however, there was increased expression of CD44, TLR2, and TLR4, which are receptors that are activated by HA. Interestingly, hyaluronidase treatment reduced expression of all three receptors and prevented the increase in proinflammatory cytokines and microglial activation. Furthermore, hyaluronidase treatment prevented the deficits in OPC maturation, myelination, and motor function (Vinukonda et al., 2016). Additional studies are needed to determine the functional importance of increased HA-receptor expression in the protective effects of hyaluronidase treatment.

6. CONCLUSIONS

Clinical and epidemiological evidence point to the negative impact of maternal/perinatal inflammation on neurodevelopment. Numerous preclinical models have been developed that recapitulate many of the neuropathological and behavioral features of neurodevelopmental disorders such as CP, SZ, and ASD. Researchers around the world have successfully used these animal models to gain insights into the effects of maternal/perinatal inflammation on critical cellular and molecular events involved in neurodevelopment. There is substantial evidence that maternal/perinatal inflammation induces neuroinflammation in offspring as indicated by increased cytokine expression and microglia/astrocyte activation. IL-1β, TNF-α, and IL-6 are likely the most documented cytokines involved, yet mounting evidence points to a critical role for IL-17A. Likewise, neurotransmitters such as serotonin and glutamate have been implicated in inflammation-mediated alterations in neurodevelopment. White matter damage seems to be most vulnerable to inflammatory insult during neurodevelopment, and this is partly a consequence of altered OPC maturation and myelination. Substantial evidence also points to altered microglia/astroglia function in the neuroinflammatory events associated with neurodevelopmental insults.

Of course there remains much to discover, but significant advancements are being made in terms of the mechanisms involved in inflammatory-mediated deficits in neurodevelopment. Importantly, these advances have led to novel, therapeutic strategies that show promise in preclinical models. It is expected therefore, that further investigation will lead to effective therapeutic options to prevent or minimize the adverse effects of maternal/perinatal inflammation on neurodevelopment; and pharmacological strategies will likely involve maternal treatment and/or treatment in the postnatal period.

REFERENCES

Alshaikh, B., Yusuf, K., Sauve, R., 2013. Neurodevelopmental outcomes of very low birth weight infants with neonatal sepsis: systematic review and meta-analysis. J. Perinatol. 33, 558–564.

American Psychiatric Association, 2013. Diagnostic and Statistical Manual of Mental Disorders, fifth ed. American Psychiatric Publishing, Arlington, VA.

Andersen, S.L., 2003. Trajectories of brain development: point of vulnerability or window of opportunity? Neurosci. Biobehav. Rev. 27, 3–18.

Arcuri, C., Mecca, C., Bianchi, R., Giambanco, I., Donato, R., 2017. The pathophysiological role of microglia in dynamic surveillance, phagocytosis and structural remodeling of the developing CNS. Front. Mol. Neurosci. 10, 191.

Arnold, T., Betsholtz, C., 2013. Correction: the importance of microglia in the development of the vasculature in the central nervous system. Vasc. Cell 5, 12.

Avendano, B.C., Montero, T.D., Chavez, C.E., von Bernhardi, R., Orellana, J.A., 2015. Prenatal exposure to inflammatory conditions increases Cx43 and Panx1 unopposed channel opening and activation of astrocytes in the offspring effect on neuronal survival. Glia 63, 2058–2072.

Awad, N., Khatib, N., Ginsberg, Y., Weiner, Z., Maravi, N., Thaler, I., Ross, M.G., Itsokovitz-Eldor, J., Beloosesky, R., 2011. N-acetyl-cysteine (NAC) attenuates LPS-induced maternal and amniotic fluid oxidative stress and inflammatory responses in the preterm gestation. Am. J. Obstet. Gynecol. 204 (450), e15–20.

Back, S.A., 2017. White matter injury in the preterm infant: pathology and mechanisms. Acta Neuropathol. 134, 331–349.

Back, S.A., Han, B.H., Luo, N.L., Chricton, C.A., Xanthoudakis, S., Tam, J., Arvin, K.L., Holtzman, D.M., 2002. Selective vulnerability of late oligodendrocyte progenitors to hypoxia-ischemia. J. Neurosci. 22, 455–463.

Banks, W.A., Robinson, S.M., 2010. Minimal penetration of lipopolysaccharide across the murine blood-brain barrier. Brain Behav. Immun. 24, 102–109.

Belanger, M., Allaman, I., Magistretti, P.J., 2011. Brain energy metabolism: focus on astrocyte-neuron metabolic cooperation. Cell Metab. 14, 724–738.

Ben Haim, L., Rowitch, D.H., 2017. Functional diversity of astrocytes in neural circuit regulation. Nat. Rev. Neurosci. 18, 31–41.

Bialas, A.R., Stevens, B., 2013. TGF-beta signaling regulates neuronal C1q expression and developmental synaptic refinement. Nat. Neurosci. 16, 1773–1782.

Bilbo, S.D., Schwarz, J.M., 2009. Early-life programming of later-life brain and behavior: a critical role for the immune system. Front. Behav. Neurosci. 3, 14.

Bjornsen, L.P., Hadera, M.G., Zhou, Y., Danbolt, N.C., Sonnewald, U., 2014. The GLT-1 (EAAT2; slc1a2) glutamate transporter is essential for glutamate homeostasis in the neocortex of the mouse. J. Neurochem. 128, 641–649.

Boppana, S.B., Ross, S.A., Fowler, K.B., 2013. Congenital cytomegalovirus infection: clinical outcome. Clin. Infect. Dis. 57 (Suppl. 4), S178–81.

Bosworth, A.P., Allen, N.J., 2017. The diverse actions of astrocytes during synaptic development. Curr. Opin. Neurobiol. 47, 38–43.

Brown, A.S., Derkits, E.J., 2010. Prenatal infection and schizophrenia: a review of epidemiologic and translational studies. Am. J. Psychiatry 167, 261–280.

Brown, A.S., Patterson, P.H., 2011. Maternal infection and schizophrenia: implications for prevention. Schizophr. Bull. 37, 284–290.

Buosi, A.S., Matias, I., Araujo, A.P., Batista, C., Gomes, F.C., 2018. Heterogeneity in synaptogenic profile of astrocytes from different brain regions. Mol. Neurobiol. 55 (1), 751–762.

Buser, J.R., Maire, J., Riddle, A., Gong, X., Nguyen, T., Nelson, K., Luo, N.L., Ren, J., Struve, J., Sherman, L.S., Miller, S.P., Chau, V., Hendson, G., Ballabh, P., Grafe, M.R., Back, S.A., 2012. Arrested preoligodendrocyte maturation contributes to myelination failure in premature infants. Ann. Neurol. 71, 93–109.

Cannon, T.D., Yolken, R., Buka, S., Torrey, E.F., Collaborative Study Group on the Perinatal Origins of Severe Psychiatric Disorders, 2008. Decreased neurotrophic response to birth hypoxia in the etiology of schizophrenia. Biol. Psychiatry 64, 797–802.

Cardoso, F.L., Herz, J., Fernandes, A., Rocha, J., Sepodes, B., Brito, M.A., McGavern, D.B., Brites, D., 2015. Systemic inflammation in early neonatal mice induces transient and lasting neurodegenerative effects. J. Neuroinflammation 12, 82.

Carmona, M.A., Murai, K.K., Wang, L., Roberts, A.J., Pasquale, E.B., 2009. Glial ephrin-A3 regulates hippocampal dendritic spine morphology and glutamate transport. Proc. Natl. Acad. Sci. U. S. A. 106, 12524–12529.

Cassella, S.N., Hemmerle, A.M., Lundgren, K.H., Kyser, T.L., Ahlbrand, R., Bronson, S.L., Richtand, N.M., Seroogy, K.B., 2016. Maternal immune activation alters glutamic acid decarboxylase-67 expression in the brains of adult rat offspring. Schizophr. Res. 171, 195–199.
Chang, E.Y., Zhang, J., Sullivan, S., Newman, R., Singh, I., 2011. N-acetylcysteine attenuates the maternal and fetal proinflammatory response to intrauterine LPS injection in an animal model for preterm birth and brain injury. J. Matern. Fetal Neonatal Med. 24, 732–740.
Chao, M.W., Chen, C.P., Yang, Y.H., Chuang, Y.C., Chu, T.Y., Tseng, C.Y., 2016. N-acetylcysteine attenuates lipopolysaccharide-induced impairment in lamination of Ctip2-and Tbr1-expressing cortical neurons in the developing rat fetal brain. Sci. Rep. 6, 32373.
Chattopadhyaya, B., Di Cristo, G., Wu, C.Z., Knott, G., Kuhlman, S., Fu, Y., Palmiter, R.D., Huang, Z.J., 2007. GAD67-mediated GABA synthesis and signaling regulate inhibitory synaptic innervation in the visual cortex. Neuron 54, 889–903.
Choi, S.S., Lee, H.J., Lim, I., Satoh, J., Kim, S.U., 2014. Human astrocytes: secretome profiles of cytokines and chemokines. PLoS One 9, e92325.
Choi, G.B., Yim, Y.S., Wong, H., Kim, S., Kim, H., Kim, S.V., Hoeffer, C.A., Littman, D.R., Huh, J.R., 2016. The maternal interleukin-17a pathway in mice promotes autism-like phenotypes in offspring. Science 351, 933–939.
Christensen, D.L., Baio, J., Van Naarden Braun, K., Bilder, D., Charles, J., Constantino, J.N., Daniels, J., Durkin, M.S., Fitzgerald, R.T., Kurzius-Spencer, M., Lee, L.C., Pettygrove, S., Robinson, C., Schulz, E., Wells, C., Wingate, M.S., Zahorodny, W., Yeargin-Allsopp, M., Centers for Disease Control and Prevention (CDC), 2016. Prevalence and characteristics of autism spectrum disorder among children aged 8 years—autism and developmental disabilities monitoring network, 11 sites, United States, 2012. MMWR Surveill. Summ. 65, 1–23.
Chua, C.O., Chahboune, H., Braun, A., Dummula, K., Chua, C.E., Yu, J., Ungvari, Z., Sherbany, A.A., Hyder, F., Ballabh, P., 2009. Consequences of intraventricular hemorrhage in a rabbit pup model. Stroke 40, 3369–3377.
Chung, W.S., Clarke, L.E., Wang, G.X., Stafford, B.K., Sher, A., Chakraborty, C., Joung, J., Foo, L.C., Thompson, A., Chen, C., Smith, S.J., Barres, B.A., 2013. Astrocytes mediate synapse elimination through MEGF10 and MERTK pathways. Nature 504, 394–400.
Clarke, L.E., Barres, B.A., 2013. Emerging roles of astrocytes in neural circuit development. Nat. Rev. Neurosci. 14, 311–321.
Colombo, E., Farina, C., 2016. Astrocytes: key regulators of neuroinflammation. Trends Immunol. 37, 608–620.
Colver, A., Fairhurst, C., Pharoah, P.O., 2014. Cerebral palsy. Lancet 383, 1240–1249.
Comim, C.M., Bussmann, R.M., Simao, S.R., Ventura, L., Freiberger, V., Patricio, J.J., Palmas, D., Mendonca, B.P., Cassol Jr., O.J., Quevedo, J., 2016. Experimental neonatal sepsis causes long-term cognitive impairment. Mol. Neurobiol. 53, 5928–5934.
Cordeiro, C.N., Tsimis, M., Burd, I., 2015. Infections and brain development. Obstet. Gynecol. Surv. 70, 644–655.
Cunningham, C.L., Martinez-Cerdeno, V., Noctor, S.C., 2013. Microglia regulate the number of neural precursor cells in the developing cerebral cortex. J. Neurosci. 33, 4216–4233.
Debnath, M., Venkatasubramanian, G., Berk, M., 2015. Fetal programming of schizophrenia: select mechanisms. Neurosci. Biobehav. Rev. 49, 90–104.
De Bundel, D., Zussy, C., Espallergues, J., Gerfen, C.R., Girault, J.A., Valjent, E., 2016. Dopamine D2 receptors gate generalization of conditioned threat responses through mTORC1 signaling in the extended amygdala. Mol. Psychiatry 21, 1545–1553.
Deng, Y., Xie, D., Fang, M., Zhu, G., Chen, C., Zeng, H., Lu, J., Charanjit, K., 2014. Astrocyte-derived proinflammatory cytokines induce hypomyelination in the periventricular white matter in the hypoxic neonatal brain. PLoS One 9, e87420.

Deshmukh, M., Johnson Jr., E.M., 1997. Programmed cell death in neurons: focus on the pathway of nerve growth factor deprivation-induced death of sympathetic neurons. Mol. Pharmacol. 51, 897–906.
Dhobale, M., Mehendale, S., Pisal, H., D'Souza, V., Joshi, S., 2012. Association of brain-derived neurotrophic factor and tyrosine kinase B receptor in pregnancy. Neuroscience 216, 31–37.
Dohare, P., Zia, M.T., Ahmed, E., Ahmed, A., Yadala, V., Schober, A.L., Ortega, J.A., Kayton, R., Ungvari, Z., Mongin, A.A., Ballabh, P., 2016. AMPA-kainate receptor inhibition promotes neurologic recovery in premature rabbits with intraventricular hemorrhage. J. Neurosci. 36, 3363–3377.
Dominguez Rubio, A.P., Sordelli, M.S., Salazar, A.I., Aisemberg, J., Bariani, M.V., Cella, M., Rosenstein, R.E., Franchi, A.M., 2014. Melatonin prevents experimental preterm labor and increases offspring survival. J. Pineal Res. 56, 154–162.
Dominguez Rubio, A.P., Correa, F., Aisemberg, J., Dorfman, D., Bariani, M.V., Rosenstein, R.E., Zorrilla Zubilete, M., Franchi, A.M., 2017. Maternal administration of melatonin exerts short- and long-term neuroprotective effects on the offspring from lipopolysaccharide-treated mice. J. Pineal Res. 63, e12439.
Du, X., Fleiss, B., Li, H., D'Angelo, B., Sun, Y., Zhu, C., Hagberg, H., Levy, O., Mallard, C., Wang, X., 2011. Systemic stimulation of TLR2 impairs neonatal mouse brain development. PLoS One 6, e19583.
Ebrahimi, M., Yamamoto, Y., Sharifi, K., Kida, H., Kagawa, Y., Yasumoto, Y., Islam, A., Miyazaki, H., Shimamoto, C., Maekawa, M., Mitsushima, D., Yoshikawa, T., Owada, Y., 2016. Astrocyte-expressed FABP7 regulates dendritic morphology and excitatory synaptic function of cortical neurons. Glia 64, 48–62.
Elovitz, M.A., Brown, A.G., Breen, K., Anton, L., Maubert, M., Burd, I., 2011. Intrauterine inflammation, insufficient to induce parturition, still evokes fetal and neonatal brain injury. Int. J. Dev. Neurosci. 29, 663–671.
Emery, B., Lu, Q.R., 2015. Transcriptional and epigenetic regulation of oligodendrocyte development and myelination in the central nervous system. Cold Spring Harb. Perspect. Biol. 7, A020461.
Fantin, A., Vieira, J.M., Gestri, G., Denti, L., Schwarz, Q., Prykhozhij, S., Peri, F., Wilson, S.W., Ruhrberg, C., 2010. Tissue macrophages act as cellular chaperones for vascular anastomosis downstream of VEGF-mediated endothelial tip cell induction. Blood 116, 829–840.
Feigenson, K.A., Kusnecov, A.W., Silverstein, S.M., 2014. Inflammation and the two-hit hypothesis of schizophrenia. Neurosci. Biobehav. Rev. 38, 72–93.
Fern, R., Moller, T., 2000. Rapid ischemic cell death in immature oligodendrocytes: a fatal glutamate release feedback loop. J. Neurosci. 20, 34–42.
Fernandez de Cossio, L., Guzman, A., van der Veldt, S., Luheshi, G.N., 2017. Prenatal infection leads to ASD-like behavior and altered synaptic pruning in the mouse offspring. Brain Behav. Immun. 63, 88–98.
Fonseca, M.I., Chu, S.H., Hernandez, M.X., Fang, M.J., Modarresi, L., Selvan, P., Macgregor, G.R., Tenner, A.J., 2017. Cell-specific deletion of C1qa identifies microglia as the dominant source of C1q in mouse brain. J. Neuroinflammation 14, 48.
Forbes, T.A., Gallo, V., 2017. All wrapped up: environmental effects on myelination. Trends Neurosci. 40, 572–587.
Garofoli, F., Lombardi, G., Orcesi, S., Pisoni, C., Mazzucchelli, I., Angelini, M., Balottin, U., Stronati, M., 2017. An Italian prospective experience on the association between congenital cytomegalovirus infection and autistic spectrum disorder. J. Autism Dev. Disord. 47, 1490–1495.
Georgiadis, P., Xu, H., Chua, C., Hu, F., Collins, L., Huynh, C., Lagamma, E.F., Ballabh, P., 2008. Characterization of acute brain injuries and neurobehavioral profiles in a rabbit model of germinal matrix hemorrhage. Stroke 39, 3378–3388.

Ginhoux, F., Greter, M., Leboeuf, M., Nandi, S., See, P., Gokhan, S., Mehler, M.F., Conway, S.J., Ng, L.G., Stanley, E.R., Samokhvalov, I.M., Merad, M., 2010. Fate mapping analysis reveals that adult microglia derive from primitive macrophages. Science 330, 841–845.

Giovanoli, S., Weber-Stadlbauer, U., Schedlowski, M., Meyer, U., Engler, H., 2016a. Prenatal immune activation causes hippocampal synaptic deficits in the absence of overt microglia anomalies. Brain Behav. Immun. 55, 25–38.

Giovanoli, S., Engler, H., Engler, A., Richetto, J., Feldon, J., Riva, M.A., Schedlowski, M., Meyer, U., 2016b. Preventive effects of minocycline in a neurodevelopmental two-hit model with relevance to schizophrenia. Transl. Psychiatry 6, e772.

Girard, S., Kadhim, H., Beaudet, N., Sarret, P., Sebire, G., 2009. Developmental motor deficits induced by combined fetal exposure to lipopolysaccharide and early neonatal hypoxia/ischemia: a novel animal model for cerebral palsy in very premature infants. Neuroscience 158, 673–682.

Goeden, N., Velasquez, J., Arnold, K.A., Chan, Y., Lund, B.T., Anderson, G.M., Bonnin, A., 2016. Maternal inflammation disrupts fetal neurodevelopment via increased placental output of serotonin to the fetal brain. J. Neurosci. 36, 6041–6049.

Gomez-Nicola, D., Perry, V.H., 2015. Microglial dynamics and role in the healthy and diseased brain: a paradigm of functional plasticity. Neuroscientist 21, 169–184.

Gomez Perdiguero, E., Klapproth, K., Schulz, C., Busch, K., Azzoni, E., Crozet, L., Garner, H., Trouillet, C., de Bruijn, M.F., Geissmann, F., Rodewald, H.R., 2015. Tissue-resident macrophages originate from yolk-sac-derived erythro-myeloid progenitors. Nature 518, 547–551.

Gottfried, C., Bambini-Junior, V., Francis, F., Riesgo, R., Savino, W., 2015. The impact of neuroimmune alterations in autism spectrum disorder. Front. Psychiatry 6, 121.

Gottschall, P.E., Komaki, G., Arimura, A., 1992. Increased circulating interleukin-1 and interleukin-6 after intracerebroventricular injection of lipopolysaccharide. Neuroendocrinology 56, 935–938.

Hagberg, H., Mallard, C., Ferriero, D.M., Vannucci, S.J., Levison, S.W., Vexler, Z.S., Gressens, P., 2015. The role of inflammation in perinatal brain injury. Nat. Rev. Neurol. 11, 192–208.

Hamilton, S.P., Rome, L.H., 1994. Stimulation of in vitro myelin synthesis by microglia. Glia 11, 326–335.

Han, M., Zhang, J.C., Yao, W., Yang, C., Ishima, T., Ren, Q., Ma, M., Dong, C., Huang, X.F., Hashimoto, K., 2016. Intake of 7,8-dihydroxyflavone during juvenile and adolescent stages prevents onset of psychosis in adult offspring after maternal immune activation. Sci. Rep. 6, 36087.

Han, M., Zhang, J.C., Huang, X.F., Hashimoto, K., 2017. Intake of 7,8-dihydroxyflavone from pregnancy to weaning prevents cognitive deficits in adult offspring after maternal immune activation. Eur. Arch. Psychiatry Clin. Neurosci. 267, 479–483.

Hardeland, R., Cardinali, D.P., Srinivasan, V., Spence, D.W., Brown, G.M., Pandi-Perumal, S.R., 2011. Melatonin—a pleiotropic, orchestrating regulator molecule. Prog. Neurobiol. 93, 350–384.

Hazlett, E.A., Buchsbaum, M.S., Zhang, J., Newmark, R.E., Glanton, C.F., Zelmanova, Y., Haznedar, M.M., Chu, K.W., Nenadic, I., Kemether, E.M., Tang, C.Y., New, A.S., Siever, L.J., 2008. Frontal-striatal-thalamic mediodorsal nucleus dysfunction in schizophrenia-spectrum patients during sensorimotor gating. Neuroimage 42, 1164–1177.

Herculano-Houzel, S., 2014. The glia/neuron ratio: how it varies uniformly across brain structures and species and what that means for brain physiology and evolution. Glia 62, 1377–1391.

Heusch, G., Musiolik, J., Gedik, N., Skyschally, A., 2011. Mitochondrial STAT3 activation and cardioprotection by ischemic postconditioning in pigs with regional myocardial ischemia/reperfusion. Circ. Res. 109, 1302–1308.

Hirvonen, M., Ojala, R., Korhonen, P., Haataja, P., Eriksson, K., Gissler, M., Luukkaala, T., Tammela, O., 2014. Cerebral palsy among children born moderately and late preterm. Pediatrics 134, e1584–93.

Hochreiter-Hufford, A., Ravichandran, K.S., 2013. Clearing the dead: apoptotic cell sensing, recognition, engulfment, and digestion. Cold Spring Harb. Perspect. Biol. 5, A008748.

Hristova, M., Rocha-Ferreira, E., Fontana, X., Thei, L., Buckle, R., Christou, M., Hompoonsup, S., Gostelow, N., Raivich, G., Peebles, D., 2016. Inhibition of signal transducer and activator of transcription 3 (STAT3) reduces neonatal hypoxic-ischaemic brain damage. J. Neurochem. 136, 981–994.

Huang, Z.J., Di Cristo, G., Ango, F., 2007. Development of GABA innervation in the cerebral and cerebellar cortices. Nat. Rev. Neurosci. 8, 673–686.

Huang, J., Zhang, L., Qu, Y., Zhou, Y., Zhu, J., Li, Y., Zhu, T., Zhao, F., Tang, J., Mu, D., 2018. Histone acetylation of oligodendrocytes protects against white matter injury induced by inflammation and hypoxia-ischemia through activation of BDNF-TrkB signaling pathway in neonatal rats. Brain Res. 1688, 33–46.

Itoh, T., Beesley, J., Itoh, A., Cohen, A.S., Kavanaugh, B., Coulter, D.A., Grinspan, J.B., Pleasure, D., 2002. AMPA glutamate receptor-mediated calcium signaling is transiently enhanced during development of oligodendrocytes. J. Neurochem. 81, 390–402.

Jantzie, L.L., Corbett, C.J., Berglass, J., Firl, D.J., Flores, J., Mannix, R., Robinson, S., 2014. Complex pattern of interaction between in utero hypoxia-ischemia and intra-amniotic inflammation disrupts brain development and motor function. J. Neuroinflammation 11, 131.

Jantzie, L.L., Winer, J.L., Maxwell, J.R., Chan, L.A., Robinson, S., 2015. Modeling encephalopathy of prematurity using prenatal hypoxia-ischemia with intra-amniotic lipopolysaccharide in rats. J. Vis. Exp. 105, e53196.

Kagawa, Y., Yasumoto, Y., Sharifi, K., Ebrahimi, M., Islam, A., Miyazaki, H., Yamamoto, Y., Sawada, T., Kishi, H., Kobayashi, S., Maekawa, M., Yoshikawa, T., Takaki, E., Nakai, A., Kogo, H., Fujimoto, T., Owada, Y., 2015. Fatty acid-binding protein 7 regulates function of caveolae in astrocytes through expression of caveolin-1. Glia 63, 780–794.

Kannan, S., Dai, H., Navath, R.S., Balakrishnan, B., Jyoti, A., Janisse, J., Romero, R., Kannan, R.M., 2012. Dendrimer-based postnatal therapy for neuroinflammation and cerebral palsy in a rabbit model. Sci. Transl. Med. 4, 130ra46.

Khakh, B.S., Sofroniew, M.V., 2015. Diversity of astrocyte functions and phenotypes in neural circuits. Nat. Neurosci. 18, 942–952.

Khandaker, G.M., Zimbron, J., Lewis, G., Jones, P.B., 2013. Prenatal maternal infection, neurodevelopment and adult schizophrenia: a systematic review of population-based studies. Psychol. Med. 43, 239–257.

Kimelberg, H.K., Nedergaard, M., 2010. Functions of astrocytes and their potential as therapeutic targets. Neurotherapeutics 7, 338–353.

Kirsten, T.B., Chaves-Kirsten, G.P., Chaible, L.M., Silva, A.C., Martins, D.O., Britto, L.R., Dagli, M.L., Torrao, A.S., Palermo-Neto, J., Bernardi, M.M., 2012. Hypoactivity of the central dopaminergic system and autistic-like behavior induced by a single early prenatal exposure to lipopolysaccharide. J. Neurosci. Res. 90, 1903–1912.

Kirsten, T.B., Chaves-Kirsten, G.P., Bernardes, S., Scavone, C., Sarkis, J.E., Bernardi, M.M., Felicio, L.F., 2015. Lipopolysaccharide exposure induces maternal hypozincemia, and prenatal zinc treatment prevents autistic-like behaviors and disturbances in the striatal dopaminergic and mTOR systems of offspring. PLoS One 10, e0134565.

Knuesel, I., Chicha, L., Britschgi, M., Schobel, S.A., Bodmer, M., Hellings, J.A., Toovey, S., Prinssen, E.P., 2014. Maternal immune activation and abnormal brain development across CNS disorders. Nat. Rev. Neurol. 10, 643–660.

Krieglstein, K., Strelau, J., Schober, A., Sullivan, A., Unsicker, K., 2002. TGF-beta and the regulation of neuron survival and death. J. Physiol. Paris 96, 25–30.
Kuban, K.C., O'Shea, T.M., Allred, E.N., Paneth, N., Hirtz, D., Fichorova, R.N., Leviton, A., ELGAN Study Investigators, 2014. Systemic inflammation and cerebral palsy risk in extremely preterm infants. J. Child Neurol. 29, 1692–1698.
Lan, K.M., Tien, L.T., Cai, Z., Lin, S., Pang, Y., Tanaka, S., Rhodes, P.G., Bhatt, A.J., Savich, R.D., Fan, L.W., 2016. Erythropoietin ameliorates neonatal hypoxia-ischemia-induced neurobehavioral deficits, neuroinflammation, and hippocampal injury in the juvenile rat. Int. J. Mol. Sci. 17, 289.
Lante, F., Meunier, J., Guiramand, J., De Jesus Ferreira, M.C., Cambonie, G., Aimar, R., Cohen-Solal, C., Maurice, T., Vignes, M., Barbanel, G., 2008. Late N-acetylcysteine treatment prevents the deficits induced in the offspring of dams exposed to an immune stress during gestation. Hippocampus 18, 602–609.
Li, Y., Liu, D.X., Li, M.Y., Qin, X.X., Fang, W.G., Zhao, W.D., Chen, Y.H., 2014. Ephrin-A3 and ephrin-A4 contribute to microglia-induced angiogenesis in brain endothelial cells. Anat. Rec. (Hoboken) 297, 1908–1918.
Li, B., Concepcion, K., Meng, X., Zhang, L., 2017. Brain-immune interactions in perinatal hypoxic-ischemic brain injury. Prog. Neurobiol.
Liesi, P., Silver, J., 1988. Is astrocyte laminin involved in axon guidance in the mammalian CNS? Dev. Biol. 130, 774–785.
Loomes, R., Hull, L., Mandy, W.P.L., 2017. What is the male-to-female ratio in autism spectrum disorder? A systematic review and meta-analysis. J. Am. Acad. Child Adolesc. Psychiatry 56, 466–474.
Lu, B., Nagappan, G., Lu, Y., 2014. BDNF and synaptic plasticity, cognitive function, and dysfunction. Handb. Exp. Pharmacol. 220, 223–250.
Ma, S., Kwon, H.J., Huang, Z., 2012. A functional requirement for astroglia in promoting blood vessel development in the early postnatal brain. PLoS One 7, e48001.
MacDowell, K.S., Munarriz-Cuezva, E., Caso, J.R., Madrigal, J.L., Zabala, A., Meana, J.J., Garcia-Bueno, B., Leza, J.C., 2017. Paliperidone reverts Toll-like receptor 3 signaling pathway activation and cognitive deficits in a maternal immune activation mouse model of schizophrenia. Neuropharmacology 116, 196–207.
Makinodan, M., Tatsumi, K., Manabe, T., Yamauchi, T., Makinodan, E., Matsuyoshi, H., Shimoda, S., Noriyama, Y., Kishimoto, T., Wanaka, A., 2008. Maternal immune activation in mice delays myelination and axonal development in the hippocampus of the offspring. J. Neurosci. Res. 86, 2190–2200.
Makinson, R., Lloyd, K., Rayasam, A., McKee, S., Brown, A., Barila, G., Grissom, N., George, R., Marini, M., Fabry, Z., Elovitz, M., Reyes, T.M., 2017. Intrauterine inflammation induces sex-specific effects on neuroinflammation, white matter, and behavior. Brain Behav. Immun. 66, 277–288.
Matcovitch-Natan, O., Winter, D.R., Giladi, A., Vargas Aguilar, S., Spinrad, A., Sarrazin, S., Ben-Yehuda, H., David, E., Zelada Gonzalez, F., Perrin, P., Keren-Shaul, H., Gury, M., Lara-Astaiso, D., Thaiss, C.A., Cohen, M., Bahar Halpern, K., Baruch, K., Deczkowska, A., Lorenzo-Vivas, E., Itzkovitz, S., Elinav, E., Sieweke, M.H., Schwartz, M., Amit, I., 2016. Microglia development follows a stepwise program to regulate brain homeostasis. Science 353, aad8670.
Maxwell, J.R., Denson, J.L., Joste, N.E., Robinson, S., Jantzie, L.L., 2015. Combined in utero hypoxia-ischemia and lipopolysaccharide administration in rats induces chorioamnionitis and a fetal inflammatory response syndrome. Placenta 36, 1378–1384.
Maynard, T.M., Sikich, L., Lieberman, J.A., Lamantia, A.S., 2001. Neural development, cell-cell signaling, and the "two-hit" hypothesis of schizophrenia. Schizophr. Bull. 27, 457–476.
Mazaheri, F., Breus, O., Durdu, S., Haas, P., Wittbrodt, J., Gilmour, D., Peri, F., 2014. Distinct roles for BAI1 and TIM-4 in the engulfment of dying neurons by microglia. Nat. Commun. 5, 4046.

McIntyre, S., Taitz, D., Keogh, J., Goldsmith, S., Badawi, N., Blair, E., 2013. A systematic review of risk factors for cerebral palsy in children born at term in developed countries. Dev. Med. Child Neurol. 55, 499–508.
Meyer, U., 2014. Prenatal poly(I:C) exposure and other developmental immune activation models in rodent systems. Biol. Psychiatry 75, 307–315.
Meyer, U., Feldon, J., 2012. To poly(I:C) or not to poly(I:C): advancing preclinical schizophrenia research through the use of prenatal immune activation models. Neuropharmacology 62, 1308–1321.
Meyer, U., Feldon, J., Schedlowski, M., Yee, B.K., 2005. Towards an immuno-precipitated neurodevelopmental animal model of schizophrenia. Neurosci. Biobehav. Rev. 29, 913–947.
Meyer, U., Nyffeler, M., Engler, A., Urwyler, A., Schedlowski, M., Knuesel, I., Yee, B.K., Feldon, J., 2006. The time of prenatal immune challenge determines the specificity of inflammation-mediated brain and behavioral pathology. J. Neurosci. 26, 4752–4762.
Millan, M.J., Fone, K., Steckler, T., Horan, W.P., 2014. Negative symptoms of schizophrenia: clinical characteristics, pathophysiological substrates, experimental models and prospects for improved treatment. Eur. Neuropsychopharmacol. 24, 645–692.
Mitew, S., Hay, C.M., Peckham, H., Xiao, J., Koenning, M., Emery, B., 2014. Mechanisms regulating the development of oligodendrocytes and central nervous system myelin. Neuroscience 276, 29–47.
Mitha, A., Foix-L'Helias, L., Arnaud, C., Marret, S., Vieux, R., Aujard, Y., Thiriez, G., Larroque, B., Cambonie, G., Burguet, A., Boileau, P., Roze, J.C., Kaminski, M., Truffert, P., Ancel, P.Y., EPIPAGE Study Group, 2013. Neonatal infection and 5-year neurodevelopmental outcome of very preterm infants. Pediatrics 132, e372–80.
Modabbernia, A., Velthorst, E., Reichenberg, A., 2017. Environmental risk factors for autism: an evidence-based review of systematic reviews and meta-analyses. Mol. Autism 8, 13.
Moller, M., Swanepoel, T., Harvey, B.H., 2015. Neurodevelopmental animal models reveal the convergent role of neurotransmitter systems, inflammation, and oxidative stress as biomarkers of schizophrenia: implications for novel drug development. ACS Chem. Neurosci. 6, 987–1016.
Mottahedin, A., Svedin, P., Nair, S., Mohn, C.J., Wang, X., Hagberg, H., Ek, J., Mallard, C., 2017. Systemic activation of Toll-like receptor 2 suppresses mitochondrial respiration and exacerbates hypoxic-ischemic injury in the developing brain. J. Cereb. Blood Flow Metab. 37, 1192–1198.
Mukerji, A., Shah, V., Shah, P.S., 2015. Periventricular/intraventricular hemorrhage and neurodevelopmental outcomes: a meta-analysis. Pediatrics 136, 1132–1143.
Nieto, R., Kukuljan, M., Silva, H., 2013. BDNF and schizophrenia: from neurodevelopment to neuronal plasticity, learning, and memory. Front. Psychiatry 4, 45.
O'Loughlin, E., Pakan, J.M.P., Yilmazer-Hanke, D., McDermott, K.W., 2017. Acute in utero exposure to lipopolysaccharide induces inflammation in the pre- and postnatal brain and alters the glial cytoarchitecture in the developing amygdala. J. Neuroinflammation 14, 212.
Ortega, S.B., Kong, X., Venkataraman, R., Savedra, A.M., Kernie, S.G., Stowe, A.M., Raman, L., 2016. Perinatal chronic hypoxia induces cortical inflammation, hypomyelination, and peripheral myelin-specific T cell autoreactivity. J. Leukoc. Biol. 99, 21–29.
O'Shea, T.M., Allred, E.N., Dammann, O., Hirtz, D., Kuban, K.C., Paneth, N., Leviton, A., ELGAN Study Investigators, 2009. The ELGAN study of the brain and related disorders in extremely low gestational age newborns. Early Hum. Dev. 85, 719–725.

Oskoui, M., Coutinho, F., Dykeman, J., Jette, N., Pringsheim, T., 2013. An update on the prevalence of cerebral palsy: a systematic review and meta-analysis. Dev. Med. Child Neurol. 55, 509–519.

Osredkar, D., Sabir, H., Falck, M., Wood, T., Maes, E., Flatebo, T., Puchades, M., Thoresen, M., 2015. Hypothermia does not reverse cellular responses caused by lipopolysaccharide in neonatal hypoxic-ischaemic brain injury. Dev. Neurosci. 37, 390–397.

Pang, Y., Dai, X., Roller, A., Carter, K., Paul, I., Bhatt, A.J., Lin, R.C., Fan, L.W., 2016. Early postnatal lipopolysaccharide exposure leads to enhanced neurogenesis and impaired communicative functions in rats. PLoS One 11, e0164403.

Paolicelli, R.C., Bolasco, G., Pagani, F., Maggi, L., Scianni, M., Panzanelli, P., Giustetto, M., Ferreira, T.A., Guiducci, E., Dumas, L., Ragozzino, D., Gross, C.T., 2011. Synaptic pruning by microglia is necessary for normal brain development. Science 333, 1456–1458.

Park, M.H., Lee, S.M., Lee, J.W., Son, D.J., Moon, D.C., Yoon, D.Y., Hong, J.T., 2006. ERK-mediated production of neurotrophic factors by astrocytes promotes neuronal stem cell differentiation by erythropoietin. Biochem. Biophys. Res. Commun. 339, 1021–1028.

Patterson, P.H., 2011. Maternal infection and immune involvement in autism. Trends Mol. Med. 17, 389–394.

Perry, W., Minassian, A., Lopez, B., Maron, L., Lincoln, A., 2007. Sensorimotor gating deficits in adults with autism. Biol. Psychiatry 61, 482–486.

Pierre, W.C., Smith, P.L., Londono, I., Chemtob, S., Mallard, C., Lodygensky, G.A., 2017. Neonatal microglia: the cornerstone of brain fate. Brain Behav. Immun. 59, 333–345.

Ray, M.T., Shannon Weickert, C., Webster, M.J., 2014. Decreased BDNF and TrkB mRNA expression in multiple cortical areas of patients with schizophrenia and mood disorders. Transl. Psychiatry 4, e389.

Reemst, K., Noctor, S.C., Lucassen, P.J., Hol, E.M., 2016. The indispensable roles of microglia and astrocytes during brain development. Front. Hum. Neurosci. 10, 566.

Reisinger, S., Khan, D., Kong, E., Berger, A., Pollak, A., Pollak, D.D., 2015. The poly(I:C)-induced maternal immune activation model in preclinical neuropsychiatric drug discovery. Pharmacol. Ther. 149, 213–226.

Robinson, S., Franic, L.A., 2001. Chemokine GRO1 and the spatial and temporal regulation of oligodendrocyte precursor proliferation. Dev. Neurosci. 23, 338–345.

Robinson, S., Tani, M., Strieter, R.M., Ransohoff, R.M., Miller, R.H., 1998. The chemokine growth-regulated oncogene-alpha promotes spinal cord oligodendrocyte precursor proliferation. J. Neurosci. 18, 10457–10463.

Rudenko, A., Tsai, L.H., 2014. Epigenetic modifications in the nervous system and their impact upon cognitive impairments. Neuropharmacology 80, 70–82.

Rumajogee, P., Bregman, T., Miller, S.P., Yager, J.Y., Fehlings, M.G., 2016. Rodent hypoxia-ischemia models for cerebral palsy research: a systematic review. Front. Neurol. 7, 57.

Sakamoto, A., Moriuchi, H., Matsuzaki, J., Motoyama, K., Moriuchi, M., 2015. Retrospective diagnosis of congenital cytomegalovirus infection in children with autism spectrum disorder but no other major neurologic deficit. Brain Dev. 37, 200–205.

Savard, A., Brochu, M.E., Chevin, M., Guiraut, C., Grbic, D., Sebire, G., 2015. Neuronal self-injury mediated by IL-1beta and MMP-9 in a cerebral palsy model of severe neonatal encephalopathy induced by immune activation plus hypoxia-ischemia. J. Neuroinflammation 12, 111.

Schafer, D.P., Lehrman, E.K., Kautzman, A.G., Koyama, R., Mardinly, A.R., Yamasaki, R., Ransohoff, R.M., Greenberg, M.E., Barres, B.A., Stevens, B., 2012. Microglia sculpt postnatal neural circuits in an activity and complement-dependent manner. Neuron 74, 691–705.

Schlapbach, L.J., Aebischer, M., Adams, M., Natalucci, G., Bonhoeffer, J., Latzin, P., Nelle, M., Bucher, H.U., Latal, B., Swiss Neonatal Network and Follow-Up Group, 2011. Impact of sepsis on neurodevelopmental outcome in a Swiss National Cohort of extremely premature infants. Pediatrics 128, e348–57.

Seleme, M.C., Kosmac, K., Jonjic, S., Britt, W.J., 2017. Tumor necrosis factor alpha-induced recruitment of inflammatory mononuclear cells leads to inflammation and altered brain development in murine cytomegalovirus-infected newborn mice. J. Virol. 91.

Semple, B.D., Blomgren, K., Gimlin, K., Ferriero, D.M., Noble-Haeusslein, L.J., 2013. Brain development in rodents and humans: identifying benchmarks of maturation and vulnerability to injury across species. Prog. Neurobiol. 106–107, 1–16.

Shigemoto-Mogami, Y., Hoshikawa, K., Goldman, J.E., Sekino, Y., Sato, K., 2014. Microglia enhance neurogenesis and oligodendrogenesis in the early postnatal subventricular zone. J. Neurosci. 34, 2231–2243.

Shrivastava, K., Llovera, G., Recasens, M., Chertoff, M., Gimenez-Llort, L., Gonzalez, B., Acarin, L., 2013. Temporal expression of cytokines and signal transducer and activator of transcription factor 3 activation after neonatal hypoxia/ischemia in mice. Dev. Neurosci. 35, 212–225.

Simeone, J.C., Ward, A.J., Rotella, P., Collins, J., Windisch, R., 2015. An evaluation of variation in published estimates of schizophrenia prevalence from 1990 horizontal line 2013: a systematic literature review. BMC Psychiatry 15, 193.

Spohr, T.C., Dezonne, R.S., Rehen, S.K., Gomes, F.C., 2014. LPA-primed astrocytes induce axonal outgrowth of cortical progenitors by activating PKA signaling pathways and modulating extracellular matrix proteins. Front. Cell. Neurosci. 8, 296.

Stevens, B., Allen, N.J., Vazquez, L.E., Howell, G.R., Christopherson, K.S., Nouri, N., Micheva, K.D., Mehalow, A.K., Huberman, A.D., Stafford, B., Sher, A., Litke, A.M., Lambris, J.D., Smith, S.J., John, S.W., Barres, B.A., 2007. The classical complement cascade mediates CNS synapse elimination. Cell 131, 1164–1178.

Stoll, B.J., Hansen, N.I., Adams-Chapman, I., Fanaroff, A.A., Hintz, S.R., Vohr, B., Higgins, R.D., National Institute of Child Health and Human Development Neonatal Research Network, 2004. Neurodevelopmental and growth impairment among extremely low-birth-weight infants with neonatal infection. JAMA 292, 2357–2365.

Swerdlow, N.R., Light, G.A., Thomas, M.L., Sprock, J., Calkins, M.E., Green, M.F., Greenwood, T.A., Gur, R.E., Gur, R.C., Lazzeroni, L.C., Nuechterlein, K.H., Radant, A.D., Seidman, L.J., Siever, L.J., Silverman, J.M., Stone, W.S., Sugar, C.A., Tsuang, D.W., Tsuang, M.T., Turetsky, B.I., Braff, D.L., 2017. Deficient prepulse inhibition in schizophrenia in a multi-site cohort: internal replication and extension. Schizophr. Res. https://doi.org/10.1016/j.schres.2017.05.013, in press.

Takahashi, M., Shirakawa, O., Toyooka, K., Kitamura, N., Hashimoto, T., Maeda, K., Koizumi, S., Wakabayashi, K., Takahashi, H., Someya, T., Nawa, H., 2000. Abnormal expression of brain-derived neurotrophic factor and its receptor in the corticolimbic system of schizophrenic patients. Mol. Psychiatry 5, 293–300.

Tonge, D.A., de Burgh, H.T., Docherty, R., Humphries, M.J., Craig, S.E., Pizzey, J., 2012. Fibronectin supports neurite outgrowth and axonal regeneration of adult brain neurons in vitro. Brain Res. 1453, 8–16.

Tsai, H.H., Frost, E., To, V., Robinson, S., Ffrench-Constant, C., Geertman, R., Ransohoff, R.M., Miller, R.H., 2002. The chemokine receptor CXCR2 controls positioning of oligodendrocyte precursors in developing spinal cord by arresting their migration. Cell 110, 373–383.

van Deijk, A.F., Camargo, N., Timmerman, J., Heistek, T., Brouwers, J.F., Mogavero, F., Mansvelder, H.D., Smit, A.B., Verheijen, M.H., 2017. Astrocyte lipid metabolism is critical for synapse development and function in vivo. Glia 65, 670–682.

Vannucci, R.C., Vannucci, S.J., 1997. A model of perinatal hypoxic-ischemic brain damage. Ann. N. Y. Acad. Sci. 835, 234–249.

Vinukonda, G., Csiszar, A., Hu, F., Dummula, K., Pandey, N.K., Zia, M.T., Ferreri, N.R., Ungvari, Z., Lagamma, E.F., Ballabh, P., 2010. Neuroprotection in a rabbit model of intraventricular haemorrhage by cyclooxygenase-2, prostanoid receptor-1 or tumour necrosis factor-alpha inhibition. Brain 133, 2264–2280.

Vinukonda, G., Dohare, P., Arshad, A., Zia, M.T., Panda, S., Korumilli, R., Kayton, R., Hascall, V.C., Lauer, M.E., Ballabh, P., 2016. Hyaluronidase and hyaluronan oligosaccharides promote neurological recovery after intraventricular hemorrhage. J. Neurosci. 36, 872–889.

van Tilborg, E., Heijnen, C.J., Benders, M.J., van Bel, F., Fleiss, B., Gressens, P., Nijboer, C.H., 2016. Impaired oligodendrocyte maturation in preterm infants: potential therapeutic targets. Prog. Neurobiol. 136, 28–49.

van Tilborg, E., Achterberg, E.J.M., van Kammen, C.M., van der Toorn, A., Groenendaal, F., Dijkhuizen, R.M., Heijnen, C.J., Vanderschuren, L., Benders, M., Nijboer, C.H.A., 2017. Combined fetal inflammation and postnatal hypoxia causes myelin deficits and autism-like behavior in a rat model of diffuse white matter injury. Glia 66, 1–16.

Vuong, H.E., Hsiao, E.Y., 2017. Emerging roles for the gut microbiome in autism spectrum disorder. Biol. Psychiatry 81, 411–423.

Wang, L.W., Lin, Y.C., Wang, S.T., Yeh, T.F., Huang, C.C., 2014. Hypoxic/ischemic and infectious events have cumulative effects on the risk of cerebral palsy in very-low-birth-weight preterm infants. Neonatology 106, 209–215.

Waterhouse, E.G., Xu, B., 2009. New insights into the role of brain-derived neurotrophic factor in synaptic plasticity. Mol. Cell. Neurosci. 42, 81–89.

Watson, R.E., DeSesso, J.M., Hurtt, M.E., Cappon, G.D., 2006. Postnatal growth and morphological development of the brain: a species comparison. Birth Defects Res. 77, 471–484.

Watzlawik, J.O., Kahoud, R.J., O'Toole, R.J., White, K.A., Ogden, A.R., Painter, M.M., Wootla, B., Papke, L.M., Denic, A., Weimer, J.M., Carey, W.A., Rodriguez, M., 2015. Abbreviated exposure to hypoxia is sufficient to induce CNS dysmyelination, modulate spinal motor neuron composition, and impair motor development in neonatal mice. PLoS One 10, e0128007.

Wong, C.S., Jow, G.M., Kaizaki, A., Fan, L.W., Tien, L.T., 2014. Melatonin ameliorates brain injury induced by systemic lipopolysaccharide in neonatal rats. Neuroscience 267, 147–156.

Xu, D.X., Wang, H., Ning, H., Zhao, L., Chen, Y.H., 2007. Maternally administered melatonin differentially regulates lipopolysaccharide-induced proinflammatory and anti-inflammatory cytokines in maternal serum, amniotic fluid, fetal liver, and fetal brain. J. Pineal Res. 43, 74–79.

Yanni, D., Korzeniewski, S.J., Allred, E.N., Fichorova, R.N., O'Shea, T.M., Kuban, K., Dammann, O., Leviton, A., 2017. Both antenatal and postnatal inflammation contribute information about the risk of brain damage in extremely preterm newborns. Pediatr. Res. 82, 691–696.

Yanowitz, T.D., Jordan, J.A., Gilmour, C.H., Towbin, R., Bowen, A., Roberts, J.M., Brozanski, B.S., 2002. Hemodynamic disturbances in premature infants born after chorioamnionitis: association with cord blood cytokine concentrations. Pediatr. Res. 51, 310–316.

Yanowitz, T.D., Potter, D.M., Bowen, A., Baker, R.W., Roberts, J.M., 2006. Variability in cerebral oxygen delivery is reduced in premature neonates exposed to chorioamnionitis. Pediatr. Res. 59, 299–304.

Yellowhair, T.R., Noor, S., Maxwell, J.R., Anstine, C.V., Oppong, A.Y., Robinson, S., Milligan, E.D., Jantzie, L.L., 2017. Preclinical chorioamnionitis dysregulates CXCL1/CXCR2 signaling throughout the placental-fetal-brain axis. Exp. Neurol.

Yeung, M.S., Zdunek, S., Bergmann, O., Bernard, S., Salehpour, M., Alkass, K., Perl, S., Tisdale, J., Possnert, G., Brundin, L., Druid, H., Frisen, J., 2014. Dynamics of oligodendrocyte generation and myelination in the human brain. Cell 159, 766–774.

Zaghloul, N., Patel, H., Ahmed, M.N., 2017. A model of periventricular leukomalacia (PVL) in neonate mice with histopathological and neurodevelopmental outcomes mimicking human PVL in neonates. PLoS One 12, e0175438.

Zalocusky, K.A., Ramakrishnan, C., Lerner, T.N., Davidson, T.J., Knutson, B., Deisseroth, K., 2016. Nucleus accumbens D2R cells signal prior outcomes and control risky decision-making. Nature 531, 642–646.

Zhang, Z., Bassam, B., Thomas, A.G., Williams, M., Liu, J., Nance, E., Rojas, C., Slusher, B.S., Kannan, S., 2016. Maternal inflammation leads to impaired glutamate homeostasis and up-regulation of glutamate carboxypeptidase II in activated microglia in the fetal/newborn rabbit brain. Neurobiol. Dis. 94, 116–128.

CHAPTER NINE

Enduring Behavioral and Brain Impacts of Prenatal Stress and Childhood Adversity and Their Potential Multigenerational Consequences

Deborah A. Cory-Slechta[1]
Department of Environmental Medicine, University of Rochester School of Medicine, Rochester, NY 14642, United States
[1]Corresponding author: e-mail address: deborah_cory-slechta@urmc.rochester.edu

Contents

Advances in Neurotoxicology, Volume 2
ISSN 2468-7480
https://doi.org/10.1016/bs.ant.2018.03.007

1. INTRODUCTION

This chapter focuses on developmental stress, specifically prenatal stress and what has been deemed early behavioral adversity in young children. It is intended as an overview that elaborates known neurobehavioral and neurobiological consequences of such stressors from human studies and animal models, and mechanisms by which such stressors can cause biological reprogramming and thus produce long-term consequences. Of increasing interest is the potential for these early stressors to alter phenotypes in subsequent generations, with both multigenerational and transgenerational effects having been reported. These latter effects can derive from epigenetic and/or cultural and behavioral transmission. This chapter also includes discussions of many of the issues in research on prenatal stress and early developmental adversity that need to be considered in interpretation of any studies. While not specifically reviewing animal models of stress, the chapter does emphasize the need for the development of animal models that more closely mimic the context of stress in the human environment.

2. DEFINING STRESS

Stress has been defined as the response of an organism to regain homeostasis following a threatening stimulus, as initially based on the work of Selye in animal models (Selye, 1956), findings that were later confirmed in human studies (Holmes and Rahe, 1967). As is evident, stress can be encountered at many points throughout the lifetime, and any given stressor may be experienced acutely and/or chronically. Stress can arise from many different types of stimuli/events that can be physical or psychological (behavioral) in nature. While some such stimuli are universally experienced as stressors, others stressors may be far more unique to an individual. The cumulative experience of stress over the lifetime is, of course, unique to each individual in terms of the nature, duration, saliency and other characteristics of that stressor.

As has been elucidated by a significant body of research, stressful experiences can have consequences for physical and/or mental health depending upon the context of the stressor, and whether it is manifest acutely and/or persists over time. Protracted or cumulative stress has been associated with multiple adverse mental health effects that include major depressive disorder and depressive symptoms, and substance and alcohol abuse (Thoits, 2010). In addition, stress has been linked to multiple physical diseases. A role of stress in cardiovascular disease is well established and thought to derive from pathogenic mechanisms that can include myocardial ischemia and inflammatory and coagulatory events (Lagraauw et al., 2015; Wirtz and von Kanel, 2017), and can be co-morbid with depression (Halaris, 2017). Stress has also been linked to the development of cancer as well as its progression, which may arise from effects of stress on DNA repair and cellular aging (Antoni et al., 2006; Volden and Conzen, 2013). Links between stress and the progression of HIV and other viral diseases are also observed (Leserman et al., 2002). Stress has also been described as contributing to neurodegenerative disorders including both Alzheimer's and Parkinson's diseases (Goldstein and Kopin, 2017; Vyas et al., 2016). Studies have now also been elucidating at the role of stress in multiple other diseases, such as diabetes control (Marcovecchio and Chiarelli, 2012), metabolic syndrome (Tamashiro et al., 2011) and inflammatory bowel disease (Brzozowski et al., 2016; Caso et al., 2008; Melinder et al., 2017).

3. BIOLOGICAL MEDIATION OF STRESS

The biological mediation of stress has been well established and includes actions of peripheral systems such as the hypothalamic–pituitary–adrenal (HPA) axis and the sympathetic adrenal medullary (SAM) system, as well as the interactions of these systems with other organs, including the central nervous system (CNS). Stressors originate in the environment and are perceived via the CNS and processed via the brain and HPA axis as recently reviewed by Herman et al. (2016). Stress results in the release of corticotropin-releasing hormone (CRH) from the periventricular nucleus of the hypothalamus that subsequently results in the release of adrenocorticotropic hormone (ACTH) from the anterior pituitary. ACTH then acts on the adrenal gland to result in the release of glucocorticoids (cortisol in humans, corticosterone in rodents). This is referred to as the rising phase of HPA axis function. The system also features a negative feedback loop to shut off the stress response that includes several different mechanisms, such as glucocorticoid-induced shut-off of neurons within the periventricular nucleus of the hypothalamus, GABAergic inhibition, and effects on the glucocorticoid receptors (mineralocorticoid and glucocorticoid) in the hippocampus. This constitute the falling phase of HPA axis function (Herman et al., 2016).

Glucocorticoids actually have multiple critical functions in physiology beyond mediation of the stress response. The HPA axis, for example, has broad interactions with the CNS, particularly the mesocorticolimbic system of the brain that includes the nucleus accumbens, frontal cortex and hippocampus, as well as limbic regions such as the amygdala. Numerous animal studies have demonstrated changes in dopaminergic function in mesocorticolimbic brain regions in response to stress (Belujon and Grace, 2015). In the context of controllable vs. uncontrollable stressors, it has been shown that the consequences of stress on dopamine systems differs in these two conditions (Abbott et al., 1984; Anisman and Matheson, 2005). Rodents exhibiting learned helplessness via exposures to uncontrollable shocks show a decrease in dopamine activity in the dopaminergic cell bodies of the ventral tegmental area, whereas dopamine activity levels of animals that received controllable shock deliveries do not differ from controls (Belujon and Grace, 2014). In part, such events reflect the presence of glucocorticoid receptors within brain, including within mesocorticolimbic regions (Sah et al., 2005; Sanchez et al., 2000).

Interactions between glucocorticoids in brain and dopamine function are also well established. Corticosterone can enhance dopamine signaling within the nucleus accumbens (Graf et al., 2013). Deletion of the glucocorticoid receptor gene in the forebrain of mice attenuated depression-like behavioral effects of chronic stress (Jacobson, 2014). Adrenalectomy of rats decreased dopamine levels in the shell of the nucleus accumbens, the region within that structure considered most involved in mediation of stress and of reward-mediated behaviors (Barrot et al., 2000). HPA axis interactions with mesocorticolimbic circuitry of the brain are considered to underlie the behavioral consequences of stress, including the psychiatric disorders, as well as behavioral and cognitive deficits. A review of human and experimental animal studies of chronic juvenile stress suggested that psychopathology reflects the number of stressful events experienced during childhood, particularly of multiple types of stressors (Watt et al., 2017) as well as reduced prefrontal cortex activity in response to such stress that would be consistent with deficits in dopamine- and glutamate-mediated executive functions that are also seen in such studies (Comeau et al., 2014; Gould et al., 2012; Han et al., 2011; Schalinski et al., 2017). Childhood trauma appears to have particularly notable effects across measures of cognition, including general intelligence, learning and memory, language/verbal skills, information processing and executive functions as noted in a recent meta-analysis (Malarbi et al., 2017), behaviors for which dopamine and glutamate in mesocorticolimbic systems are crucial. Similarly, rodents show impairments in cognitive flexibility as measured using reversal learning and set-shifting tasks that can be rescued by prevention of corticosterone synthesis, increased post-synaptic norepinephrine function, increased serotonin availability or by dopamine receptor activation (Hurtubise and Howland, 2017).

Studies in rodents indicate that glucocorticoid receptor mRNA is present at the earliest time point that was examined, gestational day 14 in the hippocamposeptal formation, and by gestational day 16 in the paraventricular nucleus of the hypothalamus. By late gestation, glucocorticoid receptor mRNA was also found in piriform cortex and dorsal endopiriform nucleus, specific nuclei within the amygdala and the suprachiasmatic nucleus (Yi et al., 1994). It is important to note, however, that these distribution patterns exhibit marked sex differences (MacLusky et al., 1996).

4. CONSIDERATIONS FOR STRESS RESEARCH

While the above described definitions and mediation of stress appear straightforward and simple, in reality, the effects of stress are highly context-dependent, differing in relation to multiple factors, all of which must be considered in any interpretation of outcomes from studies aimed at studying stress.

4.1 Demographic Features of Stress

The nature and chronicity of stress is related to socioeconomic status (SES) (Cohen et al., 2006), and this association is independent of race, age, gender or body mass. Living in conditions of poverty is associated with more exposures to stressful events, both physical and psychological, than is living in middle class conditions. These can include living in substandard housing, crowding, noise, violence, parental conflict and turmoil, lack of family and social support, and discrimination. Associations of stress with poverty have been demonstrated in both the urban and rural poor populations (Brown et al., 2016; Dubow et al., 1991; Evans and English, 2002). These differences are borne out by evaluation of markers of stress as well. For example, lower SES was associated with higher levels of both cortisol and epinephrine, and marginally with increased levels of norepinephrine in a study of men and women 21–55 years of age (Cohen et al., 2006).

Such findings likewise extend to children. Lupien et al. (2001) found that in a population of 307 children from 6 to 10 years of age that low SES was associated with higher morning salivary cortisol levels relative to children from higher SES conditions. Notably, in a population of 169 children attending a Head Start program, increased salivary cortisol levels followed by down-regulation of the increase were associated with measures of executive function and self-regulation, consistent with a relationship between reactivity of the HPA axis and cognition (Blair et al., 2005).

SES is also correlated with differences in health, i.e., health disparities that are attributed at least in part to the greater stress levels of low SES conditions. As has been repeatedly demonstrated, individuals of lower SES have the highest rates of mortality, disability and mental disorders (Dohrenwend et al., 1992; Elo and Preston, 1996; Hayward et al., 2000; House et al., 1988). These health inequalities are observed at across the lifespan, ranging from neonatal outcomes and infant mortality to diseases of aging (Glymour et al., 2014). Such disparities are not a simple dichotomy but exhibit a type of "dose–response" in relation to markers of SES in studies.

4.2 Not All Stress is Bad: Pathology vs. Resilience as an Outcome

As is increasingly being recognized, however, not all stimuli or events that might inherently thought to be stressful ultimately have such consequences. In fact, exposures to some such events can lead instead to a resilient behavioral phenotype that can better handle subsequent stressful experiences. Studies in animal models suggest a restriction of the word stress to events that are uncontrollable and unpredictable rather than events that are controllable and predictable. In cases of uncontrollable and unpredictable events, the absence of an anticipatory response defines unpredictable events and a reduced recovery defines uncontrollable events (Koolhaas et al., 2011, 2017). Further, stress, in this view, should be considered a cognitive perception of uncontrollability and/or unpredictability that produces physiological and behavioral responses. Support for this premise derives from a body of experimental research comparing these two conditions as well as assessments of putative stress markers in non-stress test conditions (Maier et al., 2006). For example, increases in corticosterone, invariably used as a validation of a stress response, can actually be as high or higher under conditions of appetitive and rewarding behaviors, e.g., sexual behavior (Bonilla-Jaime et al., 2006) and anticipation of food reward (Egan and Ulrich-Lai, 2015; Olivo et al., 2017). In experimental conditions, reinforcement (reward) procedures and extinction (removal of reward) both increased corticosterone, but it declined more rapidly under conditions of reinforcement than extinction, with similar effects seen in comparing losers vs. winners subjected to a social confrontation paradigm (Bartolomucci et al., 2003; De Boer et al., 1990), indicative of a stress response characterized by a delayed corticosterone negative feedback. Additional studies have shown that stressful stimuli, defined in the context of uncontrollability and unpredictability, increase adrenaline, but not noradrenaline (Scheurink et al., 1999). Thus, differences in both the time course of corticosterone responses as well as increases in adrenaline per se may be specific markers differentiating the stress response from simple behavioral arousal.

Indeed, it is exposures to controllable and/or predictable stressors that are considered critical to the emergence of a resilient phenotype. For example, in studies that compare responses in animals that receive physically identical stressors (shock delivery) with one group being able to terminate the stressor (controllable) while the other cannot (uncontrollable), it is only the latter group that exhibits a subsequent inability to learn to escape shocks, demonstrates anxiety-like behaviors, neophobia, and development of ulcers (Maier et al., 2006). In an elegant series of experiments, mechanisms of this aspect of stress controllability include the inhibition of limbic and brain stem reactions to a subsequent stressor. More specifically, controllable stress activates regions of the ventral medial prefrontal cortex which subsequently activate brain stem and limbic regions that can inhibit a subsequent potential stress-induced activation of these regions (Maier et al., 2006). This sequence of events requires medial prefrontal cortex activation during the stressor, a finding that has likewise been demonstrated wherein the combined activation of prefrontal cortex via administration of picrotoxin together with exposure to inescapable shock led to subsequent immunization to the effects of the uncontrollable shock (Maier et al., 2006).

An additional consideration of importance in understanding the impacts of changes in glucocorticoids in response to brain or behavioral function are numerous reports of the non-linearity of the effects of this steroid hormone. For example, a study of 20–30-year-old males and females revealed an inverted U-shaped dose–response relationship between salivary cortisol levels and recall of previously learned associations, with moderate levels of salivary cortisol resulting in the highest levels of memory recall. Similarly, in rats, U-shaped dose–effect curves were found for acquisition of water maze learning in response to cold stress, with maximal and most rapid acquisition, as well as subsequent extinction of behavior, achieved at the intermediate cold-water temperature, also associated with intermediate levels of corticosterone elevation (Salehi et al., 2010). Collectively such findings are consistent with the hypothesis (Yerkes and Dodson, 1908) that suggests that cognitive performance in complex tasks is best under conditions of optimal stress, but impaired under conditions of either low or above optimal levels of stress. Subsequent studies have examined a differential influence of stress on hippocampal function as well as other factors in explaining this non-linear relationship (Joels, 2006; Zoladz and Diamond, 2009).

5. NOT ALL STRESSORS ARE EQUAL

5.1 Acute vs. Chronic Stressor Presentation and Habituation/Adaptation

Questions with respect to stress effects have focused on both the consequences of acute stress experiences as well as repeated or chronic stress, as both occur within the human environment. It has long been known that consequences of acute and chronic stressors differ significantly in many respects, whether it be in terms of their different behavioral consequences (Katz et al., 1981), changes in cell-mediated immunity (Dhabhar and McEwen, 1997), epigenetic consequences as indicated by posttranslational histone modifications (Hunter et al., 2009) or brain neurochemistry (Roth et al., 1982). Additionally, the intensity/saliency of stress stimuli can significantly modulate the magnitude and type of response (Pitman et al., 1988), a factor critical to any comparisons across stressors (Natelson et al., 1988).

Repeated exposures to a stressor, however, can produce a rapid decline in the magnitude of the physiological response it initially produced, which has been described as habituation that results in a diminishment in the corresponding glucocorticoid response, with the rate of diminishment dependent upon stressor intensity (Konarska et al., 1990). However, as has been noted, such a phenomenon may be more appropriately called adaptation (Koolhaas et al., 2011), given that it can lead to increased predictability and controllability over the event. To avoid this problem, some studies interested in the effects of chronic stress have implemented a paradigm comprised of chronic variable stress (Roth and Katz, 1981). However, repeated presentations of a stressor, even when it is a variable pattern of stressors, may likewise lead to adaptation (Herman, 2013; Herman et al., 2016) and would thus become a more controllable event, as the organism learns that the stressor will end.

5.2 Types of Stressor

A wide array of different stressors is encountered in the human environment, and, correspondingly, a range of stressor stimuli are used in experimental animal models. These are often differentiated as physiological vs. psychological, but can include features of both dimensions. Importantly, differences in the types of stressor used can produce not only markedly different behavioral and physiological consequences but also impact the duration and the magnitude of the stress response observed. Changes in dopamine and serotonin activation differed in response to shock vs. environmental context stressors, as well as to the differences in the intensity/chronicity of the stressor (Inoue et al., 1994). Brain region-specific differences in the serotonergic response were demonstrated in a comparison of physical stressors in male rats (forced swim, tail pinch, immobilization stress, exposure to a cold environment and forced motor activity) (Kirby et al., 1997). Another study extended the observation of such differences to subsequent differences in the immune consequences of exposures to either restraint, forced swim, isolation or cold (Bowers et al., 2008). Differences in plasma ACTH norepinephrine and epinephrine concentrations were found after exposures to different stressors that include hemorrhage, i.v. insulin administration, s.c. formaldehyde injection, cold or immobilization (restraint) (Pacak et al., 1998). Collectively these and other studies argue against the initial hypothesis of stress as a unitary syndrome (Selye, 1931).

5.3 Sex Differences in Stress Outcomes and Mechanisms

Not surprisingly, based on demonstrated sex differences in brain structure and function (Bangasser and Wicks, 2017; McEwen and Milner, 2017) and of course in sex hormones and sex chromosomes, sex differences in the response to stress are well established, both in human studies and in experimental animal models (Weinstock, 2011). Such differences reflect, at least in part, sex differences in HPA axis function. Meta-analyses of studies of children ≤ 18 years of age provided evidence of alterations present even in childhood, with females showing a more variable diurnal rhythm of cortisol, and higher cortisol awakening response and a stronger cortisol response to tests of social stress than do males (Hollanders et al., 2017). In another study, differences in cortisol levels between males and females were evident at <8 years of age, with males exhibiting higher salivary cortisol at this time point, whereas the pattern reversed after 8 years of age, suggesting not only absolute differences in cortisol but differences in its trajectory with age (van der Voorn et al., 2017). In adults, stress is more prevalent in females, likely via its connection with metabolic disease (Murphy and Loria, 2017). In rodent studies, the hypothalamic–pituitary–gonadal axis is known to influence HPA axis development and function in a sex-dependent manner (Bale and Epperson, 2015; Green and McCormick, 2016). Sex differences in the neuroimmune response to stress (Bekhbat and Neigh, 2018) are thought to contribute to the increased prevalence in females of stress-related neuropsychiatric conditions.

In rodent models, sexes differ in response to social defeat stress, as do corresponding dopaminergic responses, particularly downstream of D1 receptor antagonism and/or BDNF function in the bed nucleus stria terminalis (Laman-Maharg and Trainor, 2017). Female rodents likewise show a different behavioral phenotype in response to fear conditioning, as well as in enhancement of classical eye-blink conditioning following an acute stressor than do males, with the latter effect appearing to reflect differences in the controlling circuitry (Bangasser and Wicks, 2017). Female rodents have also been shown to use more passive strategies than do males in their behavioral response to stress, including greater immobile time in the forced-swim and tail-suspension tests compared to males (Mueller and Bale, 2008).

Of note, particularly with respect to the interactions of the HPA axis with the brain mesocorticolimbic system and its relationship to executive function deficits in response to stress are the reported sex differences in brain dopamine function both in human assessments as well as in experimental animal models (Gillies et al., 2014). Human functional neuroimaging studies, for example, demonstrate sex differences in reward-related behaviors mediated by mesocorticolimbic systems. Males show greater striatal dopamine release than females (Munro et al., 2006) in PET studies. In studies examining the desire–reason dilemma, men showed a stronger reduction of activation in ventral pallidum, putamen, temporal pole and pregenual anterior cingulate during the performance than did females, as well as differences in connectivity between anteroventral prefrontal cortex and nucleus accumbens during this task (Diekhof et al., 2012). These sex differences are considered to arise from developmental programming of the ventral tegmental dopamine cell bodies early in life by gonadal hormones (Gillies et al., 2014). A marked sensitivity of the ventral tegmental area to glucocorticoids has also been reported (McArthur et al., 2005, 2007).

5.4 Implications

Such findings have multiple and significant implications. First, the observed differences in stress outcomes by context and particularly in relation to sex indicate that any mechanisms of stress consequences will also differ by sex, i.e., no single underlying mechanism is likely to accommodate both sexes. Correspondingly, mechanisms underlying stress effects may differ not only by sex but also in relation to the type, duration, intensity, etc., of the stressor. Moreover, the effects will differ depending upon the timing of the stress exposure, as early stress in particular has the capacity to program brain systems. Such findings also underscore the need to separately examine stress consequences by sex in both human and experimental animal studies. This is often limited in epidemiological studies because of sample size restrictions, but under those conditions where changes in response to stress by sex move in opposite directions, as is often the case, then combining data from both sexes would ensure the absence of an effect.

Additionally, the considerations with respect to stress as cited above have implications for the design of studies examining stress, both epidemiological and experimental. In studies relying on experimental animal models, it is increasingly important to recognize the issues raised with respect to what constitutes stress, i.e., the distinction based on restricting it to conditions that are uncontrollable/unpredictable (Koolhaas et al., 2011). This definition means that many prior studies that have used stress models not falling within this framework may need to be re-interpreted with respect to whether the paradigm used actually was consistent with stress or whether it would produce an adaptive response, knowing that even the chronic variable stressor paradigms can produce adaptation. In addition, the definition of resilience is clearly dependent upon this distinction and requires both differentiation of consequences in response to uncontrollable/unpredictable stress to controllable/predictable stress.

While not reviewing animal models of stress in this chapter, the enhanced stress effects associated with low SES suggest a need to develop animal models that more accurately reflect these conditions in the human environment. Whereas many paradigms used in animal studies lead to habituation/adaptation with repeated presentation of either homotypic or heterotypic stressors, one might actually see a sensitization to the stressors of low SES environments rather than habituation to such conditions.

5.5 Enduring Consequences of Developmental Stress

Stressors encountered during development, particularly prenatal stress and early behavioral adversity, appear to have particularly protracted consequences as attested to by an extensive literature indicating the fetal programming consequences of such exposures. Studies also suggest that such effects can be cumulative over the lifetime as an individual encounters additional stressors. Moreover, new studies suggest that the effects of such stressors may not only be multigenerational (i.e., passed on to subsequent lineages), but even transgenerational (passed from the F0 generation on to the F3 generation of females or the F2 generation in males). Collectively, these findings could contribute to the understanding of the fetal basis of disease. In addition, they suggest the importance of intervention programs not only to prevent adverse consequences of fetal programming, but also the transfer of such effects across subsequent generations.

5.6 Prenatal Stress

Prenatal stress refers to stress experienced in utero in human studies. As rodents are a altricial species, stress during the early postnatal period, considered equivalent to human third trimester, may also be used in some models in addition to gestational stress. A sizable literature now documents, both from human studies and experimental animal models, that prenatal stress has enduring consequences, both physiological and psychological that includes cardiovascular diseases and coronary heart disease, metabolic diseases such as obesity and type 2 diabetes, as well as mental health outcomes such as depression, schizophrenia and drug abuse (Braun et al., 2017; Van den Bergh et al., 2017).

Human studies of prenatal stress have often focused on such conditions as maternal depression and/or anxiety, but extend to the impact of natural disasters, recent deaths of relatives and even of poverty. As described in many other reports (Charil et al., 2010; Graignic-Philippe et al., 2014; Stavrou et al., 2017), such studies have demonstrated subsequent cognitive, behavioral and physiological consequences for offspring. Prenatal depression has been argued to increase risk in a sex-dependent manner in offspring for major depressive disorder as well as obesity/metabolic syndrome (Goldstein et al., 2016). A meta-analysis of 16 studies indicates an association of maternal stress (negative life events, anxiety/depression, bereavement, distress and job strain) with multiple atopic outcomes in offspring, including asthma, wheeze, atopic dermatitis, allergic rhinitis and IgE (Andersson et al., 2016). In another meta-analysis, maternal depression was associated with irregular fetal heart rate, increases in newborn cortisol, norepinephrine and dopamine levels, altered EEG patterns and reduced vagal tone, and increased internalizing and externalizing behaviors (Gentile, 2017).

Many (although not all (Lin et al., 2017)) prospective studies have provided supportive evidence of cognitive deficits. A meta-analysis that included >5000 children demonstrated a significant negative correlation of prenatal stress and various cognitive outcomes, among which were deficits in both verbal and nonverbal cognitive measures (Tarabulsy et al., 2014). At a somewhat later age, maternal anxiety during pregnancy was found to correlate with poorer working memory and lower grades in math (Pearson et al., 2016). Such changes have been coupled with alterations in the first 10 years of life in mental health-related outcomes, including increases in externalizing and social behavior problems and aggression, depression and anxiety in several studies (Van den Bergh et al., 2017). Notably, such effects may be particularly enduring, with such associations being seen in outcomes measured as late as 18 (Capron et al., 2015; Pearson et al., 2013) to 21 years of age (Betts et al., 2015).

Brain structure and development is also a target of human prenatal stress. In a retrospective study, individuals that were born small for gestational age and who also experienced low maternal care during childhood had smaller hippocampal volumes (Buss et al., 2007) as well as with subsequent externalizing behavior, which was mediated by reduced prefrontal cortical thickness (Sandman et al., 2015). Further, maternal anxiety was shown to produce changes in brain morphology assessed at 7 years of age that included region-specific reductions in gray matter volumes in the prefrontal cortex, premotor cortex, temporal lobe, lateral temporal cortex, postcentral gyrus and cerebellum (Buss et al., 2010) as well as with deficits in executive function (Buss et al., 2011). Higher maternal cortisol concentrations were found to be associated with a higher prevalence of child affective problems that was mediated by larger right amygdala volumes in girls but not in boys (Buss et al., 2012).

Numerous prospective imaging studies likewise point to protracted and enduring changes in brain structure. In a study of young women (mean age approximately 25 years), prenatal stress was shown to be associated with decreased gray matter volume in the medial temporal lobe and amygdala, correlating in particular with maternal depressive symptoms. In addition, alterations in connectivity of the medial temporal lobe with other regions were also found (Favaro et al., 2015). The effects in human populations are further supported by studies in experimental animals (Scheinost et al., 2017). Prenatal stress was found to increase connectivity of mesocorticolimbic regions in the right hemisphere of rat offspring in adulthood (Goelman et al., 2014), while in vivo electrophysiological recordings have demonstrated findings that include decreased synaptic activity, alterations in long-term potentiation, and decreased functional connectivity between medial prefrontal cortex and hippocampus (Barzegar et al., 2015; Negron-Oyarzo et al., 2015).

As might be expected, such findings are in accord with deficits in specific behavioral domains. A meta-analysis of 16 studies revealed medium effect sizes for an association of prenatal psychological distress with toddler (13–36 months of age) cognitive development (Kingston et al., 2015). Further, lower mental development index scores on the Bayley Scales and decreased attention and goal directedness were found in offspring of stressed mothers (Buitelaar et al., 2003; Huizink et al., 2003); lower mental development index scores were likewise reported in another retrospective cohort study (Bergman et al., 2007). High levels of maternal anxiety during pregnancy, specifically between 12 and 22 weeks of pregnancy, were associated with increased impulsivity and for males, and deficits in sustained attention measured at both 15 and 17 years of age (Van den Bergh et al., 2005, 2006). Prenatal anxiety was also found to be associated with impaired inhibitory control and working memory performance at 6–9 years of age (Buss et al., 2011), while at 24 years of age, a cognitive inflexibility was found in response to maternal major negative life events (Schwabe et al., 2012).

5.7 Childhood Adversity

Childhood adversity consists of negative experience during childhood associated with an increased risk for both adverse health and social outcomes. In humans, these would occur after birth, whereas the window in rodents is in later postnatal stages, given that the early postnatal period is considered equivalent to human third trimester. There are multiple types of such experiences that can include both psychological, psychosocial and physical conditions. Among these are caregiver and/or family maltreatment, unstable environments, deprived environments (e.g., foster homes or institutional settings), witnessing or experiencing crime or violence, and impoverished neighborhood factors. Indeed, the opportunities for adverse childhood experiences are heightened in low SES environments. Like prenatal stress, childhood adversity can lead to adverse mental and physical health consequences, with effects that are enduring, as described, along with mechanisms, in many excellent reviews (Berens et al., 2017; Bucci et al., 2016; Holz et al., 2015; Koss and Gunnar, 2017; Kuhlman et al., 2017; McLaughlin et al., 2014a; Misiak et al., 2017; Nusslock and Miller, 2016; Teicher et al., 2016; Tyrka et al., 2013).

Multiple behavioral consequences of early behavioral adversity have been observed in human studies. Numerous studies have demonstrated associations between childhood adversity and psychotic symptoms, with some studies even relating specific types of adverse experiences to particular psychotic phenotypic hallucinations (Misiak et al., 2017). In addition, cognitive deficits have been reported that can occur independently of psychosis, that include changes in cognition, memory and executive functions (Bucker et al., 2012; Perez and Widom, 1994; Vasilevski and Tucker, 2016). Children who have been neglected exhibit lower IQ scores and academic performance than typical care-raised children (Eckenrode et al., 1993; Spratt et al., 2012), and the reductions in IQ score of children raised in institutional settings are correlated with the duration of institutionalization (O'Connor et al., 2000). Additionally, studies consistently show that childhood adversity results in executive function deficits that are persistent, reflect the duration of institutionalization and do not necessarily recover post-institutionalization (Bos et al., 2009; Farah et al., 2008; Hostinar et al., 2012). Reductions in language ability also characterize childhood adversity (Windsor et al., 2011).

As might be expected based on such deficits, changes in brain structure and function are also seen in response to childhood adversity that correspond to the types of cognitive deficits cited above. These include reductions in activation of nucleus accumbens during reward anticipation and in response to a positive stimulus (Goff et al., 2013; Mehta et al., 2010). Moreover, reductions in cortical thickness are observed as is smaller total brain volume, reductions in gray matter, and reductions in sensory and association cortices as well as in cerebellum (Bauer et al., 2009; McLaughlin et al., 2014b; Mehta et al., 2009; Sheridan et al., 2012). In addition, white matter deficits, including reductions in volume of the posterior corpus callosum, have been noted (Bick et al., 2015; Sheridan et al., 2012).

It has been asserted that childhood adversity can be conceptualized as either deprivation-based or threat-based (McLaughlin et al., 2014a) and as such can be compared to animal studies based on sensory deprivation and fear learning, respectively. Environments consistent with the deprivation-based context could be considered to include poverty, institutionalization and neglect and has been compared to animal models of isolation stress (animals individually housed with no in cage enrichment). In such studies, similar reductions in brain volume have been demonstrated, as are reductions in markers of behavioral functioning such as dendritic arborization and spines. Additionally, the changes in brain structures are of greater magnitude and more persistent with increasing duration of individual housing

(Bennett et al., 1974, 1996; Diamond et al., 1975). An example of the threat context is Pavlovian fear conditioning, although other paradigms may apply as well. Effects of these stressors on brain structure are not necessarily consistent with the human observations. For example, no effects were seen in adult rats on neural progenitor cell proliferation, survival of newborn cells of brain-derived neurotrophic factor mRNA expression in response to acute restraint stress, acute or subchronic tail shock or acute, subchronic, or chronic resident-intruder stress (Hanson et al., 2011). Such findings could suggest that ideal models for mimicking threat contexts in the human environment are not fully explored or developed.

6. CUMULATIVE EFFECTS OF STRESS

In the human environment, stress exposures are generally not experienced on a single occasion, but may occur across the lifetime, with each individuals' specific experiences being unique. Thus, it is almost inevitable that stressor experience will cumulate over the lifetime. As might be expected, this type of chronic stress or cumulative adversity over life has also been shown to have detrimental consequences for both physical and mental health that depend upon the saliency and duration of the stressors. Such effects are seen even in healthy individuals. Two different hypotheses have been proposed to address consequences of developmental stress followed by adulthood stress (Paquola et al., 2017). An additive model of developmental stress and adulthood stress is postulated by the cumulative stress hypothesis, which predicts that adulthood stress response will be enhanced by prior childhood stress. The mismatch hypothesis, based on an evolutionary perspective, asserts that biological embedding to early stress alters fetal programming so that an individual exposed to stress in adulthood has coping strategies designed for a similar type of environment; increased disease risk only occurs when the developmental and adulthood environments are "mismatched" in terms of stress experience/features. Data supportive of both hypotheses have been reported, thus leaving this question as yet unanswered or in need of reformulation.

With respect to the cumulative additivity hypothesis, a study of healthy community participants, for example, found that higher cumulative adversity scores, specifically high chronic stress, correlated with reduced gray matter volume in mesocorticolimbic regions, including the medial prefrontal, anterior cingulate and insular cortical regions to a greater extent than occurred in individuals with low chronic stress (Ansell et al., 2012). Healthy individuals with high levels of chronic adversity report greater health symptoms as well as altered corticolimbic responses to acute stress (Seo et al., 2014). Some studies have examined the extent to which cumulative stress may impact aging-related alterations, including cognitive decline. In one such study, cumulative stressful experiences correlated with impaired performance in two memory tasks, particularly in elderly participants, and were accompanied by EEG changes consisting of reduced alpha- and gamma event-related synchronization which the authors interpreted as a potential basis for a failure of high-stress elderly individuals to be able to actively maintain a stimulus in working memory, as required for the memory tasks (Marshall et al., 2015).

In a study comparing the additivity vs. mismatch hypotheses, 64 individuals ranging from 14 to 26 years of age were recruited from a specialized clinic for mental health problems in younger people and subjected to an MRI and clinical assessment, and they also provided measures of both childhood and recent stress (Paquola et al., 2017). Results revealed that psychiatric symptoms were worse in individuals with higher levels of lifetime stress. The study also demonstrated a dissociation between impacted brain structures based on subjects' exposures. Specifically, subjects with mismatched childhood and adult stress exhibited reduced left hippocampal and reduced anterior cingulate cortex–ventrolateral prefrontal cortex connectivity, but greater anterior cingulate cortex–hippocampal functional connectivity as compared to individuals whose childhood and recent stress levels were matched. Thus, clinical symptoms appear to follow an additivity basis, whereas the mismatch hypothesis explained alterations in hippocampal volume and prefrontal cortex connectivity.

Experimental animal studies have addressed the hypothesis as well, with somewhat mixed results. A study in female Balb/c mice that were matched for early-life stress consisting of limited nesting material or early handling and then followed by adult stress of single housing vs. group housing produced some evidence of greater stress-related behavioral phenotype under conditions of mismatched environments (Santarelli et al., 2014). In a subsequent study, male mice were exposed again to either limited nesting material or early handling followed by either chronic social defeat stress or housing with an ovariectomized female. This resulted in a type of dissociative effect, in that HPA axis activity, anxiety-like behavior and glucocorticoid receptor expression levels in hippocampus were not altered by adversity during early life and adulthood, but rather by each experience alone (Santarelli et al., 2017). In another study, results again were somewhat equivocal, with gestational restraint stress alone influencing corticosterone, while main effects of adult stress were seen for body weight, sucrose preference measured at 6 months of age, and minimal interactions (Baker et al., 2014), whereas in male rats, memory performance in offspring that had undergone maternal separation followed by chronic unpredictable stress did not differ from controls (Zalosnik et al., 2014). Interpreting these studies is somewhat difficult however, given differences in the stress paradigms that are utilized and the extent to which these remained stressful under conditions of repeated exposures.

7. MULTIGENERATIONAL AND TRANSGENERATIONAL EFFECTS OF DEVELOPMENTAL STRESS

Significant interest has arisen in the potential for the consequences of prenatal stress to be passed across generations in either a multigenerational or transgenerational capacity (e.g., Gapp et al., 2014; Matthews and Phillips, 2012; Skinner, 2014). With exposures limited to females of the F0 generation, multigenerational consequences in females occur in the F1 (direct exposure) and germ cells of the next (F2) generations. True transgenerational effects would be seen in F3 generation where no direct exposures have occurred. In the case of male lineages, a direct effect of paternal stress in the F0 generation could produce multigenerational effects in the F1 generation, whereas a transgenerational effect would be observed in the F2 or later generation. One example of the latter is the consequences of the 1944–1945 Dutch famine, where maternal malnutrition effects were seen in the F2 generation males in the form of an increased propensity to obesity (Veenendaal et al., 2013). Although many different reports have been directed at the potential for multigenerational or transgenerational consequences of early stress, supporting evidence from human studies is very limited as of yet.

Experimental animal studies have provided some, albeit still limited, evidence of both multigenerational and transgenerational consequences of developmental stress. With respect to multigenerational transmission, paternal transmission of an F1 dysmasculinization in males following prenatal stress revealed a female-like pattern of changes in neurodevelopmental gene expression in F2 males as well as increased immobility in a tail-suspension test, but no significant impairments in a Barnes maze performance (Morgan and Bale, 2011). Abusive-type maternal care of rats was associated with alterations in DNA methylation of the BDNF gene in F2 female offspring (Roth et al., 2009). Multigenerational effects of unpredictable maternal separation with or without unpredictable maternal stress in mice were found to alter approach avoidance behavior in F2 offspring (Weiss et al., 2011). In another multigenerational study (Dias and Ressler, 2014), it was reported that exposure of F0 mice to fear conditioning based on odor stimuli increased behavioral sensitivity of both F1 and F2 offspring to the odor used for F0 conditioning and that F2 inheritance was related to parental gamete transmission.

Transgenerational studies of developmental stress are limited at present but supportive of effects that can carry across generations. Franklin et al., for example, demonstrated that the male F3 offspring of F1 male mice that had been subjected to chronic and unpredictable maternal separation showed depressive-like behavioral deficits similar to those observed in the F1 males, despite the fact that F1 males were reared in a normal environmental setting. In addition, corresponding changes in DNA methylation of the promoter region of several genes from F1-stressed males were present in the brains of F3 females, notably including the gene encoding MeCP2, a DNA binding protein implicated in Rett's syndrome and in stress (Franklin et al., 2010). Thus, this study demonstrated transgenerational impacts of early postnatal stress in both sexes. In a subsequent study, also based on early chronic and unpredictable maternal separation, alterations in social recognition memory were increased in F1 males as well as in F2 and F3 females, findings that could not be attributed to alterations in olfactory recognition. Assessment of the role of serotonin in these effects showed that F1 males had reductions in binding of the serotonin receptor agonist 8-OH-DPAT in lateral periaqueductal gray, dorsal raphe, CA1 and dentate gyrus of hippocampus and regions of the thalamus as well as increases in frontal cortex levels of serotonin. A subsequent experiment demonstrated that acute administration of 8-OH-DPAT was able to reverse deficits in social interactions observed in F3 male offspring (Franklin et al., 2011). Most recently, the serotonin-related alterations were examined using pharmacological function MRI in F1 and F2 offspring. This approach revealed a functional disconnectivity of limbic circuits across generations (Razoux et al., 2017).

In low SES environments where prenatal stress may be particularly impactful (Farah, 2017), prenatal stress may also occur across generations, consistent with the cyclicity of poverty. A series of studies in rats by Metz and colleagues (Kiss et al., 2016; McCreary et al., 2016a,b) mimics this feature of the human environment by imposing prenatal stress consisting of restraint stress and forced swimming in a semi-random sequence in an F0, F1 and F2 generation and examined multiple consequences in female F3 offspring. Under these conditions, increases in basal corticosterone levels were seen across generations, with an almost doubling between the F2 and F3 generations. Additionally, F3 generation females exhibited increases in distance traveled in an open field as well as increases in time spent in the margins of the open field, interpreted by some as an anxiety-emotional behavior. MRI analyses indicated a significant reduction in F3 cortex values, and additionally reductions in both whole brain and hippocampal volumes were observed. Based on an absolute gray volume index, F3 females also exhibited reductions in both cortical and hippocampal neuronal density.

A subsequent study by this group examined stress across the F1 through F4 generations (multigenerational stress group) as well as in the F1 group only (transgenerational) and studied behavioral laterality and brain dominance in the F4 offspring of both sexes. In a comparison of paw use as a measure of brain laterality, multigenerationally-exposed, but not transgenerationally-exposed, F4 males exhibited increase left paw preference indicative of altered brain laterality; neither multigenerational or transgenerational effects on paw preference were found in females.

Thus, while studies to date are limited, there is increasing evidence that effects of early stress may be transmitted across generations, even transgenerationally to offspring who never experienced these stressors. As noted in multiple reviews, there are two major routes by which such multigenerational and transgenerational effects and programming can occur (Bale, 2015; Gapp et al., 2014; Skinner, 2014) that may include both epigenetic and non-epigenetic passage.

8. MECHANISMS OF BIOLOGICAL PROGRAMMING OF PRENATAL STRESS AND CHILDHOOD ADVERSITY

A significant literature has addressed the biological mechanisms by which effects of developmental stress can become biologically programmed as described in numerous excellent reviews of this topic (Bale, 2015; Matthews and Phillips, 2010, 2012; Skinner, 2014). Although the HPA axis, the core stress response system of the body, and the associated glucocorticoid receptor receive the greatest level of attention, it is increasingly recognized that multiple mechanisms of programming may be involved and could even act in a synergistic capacity (Rakers et al., 2017).

8.1 Changes in HPA Axis Function

Maternal stress can also increase cortisol levels to the extent that they exceed the function of the placental enzyme 11β-hydroxysteroid dehydrogenase that normally converts cortisol to the inactive cortisone and thus protects the fetus from the higher levels of maternal glucocorticoids. Under such conditions, fetal cortisol levels in circulation increase and can bind to both glucocorticoid and mineralocorticoid receptors that are expressed in high levels in multiple fetal brain regions, including the limbic system, hypothalamus and cortex, with glucocorticoid receptor expression first seen on embryonic day 12.5 in fetal rat brain (Xiong and Zhang, 2013). Glucocorticoids can then impact many facets of brain development and function (as well as of other organ systems), including neurogenesis and gliogenesis as well as axonal and dendritic development and synaptogenesis (Matthews and Phillips, 2012) that underlie the cognitive, behavioral and morphological consequences of prenatal stress.

8.2 Catecholamine Changes

Although less consistently observed or related to increases in maternal stress and for which studies indicate little transfer of catecholamines to the fetus from the maternal blood, the release of catecholamines in response to stress can activate the sympatho-adrenomedullary system that underlies the fight or flight response to stress. Catecholamines in maternal blood result in constriction of placental blood vessels and thus reduce fetal delivery of nutrients and oxygen. However, recent studies in sheep indicate that isolation stress of ewes leads to increases in both cortisol and norepinephrine that remained elevated for >2 h and contributed to a stress-induced uterine blood flow decrease (Rakers et al., 2015). Catecholamine concentrations during midgestation have been associated with spontaneous preterm birth (Holzman et al., 2009) and recent work in sheep has shown that chronic prenatal psychosocial stress during the first and second trimesters can increase fetal catecholamine concentrations which are normally quite low (Rakers et al., 2017).

8.3 Maternal Immunity

Another postulated mechanism of fetal programming involves maternal immunity and infectious disease. Here too some inconsistencies are noted. Specifically, data showing that acute or chronic stress can trigger cytokine release in pregnant women are limited, although it has been observed, and the ability of maternal cytokines to pass through the placenta into fetal circulation is derived from animal models. Nevertheless, increases in fetal brain cytokines can interact with fetal glial cells and could also increase the release of reactive oxygen species.

8.4 Oxidative Stress and Reactive Oxygen Species

Oxidative stress and reactive oxygen species, via their ability to damage lipids proteins and DNA, may also be involved in programming. Currently, little is known about the potential for oxidative stress to alter fetal programming, although it has been linked to adverse fetal outcomes with significant long-term health implications, such as intrauterine growth restriction. In addition, whether maternal stress itself leads to increased oxidative stress during pregnancy is not known or how a maternal-fetal transfer would occur, although correlations between maternal and fetal oxidative stress markers have been reported. Experimental animal models suggest a role for this mechanism in that studies in which pregnant dams that underwent restraint stress and were also provided with an oral antioxidant (hydroxytyrosol) prevented adverse consequences such as detriments in cognitive functioning and in hippocampal neurogenesis (Zheng et al., 2015).

8.5 Neuroactive Steroids

Neuroactive steroids other than glucocorticoids have also been proposed to play a role in fetal programming (Brunton, 2015), with the primary neuroactive steroids studied being allopregnanolone and 5α,3α-tetrahydrodeoxycorticosterone (THDOC), both potent modulators of GABAergic inhibitory functions and of neuronal function. In addition, these neuroactive steroids modulate HPA axis function, particularly negative feedback to restore homeostasis. Studies show that stress can reduce circulating levels of allopregnanolone at birth which reduces its production in brain of offspring. The latter is of potential importance to the cognitive impairments and anxiety-evoking effects of early stress, as reductions in neuroactive steroids are seen in disorders such as anxiety discrepancies in the specific nature of those effects.

8.6 Tryptophan and Serotonin

Potential roles for tryptophan and the neurotransmitter serotonin, synthesized from tryptophan, have also been proposed in biological programming (St-Pierre et al., 2016). Studies in animal models have demonstrated increases in tryptophan and serotonin in fetal brain occur in response to psychosocial stress in dams, but whether similar effects occur in human stress is unknown. Its potential programming role is supported by the known requirements for serotonin in placental function and for fetal and postnatal brain development.

8.7 Gut–Brain Axis

A more recently proposed mechanism of biological programming involves the gut–brain axis (Jašarević et al., 2015b; O'Mahony et al., 2017). This particular mechanism is a relatively recent development, and as yet more hypothetical than demonstrated. Several premises exist for such studies. The gut–brain axis, which includes neuronal, immune and endocrine pathways, constitutes a bi-directional communication system between the gastrointestinal tract and the brain that involves cytokines, neurotransmitters, including serotonin, and other neuromodulators. In one scenario, passage of the fetus through the birth canal renders exposure of a presumably sterile fetal gut to the vaginal microbiota; the vaginal microbiota has been shown to be sensitive to chronic maternal stress. Whether a fetal microbiome as yet exists is controversial. In a recent study in mice, offspring of maternally stressed dams showed an altered microbiota composition that corresponded to brain region and sex-dependent disruptions of amino acid profiles (Jašarević et al., 2015a).

8.8 Epigenetic Changes and Transgenerational Effects

Many of these proposed biological programming mechanisms also ultimately lead to epigenetic changes that have been the subject of a burgeoning literature, both for effects in F1 generations, but also for transgenerational effects. F1 epigenetic changes in response to developmental stress are the subject of numerous reviews to date (Kundakovic and Jaric, 2017; Provencal and Binder, 2015; Zannas and West, 2014). Mechanisms by which transgenerational effects in particular are passed across generations is a topic of great interest. As is becoming evident, multiple mechanisms are involved (Bale, 2015; Bohacek and Mansuy, 2017; Danchin et al., 2011; Hanson and Skinner, 2016; Sharma, 2017; van Otterdijk and Michels, 2016) that include epigenetic changes to germ cells as well as social and cultural influences. Strictly speaking, germ cell transmission involves a germline (egg or sperm)-mediated transfer of imprinted epigenetic information that occurs in the absence of direct environmental exposures. Germ cell imprinting to date has been shown to occur via different epigenetic changes, including DNA methylation, changes in noncoding RNA, histone modifications as well as changes in chromatin structure.

In addition to germ cell mediated epigenetic transgenerational effects, both social and cultural influences can contribute to phenotype and be passed across generations. Germline-independent mechanisms can also occur however, with phenotypic consequences. Reported examples of this have included parental effects, such as alterations in maternal care, in the microbiota, and in odors. Moreover, genetic and non-genetic mechanisms may interact. For example, one study reported that mutant mouse dams that expressed deficits in maternal care then socially transmitted impaired maternal behavior to female offspring with effects that were observable for two generations (Curley et al., 2008).

As noted above, and when considered in the context of physiology and the interaction of systems within the body, it is likely that multiple programming mechanisms occur, and that interactive mechanisms of transmission across generations can be operative. Such possibilities have prompted reports of how experimental designs must be formulated to differentiate parental/social mechanisms from epigenetic mechanisms (Bohacek and Mansuy, 2017).

9. CONCLUSIONS AND RESEARCH NEEDS

As the above makes clear, the impacts of prenatal stress and early adversity on brain and behavior can be extensive and enduring, with potential adverse consequences in relation to both physical and mental health. Indeed, such effects are now being reported to have impacts across generations. These consequences of prenatal stress and early adversity arise from biological reprogramming that can occur through multiple mechanisms to the first-generation offspring as well as via both epigenetic and non-epigenetic mechanisms.

Both human studies and animal models have utilized a wide variety of different stress protocols. As previously noted, however, data suggest that the definition of stress should be limited to events that are uncontrollable/unpredictable. This raises the question as to what extent reported studies described as demonstrating effects of stress really reflect the consequences of stress rather than of adaptation to stress? Of course, repeated homotypic stress paradigms are perhaps most problematic in this regard. Our own studies and observations of rodents show that a second exposure to such usually highly salient stressors as restraint or forced swim results in dramatically different behavior, consistent with adaptation as has previously been noted (Molendijk and de Kloet, 2015). But even chronic variable stress has been reported to produce forms of adaptation. Such possible confounds suggest that measures consistent with stress effects rather than simply with behavioral activity (Koolhaas et al., 1997, 2011) be included in such studies to validate the assay. These same concerns arise in the context of defining resilience phenotypes and associated mechanisms.

A question that arises from experimental animal models is how relevant are some of these models to the demographics of stress? Stress differs significantly in breadth, intensity and likely duration in impoverished communities. Poverty is often defined in terms of resource deprivation, e.g., lack of access to financial, environmental, social and other resources that would provide a safe and stable life. It is difficult to see how restraint stress, one of the most widely used prenatal stress paradigms, is translationally relevant to such conditions. While such stressors as deprivation of nesting material and maternal separation may be more relevant to the human environment, they are also of limited duration in relation to human poverty where stress is typically enduring. Thus, a challenge is the development and validation of chronic stress models that can be used across generations and that do not lead to adaptation, but to the opposite, namely a behavioral sensitization and/or anticipatory response to stimuli associated with the stressor. Indeed, one might expect that with processes of stimulus and response generalization that even lower levels of relevant stimuli as well as similar stimuli would come to evoke stress responses. Similar to such a hypothesis, it was shown that exposure to the environment in which rats had been exposed to inescapable shock, i.e., the cues that were present during stress, prolonged the duration of subsequent behavioral depression and learned helplessness (Maier, 2001), and another study reported that a single exposures of rats to immobilization restraint stress markedly increased anxiety-like behavior in an elevated plus maze (Belda et al., 2008). As such studies in non-human primates that have focused on social inequality aversion may be more relevant and could be adapted for rodents.

This dynamic trajectory of the protracted effects of early stress suggests that prevention is the best route of eliminating the adverse effects on offspring and potentially on their offspring. That possibility is likely not feasible or achievable except in a minimal sense. Furthermore, some "stresses," particularly when events are controllable and/or predictable are likely to invoke resilient phenotypes and thus are beneficial. In the case of the more salient uncontrollable, unpredictable stress, and its adverse consequences, another research opportunity involves intervention strategies that could be used to reverse or mitigate these consequences. While genetic and epigenetic studies sometimes suggest induction of such changes in a therapeutic context, these possibilities are remote and unlikely for several reasons including their potential lack of specificity. Behavioral therapeutic interventions represent another approach that may offer possibilities in addition to or instead of pharmacological approaches and deserve more systematic assessment in experimental animal models.

REFERENCES

Abbott, B.B., Schoen, L.S., Badia, P., 1984. Predictable and unpredictable shock: behavioral measures of aversion and physiological measures of stress. Psychol. Bull. 96, 45–71.

Andersson, N.W., Hansen, M.V., Larsen, A.D., Hougaard, K.S., Kolstad, H.A., Schlunssen, V., 2016. Prenatal maternal stress and atopic diseases in the child: a systematic review of observational human studies. Allergy 71, 15–26.

Anisman, H., Matheson, K., 2005. Stress, depression, and anhedonia: caveats concerning animal models. Neurosci. Biobehav. Rev. 29, 525–546.

Ansell, E.B., Rando, K., Tuit, K., Guarnaccia, J., Sinha, R., 2012. Cumulative adversity and smaller gray matter volume in medial prefrontal, anterior cingulate, and insula regions. Biol. Psychiatry 72, 57–64.

Antoni, M.H., Lutgendorf, S.K., Cole, S.W., Dhabhar, F.S., Sephton, S.E., McDonald, P.G., Stefanek, M., Sood, A.K., 2006. The influence of bio-behavioural factors on tumour biology: pathways and mechanisms. Nat. Rev. Cancer 6, 240–248.

Baker, S.L., Mileva, G., Huta, V., Bielajew, C., 2014. In utero programming alters adult response to chronic mild stress: part 3 of a longitudinal study. Brain Res. 1588, 175–189.

Bale, T.L., 2015. Epigenetic and transgenerational reprogramming of brain development. Nat. Rev. Neurosci. 16, 332–344.

Bale, T.L., Epperson, C.N., 2015. Sex differences and stress across the lifespan. Nat. Neurosci. 18, 1413–1420.

Bangasser, D.A., Wicks, B., 2017. Sex-specific mechanisms for responding to stress. J. Neurosci. Res. 95, 75–82.

Barrot, M., Marinelli, M., Abrous, D.N., Rouge-Pont, F., Le Moal, M., Piazza, P.V., 2000. The dopaminergic hyper-responsiveness of the shell of the nucleus accumbens is hormone-dependent. Eur. J. Neurosci. 12, 973–979.

Bartolomucci, A., Palanza, P., Costoli, T., Savani, E., Laviola, G., Parmigiani, S., Sgoifo, A., 2003. Chronic psychosocial stress persistently alters autonomic function and physical activity in mice. Physiol. Behav. 80, 57–67.

Barzegar, M., Sajjadi, F.S., Talaei, S.A., Hamidi, G., Salami, M., 2015. Prenatal exposure to noise stress: anxiety, impaired spatial memory, and deteriorated hippocampal plasticity in postnatal life. Hippocampus 25, 187–196.

Bauer, P.M., Hanson, J.L., Pierson, R.K., Davidson, R.J., Pollak, S.D., 2009. Cerebellar volume and cognitive functioning in children who experienced early deprivation. Biol. Psychiatry 66, 1100–1106.

Bekhbat, M., Neigh, G.N., 2018. Sex differences in the neuro-immune consequences of stress: focus on depression and anxiety. Brain Behav. Immun. 67, 1–12.

Belda, X., Fuentes, S., Nadal, R., Armario, A., 2008. A single exposure to immobilization causes long-lasting pituitary-adrenal and behavioral sensitization to mild stressors. Horm. Behav. 54, 654–661.

Belujon, P., Grace, A.A., 2014. Restoring mood balance in depression: ketamine reverses deficit in dopamine-dependent synaptic plasticity. Biol. Psychiatry 76, 927–936.

Belujon, P., Grace, A.A., 2015. Regulation of dopamine system responsivity and its adaptive and pathological response to stress. Proc. Biol. Sci. 282.

Bennett, E.L., Rosenzweig, M.R., Diamond, M.C., Morimoto, H., Hebert, M., 1974. Effects of successive environments on brain measures. Physiol. Behav. 12, 621–631.

Bennett, E.L., Diamond, M.C., Krech, D., Rosenzweig, M.R., 1996. Chemical and anatomical plasticity of brain. 1964. J. Neuropsychiatry Clin. Neurosci. 8, 459–470.

Berens, A.E., Jensen, S.K.G., Nelson 3rd, C.A., 2017. Biological embedding of childhood adversity: from physiological mechanisms to clinical implications. BMC Med. 15, 135.

Bergman, K., Sarkar, P., O'Connor, T.G., Modi, N., Glover, V., 2007. Maternal stress during pregnancy predicts cognitive ability and fearfulness in infancy. J. Am. Acad. Child Adolesc. Psychiatry 46, 1454–1463.

Betts, K.S., Williams, G.M., Najman, J.M., Alati, R., 2015. The relationship between maternal depressive, anxious, and stress symptoms during pregnancy and adult offspring behavioral and emotional problems. Depress. Anxiety 32, 82–90.
Bick, J., Zhu, T., Stamoulis, C., Fox, N.A., Zeanah, C., Nelson, C.A., 2015. Effect of early institutionalization and foster care on long-term white matter development: a randomized clinical trial. JAMA Pediatr. 169, 211–219.
Blair, C., Granger, D., Peters Razza, R., 2005. Cortisol reactivity is positively related to executive function in preschool children attending head start. Child Dev. 76, 554–567.
Bohacek, J., Mansuy, I.M., 2017. A guide to designing germline-dependent epigenetic inheritance experiments in mammals. Nat. Methods 14, 243–249.
Bonilla-Jaime, H., Vazquez-Palacios, G., Arteaga-Silva, M., Retana-Marquez, S., 2006. Hormonal responses to different sexually related conditions in male rats. Horm. Behav. 49, 376–382.
Bos, K.J., Fox, N., Zeanah, C.H., Nelson III, C.A., 2009. Effects of early psychosocial deprivation on the development of memory and executive function. Front. Behav. Neurosci. 3, 16.
Bowers, S.L., Bilbo, S.D., Dhabhar, F.S., Nelson, R.J., 2008. Stressor-specific alterations in corticosterone and immune responses in mice. Brain Behav. Immun. 22, 105–113.
Braun, K., Bock, J., Wainstock, T., Matas, E., Gaisler-Salomon, I., Fegert, J., Ziegenhain, U., Segal, M., 2017. Experience-induced transgenerational (re-)programming of neuronal structure and functions: impact of stress prior and during pregnancy. Neurosci. Biobehav. Rev.
Brown, S., Fite, P.J., Poquiz, J., 2016. Moderating effects of gender on outcomes associated with stressful life events among elementary school-age youth. Child Psychiatry Hum. Dev. 47, 593–602.
Brunton, P.J., 2015. Programming the brain and behaviour by early-life stress: a focus on neuroactive steroids. J. Neuroendocrinol. 27, 468–480.
Brzozowski, B., Mazur-Bialy, A., Pajdo, R., Kwiecien, S., Bilski, J., Zwolinska-Wcislo, M., Mach, T., Brzozowski, T., 2016. Mechanisms by which stress affects the experimental and clinical inflammatory bowel disease (IBD): role of brain-gut axis. Curr. Neuropharmacol. 14, 892–900.
Bucci, M., Marques, S.S., Oh, D., Harris, N.B., 2016. Toxic stress in children and adolescents. Adv. Pediatr. Infect. Dis. 63, 403–428.
Bucker, J., Kapczinski, F., Post, R., Cereser, K.M., Szobot, C., Yatham, L.N., Kapczinski, N.S., Kauer-Sant'Anna, M., 2012. Cognitive impairment in school-aged children with early trauma. Compr. Psychiatry 53, 758–764.
Buitelaar, J.K., Huizink, A.C., Mulder, E.J., de Medina, P.G., Visser, G.H., 2003. Prenatal stress and cognitive development and temperament in infants. Neurobiol. Aging 24 (Suppl. 1), S53–60. discussion S67–S68.
Buss, C., Lord, C., Wadiwalla, M., Hellhammer, D.H., Lupien, S.J., Meaney, M.J., Pruessner, J.C., 2007. Maternal care modulates the relationship between prenatal risk and hippocampal volume in women but not in men. J. Neurosci. 27, 2592–2595.
Buss, C., Davis, E.P., Muftuler, L.T., Head, K., Sandman, C.A., 2010. High pregnancy anxiety during mid-gestation is associated with decreased gray matter density in 6-9-year-old children. Psychoneuroendocrinology 35, 141–153.
Buss, C., Davis, E.P., Hobel, C.J., Sandman, C.A., 2011. Maternal pregnancy-specific anxiety is associated with child executive function at 6-9 years age. Stress 14, 665–676.
Buss, C., Davis, E.P., Shahbaba, B., Pruessner, J.C., Head, K., Sandman, C.A., 2012. Maternal cortisol over the course of pregnancy and subsequent child amygdala and hippocampus volumes and affective problems. Proc. Natl. Acad. Sci. U. S. A. 109, E1312–9.
Capron, L.E., Glover, V., Pearson, R.M., Evans, J., O'Connor, T.G., Stein, A., Murphy, S.E., Ramchandani, P.G., 2015. Associations of maternal and paternal antenatal mood with offspring anxiety disorder at age 18 years. J. Affect. Disord. 187, 20–26.

Caso, J.R., Leza, J.C., Menchen, L., 2008. The effects of physical and psychological stress on the gastro-intestinal tract: lessons from animal models. Curr. Mol. Med. 8, 299–312.
Charil, A., Laplante, D.P., Vaillancourt, C., King, S., 2010. Prenatal stress and brain development. Brain Res. Rev. 65, 56–79.
Cohen, S., Doyle, W.J., Baum, A., 2006. Socioeconomic status is associated with stress hormones. Psychosom. Med. 68, 414–420.
Comeau, W.L., Winstanley, C.A., Weinberg, J., 2014. Prenatal alcohol exposure and adolescent stress—unmasking persistent attentional deficits in rats. Eur. J. Neurosci. 40, 3078–3095.
Curley, J.P., Champagne, F.A., Bateson, P., Keverne, E.B., 2008. Transgenerational effects of impaired maternal care on behaviour of offspring and grandoffspring. Anim. Behav. 75, 1551–1561.
Danchin, E., Charmantier, A., Champagne, F.A., Mesoudi, A., Pujol, B., Blanchet, S., 2011. Beyond DNA: integrating inclusive inheritance into an extended theory of evolution. Nat. Rev. Genet. 12, 475–486.
De Boer, S.F., Slangen, J.L., Van der Gugten, J., 1990. Plasma catecholamine and corticosterone levels during active and passive shock-prod avoidance behavior in rats: effects of chlordiazepoxide. Physiol. Behav. 47, 1089–1098.
Dhabhar, F.S., McEwen, B.S., 1997. Acute stress enhances while chronic stress suppresses cell-mediated immunity in vivo: a potential role for leukocyte trafficking. Brain Behav. Immun. 11, 286–306.
Diamond, M.C., Lindner, B., Johnson, R., Bennett, E.L., Rosenzweig, M.R., 1975. Differences in occipital cortical synapses from environmentally enriched, impoverished, and standard colony rats. J. Neurosci. Res. 1, 109–119.
Dias, B.G., Ressler, K.J., 2014. Parental olfactory experience influences behavior and neural structure in subsequent generations. Nat. Neurosci. 17, 89–96.
Diekhof, E.K., Keil, M., Obst, K.U., Henseler, I., Dechent, P., Falkai, P., Gruber, O., 2012. A functional neuroimaging study assessing gender differences in the neural mechanisms underlying the ability to resist impulsive desires. Brain Res. 1473, 63–77.
Dohrenwend, B.P., Levav, I., Shrout, P.E., Schwartz, S., Naveh, G., Link, B.G., Skodol, A.E., Stueve, A., 1992. Socioeconomic status and psychiatric disorders: the causation-selection issue. Science 255, 946–952.
Dubow, E.F., Tisak, J., Causey, D., Hryshko, A., Reid, G., 1991. A two-year longitudinal study of stressful life events, social support, and social problem-solving skills: contributions to children's behavioral and academic adjustment. Child Dev. 62, 583–599.
Eckenrode, J., Laird, M., Doris, J., 1993. School performance and disciplinary problems among abused and neglected children. Dev. Psychol. 29, 53–62.
Egan, A.E., Ulrich-Lai, Y.M., 2015. Activation of physiological stress responses by a natural reward: novel vs. repeated sucrose intake. Physiol. Behav. 150, 43–52.
Elo, I.T., Preston, S.H., 1996. Educational differentials in mortality: United States, 1979-85. Soc. Sci. Med. 42, 47–57.
Evans, G.W., English, K., 2002. The environment of poverty: multiple stressor exposure, psychophysiological stress, and socioemotional adjustment. Child Dev. 73, 1238–1248.
Farah, M.J., 2017. The neuroscience of socioeconomic status: correlates, causes, and consequences. Neuron 96, 56–71.
Farah, M.J., Betancourt, L., Shera, D.M., Savage, J.H., Giannetta, J.M., Brodsky, N.L., Malmud, E.K., Hurt, H., 2008. Environmental stimulation, parental nurturance and cognitive development in humans. Dev. Sci. 11, 793–801.
Favaro, A., Tenconi, E., Degortes, D., Manara, R., Santonastaso, P., 2015. Neural correlates of prenatal stress in young women. Psychol. Med. 45, 2533–2543.
Franklin, T.B., Russig, H., Weiss, I.C., Graff, J., Linder, N., Michalon, A., Vizi, S., Mansuy, I.M., 2010. Epigenetic transmission of the impact of early stress across generations. Biol. Psychiatry 68, 408–415.

Franklin, T.B., Linder, N., Russig, H., Thony, B., Mansuy, I.M., 2011. Influence of early stress on social abilities and serotonergic functions across generations in mice. PLoS One 6, e21842.

Gapp, K., von Ziegler, L., Tweedie-Cullen, R.Y., Mansuy, I.M., 2014. Early life epigenetic programming and transmission of stress-induced traits in mammals: how and when can environmental factors influence traits and their transgenerational inheritance? Bioessays 36, 491–502.

Gentile, S., 2017. Untreated depression during pregnancy: short- and long-term effects in offspring. A systematic review. Neuroscience 342, 154–166.

Gillies, G.E., Virdee, K., McArthur, S., Dalley, J.W., 2014. Sex-dependent diversity in ventral tegmental dopaminergic neurons and developmental programing: a molecular, cellular and behavioral analysis. Neuroscience 282, 69–85.

Glymour, M.M., Avendano, M., Kawachi, I., 2014. Socioeconomic status and health. In: Berkman, L., Kawachi, I., Glymour, M.M. (Eds.), Social Epidemiology. Oxford University Press, Oxford.

Goelman, G., Ilinca, R., Zohar, I., Weinstock, M., 2014. Functional connectivity in prenatally stressed rats with and without maternal treatment with ladostigil, a brain-selective monoamine oxidase inhibitor. Eur. J. Neurosci. 40, 2734–2743.

Goff, B., Gee, D.G., Telzer, E.H., Humphreys, K.L., Gabard-Durnam, L., Flannery, J., Tottenham, N., 2013. Reduced nucleus accumbens reactivity and adolescent depression following early-life stress. Neuroscience 249, 129–138.

Goldstein, D.S., Kopin, I.J., 2017. Linking stress, catecholamine autotoxicity, and allostatic load with neurodegenerative diseases: a focused review in memory of Richard Kvetnansky. Cell. Mol. Neurobiol.

Goldstein, J.M., Holsen, L., Huang, G., Hammond, B.D., James-Todd, T., Cherkerzian, S., Hale, T.M., Handa, R.J., 2016. Prenatal stress-immune programming of sex differences in comorbidity of depression and obesity/metabolic syndrome. Dialogues Clin. Neurosci. 18, 425–436.

Gould, F., Clarke, J., Heim, C., Harvey, P.D., Majer, M., Nemeroff, C.B., 2012. The effects of child abuse and neglect on cognitive functioning in adulthood. J. Psychiatr. Res. 46, 500–506.

Graf, E.N., Wheeler, R.A., Baker, D.A., Ebben, A.L., Hill, J.E., McReynolds, J.R., Robble, M.A., Vranjkovic, O., Wheeler, D.S., Mantsch, J.R., Gasser, P.J., 2013. Corticosterone acts in the nucleus accumbens to enhance dopamine signaling and potentiate reinstatement of cocaine seeking. J. Neurosci. 33, 11800–11810.

Graignic-Philippe, R., Dayan, J., Chokron, S., Jacquet, A.Y., Tordjman, S., 2014. Effects of prenatal stress on fetal and child development: a critical literature review. Neurosci. Biobehav. Rev. 43, 137–162.

Green, M.R., McCormick, C.M., 2016. Sex and stress steroids in adolescence: gonadal regulation of the hypothalamic-pituitary-adrenal axis in the rat. Gen. Comp. Endocrinol. 234, 110–116.

Halaris, A., 2017. Inflammation-associated co-morbidity between depression and cardiovascular disease. In: Dantzer, R., Capuron, L. (Eds.), Inflammation-Associated Depression: Evidence, Mechanisms and Implications. Springer International Publishing, Cham.

Han, X., Wang, W., Xue, X., Shao, F., Li, N., 2011. Brief social isolation in early adolescence affects reversal learning and forebrain BDNF expression in adult rats. Brain Res. Bull. 86, 173–178.

Hanson, M.A., Skinner, M.K., 2016. Developmental origins of epigenetic transgenerational inheritance. Environ. Epigenet. 2.

Hanson, N.D., Owens, M.J., Boss-Williams, K.A., Weiss, J.M., Nemeroff, C.B., 2011. Several stressors fail to reduce adult hippocampal neurogenesis. Psychoneuroendocrinology 36, 1520–1529.

Hayward, M., Miles, T., Crimmins, E., Yang, Y., 2000. The significance of socioeconomic status in explaining the racial gap in chronic health conditions. Am. Sociol. Rev. 65, 910–930.

Herman, J.P., 2013. Neural control of chronic stress adaptation. Front. Behav. Neurosci. 7, 61.

Herman, J.P., McKlveen, J.M., Ghosal, S., Kopp, B., Wulsin, A., Makinson, R., Scheimann, J., Myers, B., 2016. Regulation of the hypothalamic-pituitary-adrenocortical stress response. Compr. Physiol. 6, 603–621.

Hollanders, J.J., van der Voorn, B., Rotteveel, J., Finken, M.J.J., 2017. Is HPA axis reactivity in childhood gender-specific? A systematic review. Biol. Sex Differ. 8, 23.

Holmes, T.H., Rahe, R.H., 1967. The social readjustment rating scale. J. Psychosom. Res. 11, 213–218.

Holz, N.E., Laucht, M., Meyer-Lindenberg, A., 2015. Recent advances in understanding the neurobiology of childhood socioeconomic disadvantage. Curr. Opin. Psychiatry 28, 365–370.

Holzman, C., Senagore, P., Tian, Y., Bullen, B., Devos, E., Leece, C., Zanella, A., Fink, G., Rahbar, M.H., Sapkal, A., 2009. Maternal catecholamine levels in midpregnancy and risk of preterm delivery. Am. J. Epidemiol. 170, 1014–1024.

Hostinar, C.E., Stellern, S.A., Schaefer, C., Carlson, S.M., Gunnar, M.R., 2012. Associations between early life adversity and executive function in children adopted internationally from orphanages. Proc. Natl. Acad. Sci. U. S. A. 109 (Suppl. 2), 17208–17212.

House, J.S., Landis, K.R., Umberson, D., 1988. Social relationships and health. Science 241, 540–545.

Huizink, A.C., Robles de Medina, P.G., Mulder, E.J., Visser, G.H., Buitelaar, J.K., 2003. Stress during pregnancy is associated with developmental outcome in infancy. J. Child Psychol. Psychiatry 44, 810–818.

Hunter, R.G., McCarthy, K.J., Milne, T.A., Pfaff, D.W., McEwen, B.S., 2009. Regulation of hippocampal H3 histone methylation by acute and chronic stress. Proc. Natl. Acad. Sci. U. S. A. 106, 20912–20917.

Hurtubise, J.L., Howland, J.G., 2017. Effects of stress on behavioral flexibility in rodents. Neuroscience 345, 176–192.

Inoue, T., Tsuchiya, K., Koyama, T., 1994. Regional changes in dopamine and serotonin activation with various intensity of physical and psychological stress in the rat brain. Pharmacol. Biochem. Behav. 49, 911–920.

Jacobson, L., 2014. Forebrain glucocorticoid receptor gene deletion attenuates behavioral changes and antidepressant responsiveness during chronic stress. Brain Res. 1583, 109–121.

Jašarević, E., Howerton, C.L., Howard, C.D., Bale, T.L., 2015a. Alterations in the vaginal microbiome by maternal stress are associated with metabolic reprogramming of the offspring gut and brain. Endocrinology 156, 3265–3276.

Jašarević, E., Rodgers, A.B., Bale, T.L., 2015b. A novel role for maternal stress and microbial transmission in early life programming and neurodevelopment. Neurobiol. Stress 1, 81–88.

Joels, M., 2006. Corticosteroid effects in the brain: U-shape it. Trends Pharmacol. Sci. 27, 244–250.

Katz, R.J., Roth, K.A., Carroll, B.J., 1981. Acute and chronic stress effects on open field activity in the rat: implications for a model of depression. Neurosci. Biobehav. Rev. 5, 247–251.

Kingston, D., McDonald, S., Austin, M.P., Tough, S., 2015. Association between prenatal and postnatal psychological distress and toddler cognitive development: a systematic review. PLoS One 10, e0126929.

Kirby, L.G., Chou-Green, J.M., Davis, K., Lucki, I., 1997. The effects of different stressors on extracellular 5-hydroxytryptamine and 5-hydroxyindoleacetic acid. Brain Res. 760, 218–230.

Kiss, D., Ambeskovic, M., Montina, T., Metz, G.A., 2016. Stress transgenerationally programs metabolic pathways linked to altered mental health. Cell. Mol. Life Sci. 73, 4547–4557.

Konarska, M., Stewart, R.E., McCarty, R., 1990. Habituation and sensitization of plasma catecholamine responses to chronic intermittent stress: effects of stressor intensity. Physiol. Behav. 47, 647–652.

Koolhaas, J.M., Meerlo, P., De Boer, S.F., Strubbe, J.H., Bohus, B., 1997. The temporal dynamics of the stress response. Neurosci. Biobehav. Rev. 21, 775–782.

Koolhaas, J.M., Bartolomucci, A., Buwalda, B., de Boer, S.F., Flugge, G., Korte, S.M., Meerlo, P., Murison, R., Olivier, B., Palanza, P., Richter-Levin, G., Sgoifo, A., Steimer, T., Stiedl, O., van Dijk, G., Wohr, M., Fuchs, E., 2011. Stress revisited: a critical evaluation of the stress concept. Neurosci. Biobehav. Rev. 35, 1291–1301.

Koolhaas, J.M., De Boer, S.F., Buwalda, B., Meerlo, P., 2017. Social stress models in rodents: towards enhanced validity. Neurobiol. Stress 6, 104–112.

Koss, K.J., Gunnar, M.R., 2017. Annual research review: early adversity, the hypothalamic-pituitary-adrenocortical axis, and child psychopathology. J. Child Psychol. Psychiatry.

Kuhlman, K.R., Chiang, J.J., Horn, S., Bower, J.E., 2017. Developmental psychoneuroendocrine and psychoneuroimmune pathways from childhood adversity to disease. Neurosci. Biobehav. Rev. 80, 166–184.

Kundakovic, M., Jaric, I., 2017. The epigenetic link between prenatal adverse environments and neurodevelopmental disorders. Genes (Basel) 8, 104. https://doi.org/10.3390/genes8030104.

Lagraauw, H.M., Kuiper, J., Bot, I., 2015. Acute and chronic psychological stress as risk factors for cardiovascular disease: insights gained from epidemiological, clinical and experimental studies. Brain Behav. Immun. 50, 18–30.

Laman-Maharg, A., Trainor, B.C., 2017. Stress, sex, and motivated behaviors. J. Neurosci. Res. 95, 83–92.

Leserman, J., Petitto, J.M., Gu, H., Gaynes, B.N., Barroso, J., Golden, R.N., Perkins, D.O., Folds, J.D., Evans, D.L., 2002. Progression to AIDS, a clinical AIDS condition and mortality: psychosocial and physiological predictors. Psychol. Med. 32, 1059–1073.

Lin, Y., Xu, J., Huang, J., Jia, Y., Zhang, J., Yan, C., Zhang, J., 2017. Effects of prenatal and postnatal maternal emotional stress on toddlers' cognitive and temperamental development. J. Affect. Disord. 207, 9–17.

Lupien, S.J., King, S., Meaney, M.J., McEwen, B.S., 2001. Can poverty get under your skin? Basal cortisol levels and cognitive function in children from low and high socioeconomic status. Dev. Psychopathol. 13, 653–676.

MacLusky, N.J., Yuan, H., Elliott, J., Brown, T.J., 1996. Sex differences in corticosteroid binding in the rat brain: an in vitro autoradiographic study. Brain Res. 708, 71–81.

Maier, S.F., 2001. Exposure to the stressor environment prevents the temporal dissipation of behavioral depression/learned helplessness. Biol. Psychiatry 49, 763–773.

Maier, S.F., Amat, J., Baratta, M.V., Paul, E., Watkins, L.R., 2006. Behavioral control, the medial prefrontal cortex, and resilience. Dialogues Clin. Neurosci. 8, 397–406.

Malarbi, S., Abu-Rayya, H.M., Muscara, F., Stargatt, R., 2017. Neuropsychological functioning of childhood trauma and post-traumatic stress disorder: a meta-analysis. Neurosci. Biobehav. Rev. 72, 68–86.

Marcovecchio, M.L., Chiarelli, F., 2012. The effects of acute and chronic stress on diabetes control. Sci. Signal. 5 (247), pt10. https://doi.org/10.1126/scisignal.2003508.

Marshall, A.C., Cooper, N.R., Segrave, R., Geeraert, N., 2015. The effects of long-term stress exposure on aging cognition: a behavioral and EEG investigation. Neurobiol. Aging 36, 2136–2144.

Matthews, S.G., Phillips, D.I., 2010. Minireview: transgenerational inheritance of the stress response: a new frontier in stress research. Endocrinology 151, 7–13.

Matthews, S.G., Phillips, D.I., 2012. Transgenerational inheritance of stress pathology. Exp. Neurol. 233, 95–101.
McArthur, S., McHale, E., Dalley, J.W., Buckingham, J.C., Gillies, G.E., 2005. Altered mesencephalic dopaminergic populations in adulthood as a consequence of brief perinatal glucocorticoid exposure. J. Neuroendocrinol. 17, 475–482.
McArthur, S., McHale, E., Gillies, G.E., 2007. The size and distribution of midbrain dopaminergic populations are permanently altered by perinatal glucocorticoid exposure in a sex- region- and time-specific manner. Neuropsychopharmacology 32, 1462–1476.
McCreary, J.K., Erickson, Z.T., Metz, G.A., 2016a. Environmental enrichment mitigates the impact of ancestral stress on motor skill and corticospinal tract plasticity. Neurosci. Lett. 632, 181–186.
McCreary, J.K., Truica, L.S., Friesen, B., Yao, Y., Olson, D.M., Kovalchuk, I., Cross, A.R., Metz, G.A., 2016b. Altered brain morphology and functional connectivity reflect a vulnerable affective state after cumulative multigenerational stress in rats. Neuroscience 330, 79–89.
McEwen, B.S., Milner, T.A., 2017. Understanding the broad influence of sex hormones and sex differences in the brain. J. Neurosci. Res. 95, 24–39.
McLaughlin, K.A., Sheridan, M.A., Lambert, H.K., 2014a. Childhood adversity and neural development: deprivation and threat as distinct dimensions of early experience. Neurosci. Biobehav. Rev. 47, 578–591.
McLaughlin, K.A., Sheridan, M.A., Winter, W., Fox, N.A., Zeanah, C.H., Nelson, C.A., 2014b. Widespread reductions in cortical thickness following severe early-life deprivation: a neurodevelopmental pathway to attention-deficit/hyperactivity disorder. Biol. Psychiatry 76, 629–638.
Mehta, M.A., Golembo, N.I., Nosarti, C., Colvert, E., Mota, A., Williams, S.C., Rutter, M., Sonuga-Barke, E.J., 2009. Amygdala, hippocampal and corpus callosum size following severe early institutional deprivation: the English and Romanian Adoptees study pilot. J. Child Psychol. Psychiatry 50, 943–951.
Mehta, M.A., Gore-Langton, E., Golembo, N., Colvert, E., Williams, S.C., Sonuga-Barke, E., 2010. Hyporesponsive reward anticipation in the basal ganglia following severe institutional deprivation early in life. J. Cogn. Neurosci. 22, 2316–2325.
Melinder, C., Hiyoshi, A., Fall, K., Halfvarson, J., Montgomery, S., 2017. Stress resilience and the risk of inflammatory bowel disease: a cohort study of men living in Sweden. BMJ Open 7 (1), e014315. https://doi.org/10.1136/bmjopen-2016-014315M.
Misiak, B., Krefft, M., Bielawski, T., Moustafa, A.A., Sasiadek, M.M., Frydecka, D., 2017. Toward a unified theory of childhood trauma and psychosis: a comprehensive review of epidemiological, clinical, neuropsychological and biological findings. Neurosci. Biobehav. Rev. 75, 393–406.
Molendijk, M.L., de Kloet, E.R., 2015. Immobility in the forced swim test is adaptive and does not reflect depression. Psychoneuroendocrinology 62, 389–391.
Morgan, C.P., Bale, T.L., 2011. Early prenatal stress epigenetically programs dysmasculinization in second-generation offspring via the paternal lineage. J. Neurosci. 31, 11748–11755.
Mueller, B.R., Bale, T.L., 2008. Sex-specific programming of offspring emotionality after stress early in pregnancy. J. Neurosci. 28, 9055–9065.
Munro, C.A., McCaul, M.E., Wong, D.F., Oswald, L.M., Zhou, Y., Brasic, J., Kuwabara, H., Kumar, A., Alexander, M., Ye, W., Wand, G.S., 2006. Sex differences in striatal dopamine release in healthy adults. Biol. Psychiatry 59, 966–974.
Murphy, M.O., Loria, A.S., 2017. Sex-specific effects of stress on metabolic and cardiovascular disease: are women at higher risk? Am. J. Physiol. Regul. Integr. Comp. Physiol. 313, R1–R9.
Natelson, B.H., Ottenweller, J.E., Cook, J.A., Pitman, D., McCarty, R., Tapp, W.N., 1988. Effect of stressor intensity on habituation of the adrenocortical stress response. Physiol. Behav. 43, 41–46.

Negron-Oyarzo, I., Neira, D., Espinosa, N., Fuentealba, P., Aboitiz, F., 2015. Prenatal stress produces persistence of remote memory and disrupts functional connectivity in the hippocampal-prefrontal cortex axis. Cereb. Cortex 25, 3132–3143.

Nusslock, R., Miller, G.E., 2016. Early-life adversity and physical and emotional health across the lifespan: a neuroimmune network hypothesis. Biol. Psychiatry 80, 23–32.

O'Connor, T.G., Rutter, M., Beckett, C., Keaveney, L., Kreppner, J.M., 2000. The effects of global severe privation on cognitive competence: extension and longitudinal follow-up. English and Romanian Adoptees Study Team. Child Dev. 71, 376–390.

Olivo, D., Caba, M., Gonzalez-Lima, F., Rodriguez-Landa, J.F., Corona-Morales, A.A., 2017. Metabolic activation of amygdala, lateral septum and accumbens circuits during food anticipatory behavior. Behav. Brain Res. 316, 261–270.

O'Mahony, S.M., Clarke, G., Dinan, T.G., Cryan, J.F., 2017. Early-life adversity and brain development: is the microbiome a missing piece of the puzzle? Neuroscience 342, 37–54.

Pacak, K., Palkovits, M., Yadid, G., Kvetnansky, R., Kopin, I.J., Goldstein, D.S., 1998. Heterogeneous neurochemical responses to different stressors: a test of Selye's doctrine of nonspecificity. Am. J. Physiol. 275, R1247–55.

Paquola, C., Bennett, M.R., Hatton, S.N., Hermens, D.F., Lagopoulos, J., 2017. Utility of the cumulative stress and mismatch hypotheses in understanding the neurobiological impacts of childhood abuse and recent stress in youth with emerging mental disorder. Hum. Brain Mapp. 38, 2709–2721.

Pearson, R.M., Evans, J., Kounali, D., Lewis, G., Heron, J., Ramchandani, P.G., O'Connor, T.G., Stein, A., 2013. Maternal depression during pregnancy and the postnatal period: risks and possible mechanisms for offspring depression at age 18 years. JAMA Psychiatry 70, 1312–1319.

Pearson, R.M., Bornstein, M.H., Cordero, M., Scerif, G., Mahedy, L., Evans, J., Abioye, A., Stein, A., 2016. Maternal perinatal mental health and offspring academic achievement at age 16: the mediating role of childhood executive function. J. Child Psychol. Psychiatry 57, 491–501.

Perez, C.M., Widom, C.S., 1994. Childhood victimization and long-term intellectual and academic outcomes. Child Abuse Negl. 18, 617–633.

Pitman, D.L., Ottenweller, J.E., Natelson, B.H., 1988. Plasma corticosterone levels during repeated presentation of two intensities of restraint stress: chronic stress and habituation. Physiol. Behav. 43, 47–55.

Provencal, N., Binder, E.B., 2015. The effects of early life stress on the epigenome: from the womb to adulthood and even before. Exp. Neurol. 268, 10–20.

Rakers, F., Bischoff, S., Schiffner, R., Haase, M., Rupprecht, S., Kiehntopf, M., Kuhn-Velten, W.N., Schubert, H., Witte, O.W., Nijland, M.J., Nathanielsz, P.W., Schwab, M., 2015. Role of catecholamines in maternal-fetal stress transfer in sheep. Am. J. Obstet. Gynecol. 213 (684), e1–9.

Rakers, F., Rupprecht, S., Dreiling, M., Bergmeier, C., Witte, O.W., Schwab, M., 2017. Transfer of maternal psychosocial stress to the fetus. Neurosci. Biobehav. Rev.

Razoux, F., Russig, H., Mueggler, T., Baltes, C., Dikaiou, K., Rudin, M., Mansuy, I.M., 2017. Transgenerational disruption of functional 5-HT1AR-induced connectivity in the adult mouse brain by traumatic stress in early life. Mol. Psychiatry 22, 519–526.

Roth, K.A., Katz, R.J., 1981. Further studies on a novel animal model of depression: therapeutic effects of a tricyclic antidepressant. Neurosci. Biobehav. Rev. 5, 253–258.

Roth, K.A., Mefford, I.M., Barchas, J.D., 1982. Epinephrine, norepinephrine, dopamine and serotonin: differential effects of acute and chronic stress on regional brain amines. Brain Res. 239, 417–424.

Roth, T.L., Lubin, F.D., Funk, A.J., Sweatt, J.D., 2009. Lasting epigenetic influence of early-life adversity on the BDNF gene. Biol. Psychiatry 65, 760–769.

Sah, R., Pritchard, L.M., Richtand, N.M., Ahlbrand, R., Eaton, K., Sallee, F.R., Herman, J.P., 2005. Expression of the glucocorticoid-induced receptor mRNA in rat brain. Neuroscience 133, 281–292.
Salehi, B., Cordero, M.I., Sandi, C., 2010. Learning under stress: the inverted-U-shape function revisited. Learn. Mem. 17, 522–530.
Sanchez, M.M., Young, L.J., Plotsky, P.M., Insel, T.R., 2000. Distribution of corticosteroid receptors in the rhesus brain: relative absence of glucocorticoid receptors in the hippocampal formation. J. Neurosci. 20, 4657–4668.
Sandman, C.A., Buss, C., Head, K., Davis, E.P., 2015. Fetal exposure to maternal depressive symptoms is associated with cortical thickness in late childhood. Biol. Psychiatry 77, 324–334.
Santarelli, S., Lesuis, S.L., Wang, X.D., Wagner, K.V., Hartmann, J., Labermaier, C., Scharf, S.H., Muller, M.B., Holsboer, F., Schmidt, M.V., 2014. Evidence supporting the match/mismatch hypothesis of psychiatric disorders. Eur. Neuropsychopharmacol. 24, 907–918.
Santarelli, S., Zimmermann, C., Kalideris, G., Lesuis, S.L., Arloth, J., Uribe, A., Dournes, C., Balsevich, G., Hartmann, J., Masana, M., Binder, E.B., Spengler, D., Schmidt, M.V., 2017. An adverse early life environment can enhance stress resilience in adulthood. Psychoneuroendocrinology 78, 213–221.
Schalinski, I., Teicher, M.H., Carolus, A.M., Rockstroh, B., 2017. Defining the impact of childhood adversities on cognitive deficits in psychosis: an exploratory analysis. Schizophr. Res.
Scheinost, D., Sinha, R., Cross, S.N., Kwon, S.H., Sze, G., Constable, R.T., Ment, L.R., 2017. Does prenatal stress alter the developing connectome? Pediatr. Res. 81, 214–226.
Scheurink, A.J., Ammar, A.A., Benthem, B., van Dijk, G., Sodersten, P.A., 1999. Exercise and the regulation of energy intake. Int. J. Obes. Relat. Metab. Disord. 23 (Suppl. 3), S1–6.
Schwabe, L., Bohbot, V.D., Wolf, O.T., 2012. Prenatal stress changes learning strategies in adulthood. Hippocampus 22, 2136–2143.
Selye, H., 1931. A syndrome produced by diverse nocuous agents. Nature 138, 32.
Selye, H., 1956. The stress-concept as it presents itself in 1956. Antibiot. Chemother. 3, 1–17.
Seo, D., Tsou, K.A., Ansell, E.B., Potenza, M.N., Sinha, R., 2014. Cumulative adversity sensitizes neural response to acute stress: association with health symptoms. Neuropsychopharmacology 39, 670–680.
Sharma, A., 2017. Transgenerational epigenetics: integrating soma to germline communication with gametic inheritance. Mech. Ageing Dev. 163, 15–22.
Sheridan, M.A., Fox, N.A., Zeanah, C.H., McLaughlin, K.A., Nelson 3rd, C.A., 2012. Variation in neural development as a result of exposure to institutionalization early in childhood. Proc. Natl. Acad. Sci. U. S. A. 109, 12927–12932.
Skinner, M.K., 2014. Environmental stress and epigenetic transgenerational inheritance. BMC Med. 12, 153.
Spratt, E.G., Friedenberg, S.L., Swenson, C.C., Larosa, A., De Bellis, M.D., Macias, M.M., Summer, A.P., Hulsey, T.C., Runyan, D.K., Brady, K.T., 2012. The effects of early neglect on cognitive, language, and behavioral functioning in childhood. Psychology (Irvine) 3, 175–182.
Stavrou, S., Nicolaides, N.C., Critselis, E., Darviri, C., Charmandari, E., Chrousos, G.P., 2017. Paediatric stress: from neuroendocrinology to contemporary disorders. Eur. J. Clin. Investig. 47, 262–269.
St-Pierre, J., Laurent, L., King, S., Vaillancourt, C., 2016. Effects of prenatal maternal stress on serotonin and fetal development. Placenta 48 (Suppl. 1), S66–S71.
Tamashiro, K.L., Sakai, R.R., Shively, C.A., Karatsoreos, I.N., Reagan, L.P., 2011. Chronic stress, metabolism, and metabolic syndrome. Stress 14, 468–474.

Tarabulsy, G.M., Pearson, J., Vaillancourt-Morel, M.P., Bussieres, E.L., Madigan, S., Lemelin, J.P., Duchesneau, A.A., Hatier, D.E., Royer, F., 2014. Meta-analytic findings of the relation between maternal prenatal stress and anxiety and child cognitive outcome. J. Dev. Behav. Pediatr. 35, 38–43.
Teicher, M.H., Samson, J.A., Anderson, C.M., Ohashi, K., 2016. The effects of childhood maltreatment on brain structure, function and connectivity. Nat. Rev. Neurosci. 17, 652–666.
Thoits, P.A., 2010. Stress and health: major findings and policy implications. J. Health Soc. Behav. 51 (Suppl), S41–53.
Tyrka, A.R., Burgers, D.E., Philip, N.S., Price, L.H., Carpenter, L.L., 2013. The neurobiological correlates of childhood adversity and implications for treatment. Acta Psychiatr. Scand. 128, 434–447.
Van den Bergh, B.R., Mennes, M., Oosterlaan, J., Stevens, V., Stiers, P., Marcoen, A., Lagae, L., 2005. High antenatal maternal anxiety is related to impulsivity during performance on cognitive tasks in 14- and 15-year-olds. Neurosci. Biobehav. Rev. 29, 259–269.
van den Bergh, B.R., Mennes, M., Stevens, V., van der Meere, J., Borger, N., Stiers, P., Marcoen, A., Lagae, L., 2006. ADHD deficit as measured in adolescent boys with a continuous performance task is related to antenatal maternal anxiety. Pediatr. Res. 59, 78–82.
Van den Bergh, B.R.H., van den Heuvel, M.I., Lahti, M., Braeken, M., de Rooij, S.R., Entringer, S., Hoyer, D., Roseboom, T., Raikkonen, K., King, S., Schwab, M., 2017. Prenatal developmental origins of behavior and mental health: the influence of maternal stress in pregnancy. Neurosci. Biobehav. Rev.
van der Voorn, B., Hollanders, J.J., Ket, J.C.F., Rotteveel, J., Finken, M.J.J., 2017. Gender-specific differences in hypothalamus-pituitary-adrenal axis activity during childhood: a systematic review and meta-analysis. Biol. Sex Differ. 8, 3.
van Otterdijk, S.D., Michels, K.B., 2016. Transgenerational epigenetic inheritance in mammals: how good is the evidence? FASEB J. 30, 2457–2465.
Vasilevski, V., Tucker, A., 2016. Wide-ranging cognitive deficits in adolescents following early life maltreatment. Neuropsychology 30, 239–246.
Veenendaal, M.V., Painter, R.C., de Rooij, S.R., Bossuyt, P.M., van der Post, J.A., Gluckman, P.D., Hanson, M.A., Roseboom, T.J., 2013. Transgenerational effects of prenatal exposure to the 1944-45 Dutch famine. BJOG 120, 548–553.
Volden, P.A., Conzen, S.D., 2013. The influence of glucocorticoid signaling on tumor progression. Brain Behav. Immun. 30 (Suppl), S26–31.
Vyas, S., Rodrigues, A.J., Silva, J.M., Tronche, F., Almeida, O.F., Sousa, N., Sotiropoulos, I., 2016. Chronic stress and glucocorticoids: from neuronal plasticity to neurodegeneration. Neural Plast. 2016, 6391686.
Watt, M.J., Weber, M.A., Davies, S.R., Forster, G.L., 2017. Impact ofjuvenile chronic stress on adult cortico-accumbal function: implications for cognition and addiction. Prog. Neuropsychopharmacol. Biol. Psychiatry 79, 136–154.
Weinstock, M., 2011. Sex-dependent changes induced by prenatal stress in cortical and hippocampal morphology and behaviour in rats: an update. Stress 14, 604–613.
Weiss, I.C., Franklin, T.B., Vizi, S., Mansuy, I.M., 2011. Inheritable effect of unpredictable maternal separation on behavioral responses in mice. Front. Behav. Neurosci. 5, 3.
Windsor, J., Benigno, J.P., Wing, C.A., Carroll, P.J., Koga, S.F., Nelson 3rd, C.A., Fox, N.A., Zeanah, C.H., 2011. Effect of foster care on young children's language learning. Child Dev. 82, 1040–1046.
Wirtz, P.H., von Kanel, R., 2017. Psychological stress, inflammation, and coronary heart disease. Curr. Cardiol. Rep. 19, 111.
Xiong, F., Zhang, L., 2013. Role of the hypothalamic-pituitary-adrenal axis in developmental programming of health and disease. Front. Neuroendocrinol. 34, 27–46.

Yerkes, R.M., Dodson, J.D., 1908. The relation of strength of stimulus to rapidity of habit-formation. J. Comp. Neurol. Psychol. 18, 459–482.
Yi, S.J., Masters, J.N., Baram, T.Z., 1994. Glucocorticoid receptor mRNA ontogeny in the fetal and postnatal rat forebrain. Mol. Cell. Neurosci. 5, 385–393.
Zalosnik, M.I., Pollano, A., Trujillo, V., Suarez, M.M., Durando, P.E., 2014. Effect of maternal separation and chronic stress on hippocampal-dependent memory in young adult rats: evidence for the match-mismatch hypothesis. Stress 17, 445–450.
Zannas, A.S., West, A.E., 2014. Epigenetics and the regulation of stress vulnerability and resilience. Neuroscience 264, 157–170.
Zheng, A., Li, H., Cao, K., Xu, J., Zou, X., Li, Y., Chen, C., Liu, J., Feng, Z., 2015. Maternal hydroxytyrosol administration improves neurogenesis and cognitive function in prenatally stressed offspring. J. Nutr. Biochem. 26, 190–199.
Zoladz, P.R., Diamond, D.M., 2009. Linear and non-linear dose-response functions reveal a hormetic relationship between stress and learning. Dose Response 7, 132–148.

FSC
www.fsc.org
MIX
Papier | Fördert gute Waldnutzung
FSC® C083411

Kolibri 360 GmbH
Ferdinand-Jühlke-Straße 7
99095 Erfurt, Deutschland
produktsicherheit@kolibri360.de